Student Solutions Manual

Essential University Physics
Second Edition

Volume 1: Chapters 1–19

Richard Wolfson
Middlebury College

Brett Kraabel
PhD, Physics, University of California, Santa Barbara

Michael Schirber
PhD, Astrophysics, Ohio State University

Addison-Wesley

Boston Columbus Indianapolis New York San Francisco Upper Saddle River
Amsterdam Cape Town Dubai London Madrid Milan Munich Paris Montréal Toronto
Delhi Mexico City São Paulo Sydney Hong Kong Seoul Singapore Taipei Tokyo

Executive Editor: Nancy Whilton

Project Editor: Martha Steele

Director of Development: Michael Gillespie

Development Editor: Ashley Eklund

Editorial Assistant: Peter Alston

Managing Editor: Corinne Benson

Production Supervisor: Beth Collins

Production Management and Compositor: PreMediaGlobal

Manufacturing Buyer: Jeffrey Sargent

Senior Marketing Manager: Kerry Chapman

Cover Photo Credit: Andrew Lambert Photography / Science Photo Library

3 4 5 6 7 8 9 10—SCI—16 15 14 13 12

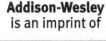

Addison-Wesley
is an imprint of

www.pearsonhighered.com

ISBN 10: 0-3217-1203-X
ISBN 13: 978-0-3217-1203-5

CONTENTS

Volume 1 contains chapters 1–19

Volume 2 contains chapters 20–39

Preface   iv

Chapter 1   Doing Physics   1-1

PART ONE   Mechanics
Chapter 2   Motion in a Straight Line   2-1
Chapter 3   Motion in Two and Three Dimensions   3-1
Chapter 4   Force and Motion   4-1
Chapter 5   Using Newton's Laws   5-1
Chapter 6   Work, Energy, and Power   6-1
Chapter 7   Conservation of Energy   7-1
Chapter 8   Gravity   8-1
Chapter 9   Systems of Particles   9-1
Chapter 10   Rotational Motion   10-1
Chapter 11   Rotational Vectors and Angular Momentum   11-1
Chapter 12   Static Equilibrium   12-1

PART TWO   Oscillations, Waves, and Fluids
Chapter 13   Oscillatory Motion   13-1
Chapter 14   Wave Motion   14-1
Chapter 15   Fluid Motion   15-1

PART THREE   Thermodynamics
Chapter 16   Temperature and Heat   16-1
Chapter 17   The Thermal Behavior of Matter   17-1
Chapter 18   Heat, Work, and the First Law of Thermodynamics   18-1
Chapter 19   The Second law of Thermodynamics   19-1

PREFACE

This *Student Solutions Manual* to *Essential University Physics,* Second Edition, by Richard Wolfson, is designed to increase your skill and confidence in solving physics problems—the key to success in your physics course. It coaches you through a range of helpful and effective problem-solving techniques and provides solutions to all the odd-numbered problems in your text. By working carefully through these techniques and solutions, you will master the proven IDEA four-step problem-solving approach used in the text (Interpret, Develop, Evaluate, Assess) and learn to successfully

- interpret problems and identify the key physics concepts involved.
- develop a plan and draw figures for your solution.
- evaluate any mathematical expressions.
- assess your solution to check that it makes sense and see how it adds to your broader understanding of physics.

Do your best to solve each problem *before* reading the solution. When you need to refer to the solution, focus on the reasoning—make sure you understand how and why each step is taken. By pushing yourself, you'll develop and hone your problem-solving skills. Don't fall into the trap of just passively reading the solutions.

Through multiple reviews involving many instructors, we have made every effort to ensure these solutions are accurate and correct. If you find any errors or ambiguities, we would be very grateful to hear from you. Please ask your professor to contact her/his Pearson Education/Addison-Wesley sales representative, who will pass on any errors to the appropriate person.

DOING PHYSICS

EXERCISES

Section 1.2 Measurements and Units

11. **INTERPRET** This problem involves comparing the sizes of two objects (a hydrogen atom and a proton), where the size of each object is expressed in different multiples of the meter (i.e., a distance).

DEVELOP Before any comparison can be made, the quantity of interest must first be expressed in the same units. From Table 1.1, we see that a nanometer (nm) is 10^{-9} m, and a femtometer (fm) is 10^{-15} m. Expressed mathematically, these relations are 1 nm = 10^{-9} m and 1 fm = 10^{-15} m, or 1 = (10^{-9} m)/nm = (10^{-15} m)/fm.

EVALUATE Using these conversion factors, the diameter of a hydrogen atom is $d_H = (0.1 \text{ nm})(10^{-9} \text{ m/nm}) = 10^{-10}$ m, and the diameter of a proton is $d_p = (1 \text{ fm})(10^{-15} \text{ m/fm}) = 10^{-15}$ m. Therefore, the ratio of the diameters of a hydrogen atom to a proton (its nucleus) is

$$\frac{d_H}{d_p} = \frac{10^{-10} \text{ m}}{10^{-15} \text{ m}} = 10^5$$

ASSESS The hydrogen atom is about 100,000 times larger than its nucleus. To get a feel for this difference in size, consider the diameter of the Earth (~ 10^7 m). If a hydrogen atom were the size of the Earth, the proton would have a size of (10^7 m)/10^5 = 10^2 m, which is the size of a football field!

13. **INTERPRET** This problem involves inverting the definition of the second to find the period of the given ^{133}Cs radiation. Note that the "period" in this context is a length of time, so our result should be in units of time.

DEVELOP By definition, 1 s $\equiv 9,192,631,770T$, where T is the period of the radiation corresponding to the transition between the two hyperfine levels of the ^{133}Cs ground state. Because this is a definition, the "second" is actually given to infinite precision, so we can write 1.000000000 s $\equiv 9,192,631,770T$ periods of ^{133}Cs radiation.

EVALUATE One period T of cesium radiation is thus

$$T = \frac{1.000000000 \text{ s}}{9,192,631,770} = 1.087827757 \times 10^{-10} \text{ s} = 108.7827757 \text{ ps}$$

ASSESS Because one nanosecond corresponds to about 9 periods of the cesium radiation, each period is about $\frac{1}{9}$ of a nanosecond. Note that there exists an alternative definition based on the frequency of the cesium-133 hyperfine transition, which is the reciprocal of the period.

15. **INTERPRET** For this problem, we need to divide a 1-cm length by the given diameter of the hydrogen atom to find how many hydrogen atoms we need to place side-by-side to span this length.

DEVELOP We first express the quantities of interest (diameter of a hydrogen atom and 1-cm line) in the same units. Since a nanometer is 10^{-9} m (Table 1.1), we see that $d_H = 0.1$ nm $= 10^{-10}$ m. In addition, 1 cm $= 10^{-2}$ m.

EVALUATE The desired number of atoms n is the length of the line divided by the diameter of a single atom:

$$n = \frac{10^{-2} \text{ m}}{10^{-10} \text{ m}} = 10^8$$

ASSESS If 1 cm corresponds to 10^8 hydrogen atoms, then each atom would correspond to 10^{-8} cm $= 10^{-10}$ m $= 0.1$ nm, which agrees with the diameter given for the hydrogen atom.

17.   **INTERPRET**   For this problem, we are looking for an angle subtended by a circular arc, so we will use the definition of angle in radians (see Fig. 1.2).

DEVELOP   From Fig. 1.2, we see that the angle in radians is the circular arc length s divided by the radius r, or $\theta = s/r$.

EVALUATE   Inserting the known quantities (s = 2.1 km, r = 3.4 km), we find the angle subtended is

$$\theta = \frac{s}{r} = \frac{2.1 \text{ km}}{3.4 \text{ km}} = 0.62 \text{ rad}$$

Using the fact that π rad $= 180°$, the result can be expressed as

$$\theta = 0.62 \text{ rad} = (0.62 \text{ rad})\left(\frac{180°}{\pi \text{ rad}}\right) \approx 35°$$

ASSESS   Because a complete circular revolution is $360°$, $35°$ is roughly 1/10 of a circle. The circumference of a circle of radius $r = 3.4$ km is $C = 2\pi(3.4 \text{ km}) = 21.4$ km. Therefore, we expect the jetliner to fly approximately 1/10 of C, or 2.1 km, which agrees with the problem statement.

19.   **INTERPRET**   For this problem, we must convert the weight of the letter in ounces to its mass in grams.

DEVELOP   Two different units for mass appear in the problem—ounces and grams. The conversion from ounces to grams is given in Appendix C (1 oz = weight of 0.02835 kg).

EVALUATE   The maximum weight of the letter is 1 oz. Using the conversion factor above, we see that this corresponds to a weight of 0.02835 kg, or 28.35 g. Because the weight in oz is given to a single significant figure, we must round our answer to a single significant figure, which gives 30 g (or 3×10^1 g).

ASSESS   The conversion factor between oz and g may be obtained based on some easily remembered conversion factors between the metric and English systems (e.g., 1 lb = weight of 0.454 kg, and 1 lb = 16 oz).

21.   **INTERPRET**   For this problem, we must express a given volume (1 m^3) in units of cm^3.

DEVELOP   Because volume has dimension of (length)^3, the problem reduces to converting m to cm. The conversion equation is 1 m = 100 cm, so the conversion factor to convert m to cm is 1 = (100 cm)/(1 m).

EVALUATE   Using this conversion factor, we obtain

$$1 \text{ m}^3 = \left(1 \text{ m}\right)^3 \overbrace{\left(\frac{100 \text{ cm}}{1 \text{ m}}\right)}^{=1}{}^3 = 10^6 \text{ cm}^3$$

ASSESS   Another way to remember this relationship is to note that 1 m^3 = 1000 liters, and 1 liter = 1000 cm^3, so 1 m^3 = 1000 × 1000 cm^3 = 10^6 cm^3.

23.   **INTERPRET**   For this problem, we have to convert units of volume and units of area. We are told the coverage of the paint (in English units) is 350 ft^2/gal.

DEVELOP   From Appendix C, we find the following conversion equations:

$$1 \text{ gal} = 3.785 \times 10^{-3} \text{ m}^3 = 3.785 \text{ L}$$
$$1 \text{ ft}^2 = 9.290 \times 10^{-2} \text{ m}^2$$

Thus, the conversion factors are 1 = 3.785 L/gal and 1 = 0.09290 m^2/ft^2.

EVALUATE   Combining the two conversion factors, we have

$$350 \text{ ft}^2/\text{gal} = \left(350\frac{\text{ft}^2}{\text{gal}}\right)\left(\frac{1 \text{ gal}}{3.785 \text{ L}}\right)\left(\frac{9.290 \times 10^{-2} \text{ m}^2}{1 \text{ ft}^2}\right) = 8.6 \text{ m}^2/\text{L}$$

ASSESS   Dividing this result by 350 gives 1 ft^2/gal = 0.025 m^2/L.

25. **INTERPRET**   This problem will require us to convert units of both length and time. Specifically, we have to convert km to m and h to s.

    DEVELOP   From Appendix C, we find the following conversion equations, which we convert into conversion factors by dividing through by km and h, respectively:

    $$1 \text{ km} \equiv 1000 \text{ m} \Rightarrow 1 = 1 \text{ km}/(1000 \text{ m})$$
    $$1 \text{ h} \equiv 3600 \text{ s} \Rightarrow 1 = 3600 \text{ s}/\text{h}$$

    EVALUATE   Combining the two conversion factors, we have

    $$1 \text{ m/s} = \left(1\frac{\not{m}}{\not{s}}\right)\overbrace{\left(\frac{3600 \not{s}}{1 \text{ h}}\right)}^{=1}\overbrace{\left(\frac{1 \text{ km}}{1000 \not{m}}\right)}^{=1} = 3.6 \text{ km/h}$$

    ASSESS   The units of the results are distance per unit time, as expected for a speed. Because one km is *defined* as 1000 m, and 1 h is *defined* as 3600 s, the conversion factor is exact. Thus, if you walk 1 meter in one second, you can walk a distance of 3.6 km in one hour (assuming you don't get tired and slow down).

27. **INTERPRET**   This problem involves converting radians to degrees.

    DEVELOP   An angle in radians is the circular arc length s divided by the radius r, or $\theta = s/r$. Because a complete revolution (360°) is *defined* as 2π radians, we have $360° \equiv 2\pi$ rad, or $1 = 360°/(2\pi)$.

    EVALUATE   Using the conversion factor above, we find that

    $$1 \text{ rad} = (1 \text{ rad})\overbrace{\left(\frac{360°}{2\pi \text{ rad}}\right)}^{=1} = 57.3°$$

    ASSESS   Because this conversion involves a definition, we know the quantities involved to infinite precision, so we could report the result to as many significant figures as desired. The result is that 1 rad is approximatley one-sixth of a complete revolution, so 6 rad corresponds approximately to one revolution.

Section 1.3 Working with Numbers

29. **INTERPRET**   We interpret this as a problem involving the conversion of time to different units.

    DEVELOP   With reference to Table 1.1 for SI prefixes, we have $1 \text{ ms} = 10^{-3}\text{s}$.

    EVALUATE   Using the above conversion factor, we obtain

    $$\frac{4.23103 \text{ m/s}}{0.57 \text{ ms}}\left(\frac{1 \text{ ms}}{10^{-3}\text{s}}\right) = 7.4 \text{ m/s}^2$$

    ASSESS   Acceleration is the physical quantity with such units. An average acceleration of 7.4 m/s^2 changes the speed of an object by 4.23103 m/s in 0.57 ms. Note that we only kept two significant figures in the answer, since that was the number of significant figures in the time quantity (see Section 1.3).

31. **INTERPRET**   This problem asks that we take the cube root of a number in scientific notation without a calculator.

    DEVELOP   As shown in the Tactics 1.1 box in the text, we can take the cube root by multiplying the exponent by 1/3.

    EVALUATE   We know that the cube root of 64 is 4. So let's rewrite the given number as 64×10^{18}. We now can calculate the cube root more easily:

    $$\sqrt[3]{6.4 \times 10^{19}} = \left(64 \times 10^{18}\right)^{1/3} = \left(64\right)^{1/3} \times 10^{18/3} = 4 \times 10^6$$

    ASSESS   We can check our work by cubing 4×10^6 and verifying that it indeed equals 6.4×10^{19}.

33. **INTERPRET**   This problem involves adding two distances that are given in different units. Therefore, before performing the sum, we must express both distances in the same units. We will choose to convert cm to m, then sum the result with 41 m.

    DEVELOP   From Table 1.1, we see that $1 \text{ cm} = 10^{-2}\text{ m}$, or $1 = 10^{-2}\text{ m/cm}$.

    EVALUATE   An airplane of initial length $L_0 = 41$ m is increased by $\Delta L = 3.6$ cm, so the final length L is

$$L = L_0 + \Delta L = 41 \text{ m} + \left(3.6 \ \cancel{\text{cm}}\right)\overbrace{\left(\frac{10^{-2} \text{ m}}{\cancel{\text{cm}}}\right)}^{=1} = 41 \text{ m} + 0.036 \text{ m} = 41.036 \text{ m}$$

However, because the data are given to two significant figures, we must round the result to two significant figures, so $L = 41$ m is the final result.

ASSESS To two significant figures, the result remains unchanged. In this context, 41 m means a length greater than or equal to 40.5 m, but less than 41.5 m, and 41 m + 0.036 m = 41.036 m satisfies this condition.

PROBLEMS

35. **INTERPRET** This problem involves exploring the numerical precision of results by evaluating $\left(\sqrt{3}\right)^3$ using two different approaches: the first involves retaining only 3 significant figures in the intermediate step, and the second involves retaining 4 significant figures in the intermediate step.

DEVELOP For part (a), retain 3 significant figures in calculating $\sqrt{3}$ and in part (b) retain 4 significant figures in the same calculation. Cube the results and round to 3 significant figures for the final answer and compare the results of (a) and (b).

EVALUATE (a) $\sqrt{3} \approx 1.73$ (to 3 significant figures), so $(1.73)^3 = 5.177 \approx 5.18$, to 3 significant figures.
(b) To 4 significant figures, $\sqrt{3} \approx 1.732$ so $(1.732)^3 \approx 5.1957$, or 5.20 to 3 significant figures.

ASSESS With a calculator, we find $\left(\sqrt{3}\right)^3 = 3^{3/2} = 5.196....$ This example shows that it is important to carry intermediate calculations to more digits than the desired accuracy for the final answer. Rounding of intermediate results could affect the final answer.

37. **INTERPRET** This problem requires us to make a rough, "order-of-magnitude" estimate, instead of a finding a precise numerical answer. Thus, although answers may vary, they should be within an order of magnitude of the given answer.

DEVELOP For problems that involve rough estimates, various assumptions typically need to be made. Such assumptions must be physically motivated with reasonable order-of-magnitude estimates. We shall assume that there are 300 million people in the United States, and that each person drinks 250 ml = 0.250 kg of milk per day.

EVALUATE Based on the assumption above, the amount consumed per year would be approximately

$$(300 \times 10^6)(365 \ \cancel{d}/\text{y})(0.250 \text{ kg}/\cancel{d}) \approx 3 \times 10^{10} \text{ kg/y}$$

Dividing this by the average annual production of one cow, we estimate the number of cows needed to be

$$N = \frac{3 \times 10^{10} \text{ kg/y}}{10^4 \text{ kg/y}} = 3 \times 10^6$$

ASSESS There are currently approximately 9 million milk cows in the United States. Based on our estimate, we do not expect any shortage of milk in the near future.

39. **INTERPRET** This problem calls for a rough estimate. The quantities of interest here are the total rate of energy consumption (i.e., power consumption) and the area needed for solar cells. The electrical power consumed by the entire population of the United States, divided by the power converted by one square meter of solar cells, is the area required by this question.

DEVELOP For problems that involve rough estimates, various assumptions typically need to be made. Such assumptions must be physically motivated with reasonable order-of-magnitude estimates. Note that we will be required to convert all data regarding power to kW, and that we must express length in common units.

EVALUATE Assume there are about 300 million people in the United States. As the problem states, the average power consumption per person person is 1.5 kW, so the total power consumption P_{tot} is

$$P_{\text{tot}} = (300 \times 10^6 \text{ people}) \times 1.5 \text{ kW/person} = 4.5 \times 10^8 \text{ kW}$$

For a solar cell with 20% efficiency in converting sunlight to electrical power, the power-yield P_{sc} per unit area A is $P_{\text{sc}}/A = (0.20)(300 \ \cancel{\text{W}}/\text{m}^2)(10^{-3} \text{ kW}/\cancel{\text{W}}) = 0.060 \text{ kW/m}^2$. Therefore, the total area A_{tot} needed is

$$A_{\text{tot}} = \frac{P_{\text{tot}}}{P_{\text{sc}}/A} = \frac{4.5 \times 10^8 \text{ kW}}{0.060 \text{ kW/m}^2} = 7.5 \times 10^9 \text{ m}^2$$

The land area of the continental United States A_{US} can be approximated as the area of a rectangle the size of the distance from New York to Los Angeles by the distance from New York to Miami, or $A_{US} \approx (5000 \text{ km}) \times (2000 \text{ km}) = 10^7 \text{ km}^2$. From Table 1.1, we know that 1 km = 10^3 m, or 1 = 10^{-3} km/m. Then the fraction of area to be covered by solar cells would be

$$\frac{A_{tot}}{A_{US}} \approx \left(\frac{7.5 \times 10^9 \text{ m}^2}{10^7 \text{ km}^2} \right) \overbrace{\left(\frac{10^{-3} \text{ km}}{1 \text{ m}} \right)}^{=1}^2 = 7.5 \times 10^{-4}$$

or, to one significant figure, approximately 0.08%.

ASSESS   This represents only a small fraction of land to be used for solar cells. The area A_{tot} is less than to the fraction of land now covered by airports.

41.   **INTERPRET**   This problem calls for a rough estimate instead of a precise numerical answer. Answers will differ depending on the assumptions used, so each assumption should be explained. We are asked to estimate the rate (volume per unit time) at which water flows over the Niagara Falls, and the time it would take for Lake Eerie to rise 1 m if the falls were shut off.

DEVELOP   For problems that involve rough estimates, various assumptions typically need to be made. Such assumptions must be physically motivated and explained.

(**a**) The discharge of Niagara Falls (i.e., the volume V flowing per unit time over the falls) is the current speed multiplied by the cross-sectional area of the channel, $D = vA$ (see Section 15.4). A road map of Niagara Falls shows that the river is about 1-km wide at the falls. The cross-sectional area A at the falls is the river width multiplied by the average depth, which is probably on the order of 1 m. The current speed may be estimated to be approximatley 3 m/s.

(**b**) To estimate the time it takes for Lake Erie to rise by 1 m, we need to know its surface area. From a road map of the area, we find that Lake Erie is about 375 km long by 75 km wide.

EVALUATE   (**a**) Based on the these assumptions, the volume V of water going over Niagara Falls each second is

$$V = D(1 \text{ s}) = vA(1 \text{ s}) = (3 \text{ m/s})(1 \text{ km}) \overbrace{\left(\frac{10^3 \text{ m}}{\text{km}} \right)}^{=1} (1 \text{ m})(1 \text{ s}) = 3 \times 10^3 \text{ m}^3$$

(b) To find the time t it would take Lake Erie to rise 1 m if the falls were blocked, we divide the discharge obtained for part (a) into the volume V_L of a 1-m layer of water covering Lake Erie. This gives

$$t = \frac{V_L}{D} = \frac{(375 \text{ km})(75 \text{ km})(1 \text{ m})}{3 \times 10^3 \text{ m}^3/\text{s}} \overbrace{\left(\frac{10^3 \text{ m}}{\text{km}} \right)}^{=1}^2$$

$$= (9 \times 10^6 \text{ s}) \left(\frac{1 \text{ h}}{3600 \text{ s}} \right) \left(\frac{1 \text{ d}}{24 \text{ h}} \right)$$

$$= 100 \text{ days}$$

ASSESS   Checking our estimate of the discharge against public information, we find that the average discharge of the Niagara Falls is roughly 2 × 10^5 ft³/s. Converting this to m³/s gives

$$2 \times 10^5 \text{ ft}^3/\text{s} = (2 \times 10^5 \text{ ft}^3/\text{s}) \overbrace{\left(\frac{0.3048 \text{ m}}{\text{ft}} \right)}^{=1}^3 = 6 \times 10^3 \text{ m}^3/\text{s}$$

which is within a factor of 2 or our very rough estimation. Not bad!

43.   **INTERPRET**   We can estimate the number of hairs in a braid by dividing the cross-sectional area of a braid by the cross-sectional area of a hair.

DEVELOP   We are not given the area of a braid, but we can assume it has a diameter of about an inch or so, which in centimeters is $d_B \approx 3 \text{ cm}$. We're told that the hair has $d_h \approx 100 \mu\text{m}$. The cross-sectional area of each is given by the formula $A = \pi r^2 = \pi (d/2)^2$.

EVALUATE   The number of hairs in a braid is roughly

$$N \approx \frac{A_b}{A_h} = \left(\frac{d_b}{d_h}\right)^2 \approx \left(\frac{3 \times 10^{-2} \text{m}}{100 \times 10^{-6} \text{m}}\right)^2 = 9 \times 10^4 \sim 10^5$$

ASSESS   We can compare this to the number of hairs on the typical person's head, which is often quoted as being around 100,000. Assuming the braid is holding all of the hair from a person's head, our answer is pretty good.

45.   **INTERPRET**   This problem calls for a rough estimate instead of a precise numerical answer. The quantity of interest here is the thickness of the bubble made from the bubble gum.

DEVELOP   For problems that involve rough estimates, various assumptions typically need to be made. Such assumptions must be physically motivated and explained. The volume V of the *bubble* (a thin spherical shell) is $V = 4\pi r^2 d$ where r is the radius and $d \ll r$ is the thickness. The mass m of such a bubble is the volume multiplied by the density, and this mass must equate with the mass (i.e., 8 g) of the wad of bubble gum. Thus, we can equate these two expressions for the mass of the bubble gum and solve for the thickness d of the bubble. Recall that the radius $r = D/2 = 5$ cm, where D is the diameter ($D = 10$ cm).

EVALUATE   With these assumptions, the thickness of the bubble is

$$m = V\rho = 4\pi r^2 d\rho$$

$$d = \frac{m}{4\pi r^2 \rho} = \frac{8 \text{ g}}{4\pi (5 \text{ cm})^2 (1 \text{ g/cm}^3)} = 0.025 \text{ cm}$$

ASSESS   The thickness of the bubble is very small. But our estimate is reasonable. Four layers of bubble-gum bubble would be about 1-mm thick. Note that this solution assumes that the bubble gum density does not change in going from the wad in the package to the chewed gum in the mouth. Is this assumption reasonable?

47.   **INTERPRET**   This is a problem that calls for a rough estimate, instead of a precise numerical answer. The quantities of interest here are the size of the electronic components on a PC chip, and the number of calculations that can be performed each second.

DEVELOP   The area of each component is the area of the chip divided by the number of components. We'll take the square root of that to get the component size $d_{comp} = \sqrt{A_{comp}}$. For part (b), to estimate the number of calculations performed per second, we take the inverse of how long it takes to do one calculation. We are told that one calculation requires that an electric impulse traverse 10^4 components each one million times. The time it takes to traverse one component is the distance across one component divided by the velocity, $t_{comp} = d_{comp}/v$. We assume the velocity is approximately the speed of light.

EVALUATE   (a) The area of each component is

$$A_{comp} = \frac{A_{chip}}{N_{comp}} = \frac{(4 \text{ mm})^2}{10^9} = 1.6 \times 10^{-14} \text{m}^2$$

Assuming the component is square

$$d_{comp} = \sqrt{A_{comp}} = \sqrt{1.6 \times 10^{-14} \text{m}^2} = 1.3 \times 10^{-7} \text{m} \sim 0.1 \ \mu\text{m}$$

(b) The time to do one calculation is the number of components to be traversed, multiplied by the number of traversals, and then multiplied by the time to traverse one component

$$t_{calc} = \left(10^4\right)\left(10^6\right) t_{comp} = 10^{10} \frac{d_{comp}}{v} = 10^{10} \frac{1.3 \times 10^{-7} \text{m}}{3 \times 10^8 \text{m/s}} = 4.2 \times 10^{-6} \text{s}$$

This means the chip can do 2×10^5 calculations per second.

ASSESS   The number of calculations per second is often referred to as FLOPS (Floating Point Operations Per Second). The performance of the above chip is 200,000 FLOPS. Modern supercomputers use parallel computing to perform at the level of more than a trillion FLOPS (or TFLOPS).

49.   **INTERPRET**   This problem involves estimating uncertainty given the number of significant figures.

DEVELOP   Because the value 3.6 can be used to represent any number between 3.55 and 3.65, rounding to two significant figures, we see that the uncertainty in the first decimal place is ±0.05 Therefore, the percent uncertainty Δ in a one-decimal-place number N is

$$\Delta = 100 \left(\frac{\pm 0.05}{N} \right) \%$$

which decreases as N increases.

EVALUATE   For the numbers given, the percent uncertainty is **(a)** $\Delta = 100(\pm 0.05/1.1)\% \approx \pm 5\%$; **(b)** $\Delta = 100(\pm 0.05/5.0)\% \approx \pm 1\%$; and **(c)** $\Delta = 100(\pm 0.05/9.9)\% \approx \pm 0.5\%$.

ASSESS   Our result indicates that, for a one-decimal place number N, whereas the uncertainty in the first decimal place is ± 0.05 independent of N, the percent uncertainty Δ becomes smaller for larger N.

51.   **INTERPRET**   This problem involves comparing the cost of gasoline in the U.S. and Canada. To do this we need to have both prices in the same currency and for the same volume of gas.

DEVELOP   Let's change the Canadian pricing into its equivalent American pricing. Therefore we take the given cost in Canadian dollars per liter, and multiply by 3.785 liters per gallon (from Appendix C) and also by 87¢ per Canadian dollar.

EVALUATE   To differentiate the two currencies, we will denote the Canadian dollar with a "hat": $\hat{\$}$. Converting the Canadian cost gives

$$\hat{\$}0.94/L \left(\frac{3.785 \text{ L}}{1 \text{ gal}} \right) \left(\frac{\$0.87}{\hat{\$}1.00} \right) = \$3.09/\text{gal}$$

This is 12¢ more than the cost in the U.S., so it would be better to buy the gas before crossing the border.

ASSESS   It certainly pays to know how to do a unit conversion!

53.   **INTERPRET**   This problem involves converting between different units and retaining the correct number of significant figures.

DEVELOP   First convert the values given in mm to m using the conversion $1 \text{ mm} = 10^{-3}$ m, or $1 = 10^{-3}$ m/mm, then apply the rules for significant figures given in the chapter: When multiplying or dividing, report your answer with the number of significant figures in the least-precise factor. When adding or subtracting, report your answer to the least-precise decimal place.

EVALUATE   **(a)** To the least-precise decimal place, we have

$$1.0 \text{ m} + \left(1 \text{ m\!m}\right) \left(\frac{10^{-3} \text{ m}}{\text{m\!m}} \right) = 1 \text{ m} + 0.001 \text{ m} = 1.001 \text{ m} = 1.0 \text{ m}$$

(b) The least-precise factor, 1 m, has one significant figure, so we find

$$(1.0 \text{ m})\left(1 \text{ m\!m}\right) \left(\frac{10^{-3} \text{ m}}{\text{m\!m}} \right) = 0.001 \text{ m}^2$$

(c) The second decimal place in 1.0 m is the least precise so

$$1.0 \text{ m} - \left(999 \text{ m\!m}\right) \overbrace{\left(\frac{10^{-3} \text{ m}}{\text{m\!m}} \right)}^{=1} = 1.0 \text{ m} - 0.999 \text{ m} = 0.001 \text{ m} = 0.0 \text{ m}$$

(d) Again, 1.0 m has two significant figures, so we find

$$\left(\frac{1.0 \text{ m}}{999 \text{ m\!m}} \right) \left(\frac{10^3 \text{ m\!m}}{\text{m}} \right) = \frac{1.0}{0.999} = 1.001001\ldots \; = 1.0 \,.$$

ASSESS   It is important to note that sometimes the answer you get, to the right precision, is unchanged! It might help to think of this in terms of money: If you're a millionaire, and you drop a nickel down a storm drain, you're still a millionaire.

55.   INTERPRET   We have to adjust the cost per bag of coffee to include the shipping costs.

DEVELOP   To find the final cost per bag, calculate the total cost for the 6 bags of coffee and then divide by 6.

EVALUATE   The total purchase is 6 bags plus shipping

$$6($8.95) + $6.90 = $60.60$$

Dividing by the number of bags gives $10.10 per bag.

ASSESS   If the same shipping costs applied to one bag of coffee, then the price per bag would be $15.85. So sometimes it pays to buy in bulk.

57.   INTERPRET   We can use the given cell size and atom size to estimate the number of atoms that fit in a cell and then compare that to the number of cells in a body.

DEVELOP   By dividing the diameter of a cell by the diameter of an atom, we can say about how many atoms span the length of a cell. But since cells and atoms are three dimensional, we will cube this number to get the number of atoms in a cell.

EVALUATE   The number of atoms in a cell is just the cube of the ratio of diameters

$$N \approx \left(\frac{d_{\text{cell}}}{d_{\text{atom}}} \right)^3 = \left(\frac{10 \ \mu\text{m}}{0.1 \ \text{nm}} \right)^3 = \left(10^5 \right)^3 = 10^{15}$$

We're told the body has about 10^{14} cells, which is technically less than our calculation for the number of atoms in a cell, but these are rough estimates. It's safer to say that the two quantities are about the same.
The answer is (c).

ASSESS   Does this make sense? A human being is roughly 2 meters tall. This is roughly 100,000 times bigger than the diameter of a cell. And we find out here that a cell is roughly 100,000 times bigger than an atom. So yes, it's reasonable that the number of cells in the body is about the same as the number of atoms in the cell.

59.   INTERPRET   The problem asks us to estimate the mass of a cell. We can use the previous answer for the volume and make some approximation for the cell's density.

DEVELOP   Since living things are mostly water, we can assume that the cell has a density roughly equal to that of water ($\rho = 1 \ \text{g/cm}^3$).

EVALUATE   Using the volume from Problem 1.58 and the density of water, the cell mass is

$$m = \rho V = \left(\frac{10^{-3} \text{kg}}{(10^{-2} \text{m})^3} \right) \left(5 \times 10^{-16} \text{m}^3 \right) = 5 \times 10^{-13} \text{kg}$$

The closest answer is (c).

ASSESS   If we multiply this by the number of cells in the body (10^{14}), we get 50 kg, which is roughly what a human weighs.

MOTION IN A STRAIGHT LINE

2

EXERCISES

Section 2.1 Average Motion

13. **INTERPRET** We need to find the average runner speed, and use that to find how long it takes them to run the additional distance.

    DEVELOP The average speed is $\bar{v} = \Delta x / \Delta t$ (Equation 2.1). Looking ahead to part (b), we will express this answer in terms of yards per minute. That means converting miles to yards and hours to minutes. A mile is 1760 yards (see Appendix C). Once we know the average speed, we will use it to determine how long ($\Delta t = \Delta x / \bar{v}$) it would take a top runner to go the extra mile and 385 yards that was added to the marathon in 1908.

    EVALUATE (a) First converting the marathon distance to yards and time to seconds

    $$\Delta x = 26 \text{ mi} \left(\frac{1760 \text{ yd}}{1 \text{ mi}} \right) + 385 \text{ yd} = 46{,}145 \text{ yd}$$

    $$\Delta t = 2 \text{ h} \left(\frac{60 \text{ min}}{1 \text{ h}} \right) + 4 \text{ min} = 124 \text{ min}$$

    Dividing these quantities, the average velocity is $\bar{v} = 372 \text{ yd/min}$.

    (b) The extra mile and 385 yards is equal to 2145 yd. The time to run this is

    $$\Delta t = \frac{\Delta x}{\bar{v}} = \frac{2145 \text{ yd}}{372 \text{ yd/min}} = 5.77 \text{ min} = 5 \text{ min, } 46 \text{ s}$$

    ASSESS The average speed that we calculated is equivalent to about 13 mi/h, which means top runners can run 26 mi marathons in roughly 2 hours. The extra distance is about 5% of the total distance, and correspondingly the extra time is about 5% of the total time, as it should be.

15. **INTERPRET** This problem asks for the time it will take a light signal to reach us from the edge of our solar system.

    DEVELOP The time is just the distance divided by the speed: $\Delta t = \Delta x / v$. The speed of light is $3.00 \times 10^8 \text{ m/s}$ (recall Section 1.2).

    EVALUATE Using the above equation

    $$\Delta t = \frac{\Delta x}{v} = \frac{\left(14 \times 10^9 \text{ mi} \right)}{\left(3.00 \times 10^8 \text{ m/s} \right)} \left(\frac{1609 \text{ m}}{1 \text{ mi}} \right) = 7.5 \times 10^4 \text{ s} = 21 \text{ h}$$

    ASSESS It takes light from the Sun 8.3 minutes to reach Earth. This means that the Voyager spacecraft will be 150 times further from us than the Sun.

17. **INTERPRET** The problem asks for the Earth's speed around the Sun. We'll use the fact that the Earth completes a full revolution in a year.

    DEVELOP The distance the Earth travels is approximately equal to the circumference ($2\pi r$) of a circle with radius equal to $1.5 \times 10^8 \text{ km}$. It takes a year, or roughly $\pi \times 10^7 \text{ s}$, to complete this orbit.

    EVALUATE (a) The average velocity in m/s is

    $$\bar{v} = \frac{2\pi r}{\Delta t} = \frac{2\pi \left(1.5 \times 10^{11} \text{ m} \right)}{\pi \times 10^7 \text{ s}} = 3.0 \times 10^4 \text{ m/s}$$

(b) Using $1609 \text{ m} = 1 \text{ mi}$ gives $\overline{v} = 19 \text{ mi/s}$.

ASSESS   It's interesting that the Earth's orbital speed is $1/10^4$ of the speed of light.

Section 2.2 Instantaneous Velocity

19.   **INTERPRET**   This problem asks us to plot the average and instantaneous velocities from the information in the text regarding the trip from Houston to Des Moines. The problem statement does not give us the times for the intermediate flights, nor the length of the layover in Kansas City, so we will have to assign these values ourselves.

DEVELOP   We can use Equation 2.1, $\overline{v} = \Delta x / \Delta t$, to calculate the average velocities. Furthermore, because each segment of the trip involves a constant velocity, the instantaneous velocity is equivalent to the average velocity, so we can apply Equation 2.1 to these segments also. To calculate the Δ-values, we subtract the initial value from the final value (e.g., for the first segment from Houston to Minneapolis, $\Delta x = x - x_0 = 700 \text{ km} - (-1000 \text{ km}) = 1700 \text{ km}$.

EVALUATE   See the figure below, on which is labeled the coordinates for each point and the velocities for each segment. The average velocity for the overall trip is labeled $\overline{v}$.

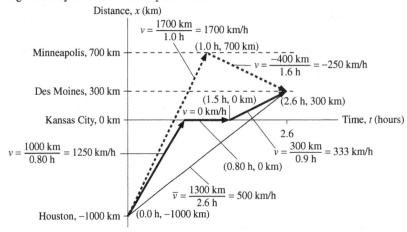

ASSESS   Although none of instantaneous velocities are equivalent to the average velocity, they arrive at the same point as if you traveled at the average velocity for the entire length of the trip.

21.   **INTERPRET**   This problem involves using calculus to express velocity given position as a function of time. We must also understand that zero velocity occurs where the slope (i.e., the derivative) of the plot is zero.

DEVELOP   The instantaneous velocity $v(t)$ can be obtained by taking the derivative of $y(t)$. The derivative of a function of the form bt^n can be obtained by using Equation 2.3.

EVALUATE   (a) The instantaneous velocity as a function of time is

$$v = \frac{dy}{dt} = b - 2ct$$

(b) By using the general expression for velocity, we find that it goes to zero at

$$v = 0 = b - 2ct$$

$$t = \frac{b}{2c} = \frac{82 \text{ m/s}}{4.9 \text{ m/s}^2} = 8.4 \text{ s}$$

ASSESS   From part (a), we see that at $t = 0$, the velocity is 82 m/s. This velocity decreases as time progresses due to the term $-2ct$, until the velocity reverses and the rocket falls back to Earth. Note also that the units for part (b) come out to be s, as expected for a time.

Section 2.3 Acceleration

23.   **INTERPRET**   The object of interest is the subway train that undergoes acceleration from rest, followed by deceleration through braking. The kinematics are one-dimensional, and we are asked to find the average acceleration over the braking period.

DEVELOP   The average acceleration over a time interval Δt is given by Equation 2.4: $\overline{a} = \Delta v / \Delta t$.

EVALUATE   Over a time interval $\Delta t = t_2 - t_1 = 48\,\text{s}$, the velocity of the train (along a straight track) changes from $v_1 = 0$ (starting at rest) to $v_2 = 17\,\text{m/s}$. The change in velocity is thus $\Delta v = v_2 - v_1 = 17\,\text{m/s} - 0.0\,\text{m/s} = 17\,\text{m/s}$. Thus, the average acceleration is

$$\bar{a} = \frac{\Delta v}{\Delta t} = \frac{17\,\text{m/s}}{48\,\text{s}} = 0.35\,\text{m/s}^2$$

ASSESS   We find that the average acceleration only depends on the change of velocity between the starting point and the end point; the intermediate velocity is irrelevant.

25.   **INTERPRET**   For this problem, the motion can be divided into two stages: (i) free fall, and (ii) stopping after striking the ground. We need to find the average acceleration for both stages.

DEVELOP   We chose a coordinate system in which the positive direction is that of the egg's velocity. For stage (i), the initial velocity is $v_1^i = 0.0\,\text{m/s}$, and the final velocity is $v_2^i = 11.0\,\text{m/s}$, so the change in velocity is $\Delta v^i = v_2^i - v_1^i = 11.0\,\text{m/s} - 0.0\,\text{m/s} = 11.0\,\text{m/s}$. The time interval for this stage is $\Delta t^i = 1.12\,\text{s}$. For the second stage, the initial velocity is $v_1^{ii} = 11.0\,\text{m/s}$, the final velocity is $v_2^{ii} = 0.0\,\text{m/s}$, so the change in velocity is $\Delta v^{ii} = v_2^{ii} - v_1^{ii} = 0\,\text{m/s} - 11\,\text{m/s} = -11.0\,\text{m/s}$. The time interval for the second stage is $\Delta t^{ii} = 0.131\,\text{s}$. Insert these values into Equation 2.4, $\bar{a} = \Delta v/\Delta t$, to find the average acceleration for each stage.

EVALUATE   While undergoing free fall (stage i), the average acceleration is

$$\bar{a}^i = \frac{\Delta v^i}{\Delta t^i} = \frac{11.0\,\text{m/s}}{1.12\,\text{s}} = 9.82\,\text{m/s}^2$$

For the stage ii, where the egg breaks on the ground, the average acceleration is

$$\bar{a}^{ii} = \frac{\Delta v^{ii}}{\Delta t^{ii}} = \frac{-11.0\,\text{m/s}}{0.131\,\text{s}} = -84.0\,\text{m/s}^2$$

ASSESS   For stage i, the acceleration is that due to gravity, and is directed downward toward the Earth. It is in the same direction as the velocity so the velocity increases during this stage. For stage ii, the acceleration is in the opposite direction (i.e., upward away from the Earth) so the velocity decreases during this stage.

27.   **INTERPRET**   The object of interest is the car, which we assume undergoes constant acceleration. The kinematics are one-dimensional.

DEVELOP   We first convert the units km/h to m/s, using the conversion factor

$$1\,\text{km/h} = \left(1\frac{\text{km}}{\text{h}}\right)\left(\frac{1000\,\text{m}}{1\,\text{km}}\right)\left(\frac{1\,\text{h}}{3600\,\text{s}}\right) = 0.278\,\text{m/s}$$

and then use Equation 2.4, $\bar{a} = \Delta v/\Delta t$, to find the average acceleration.

EVALUATE   The speed of the car at 16 s is 1000 km/h, or 278 m/s. Therefore, the average acceleration is

$$\bar{a} = \frac{v_2 - v_1}{t_2 - t_1} = \frac{(278\,\text{m/s}) - (0)}{16\,\text{s} - 0\,\text{s}} = 17\,\text{m/s}^2$$

ASSESS   The magnitude of the average acceleration is about $1.8g$, where $g = 9.8\,\text{m/s}^2$ is the gravitational acceleration. An object undergoing free fall attains only a speed of 157 m/s after 16.0 s, compared to 278 m/s for the supersonic car. Given the supersonic nature of the vehicle, the value of a is completely reasonable.

Section 2.4 Constant Acceleration

29.   **INTERPRET**   The problem is designed to establish a connection between the equation for displacement and the equation for velocity in one-dimensional kinematics.

DEVELOP   Recall that the derivative of position with respect to time dx/dt is the instantaneous velocity (see Equation 2.2b, $dx/dt = v$). Thus, by differentiating the displacement $x(t)$ given in Equation 2.10 with respect to t, we obtain the corresponding velocity $v(t)$. We can use Equation 2.3 for evaluating the derivatives.

EVALUATE Differentiating Equation 2.10, we obtain

$$\frac{dx}{dt} = \frac{d}{dt}\left(x_0 + v_0 t + \frac{1}{2}at^2\right) = 0 + v_0 + \frac{1}{2}a \cdot (2t)$$

$$v = v_0 + at$$

which is Equation 2.7. Notice that we have used Equation 2.2b and that we have used the fact that the derivative (i.e., the change in) the initial position x_0 with respect to time is zero, or $dx_0/dt = 0$.

ASSESS Both Equations 2.7 and 2.10 describe one-dimensional kinematics with constant acceleration a, but whereas Equation 2.10 gives the displacement, Equation 2.7 gives the final velocity.

31. **INTERPRET** This is a one-dimensional kinematics problem with constant acceleration. We are asked to find the acceleration and the assent time for a rocket given its speed and the distance it travels.

    DEVELOP The three quantities of interest; displacement, velocity, and acceleration, are related by Equation 2.11, $v^2 = v_0^2 + 2a(x - x_0)$. Solve this equation for acceleration for part (a). Once the acceleration is known, the time elapsed for the ascent can be calculated by using Equation 2.7, $v = v_0 + at$.

    EVALUATE (a) Taking x to indicate the upward direction, we know that $x - x_0 = 85$ km $= 85,000$ m, $v_0 = 0$ (the rocket starts from rest), and $v = 2.8$ km/s $= 2800$ m/s. Therefore, from Equation 2.11, the acceleration is

    $$v^2 = v_0^2 + 2a(x - x_0)$$

    $$a = \frac{v^2 - v_0^2}{2(x - x_0)} = \frac{(2800 \text{ m/s})^2 - (0 \text{ m/s})^2}{2(85,000 \text{ m})} = 46 \text{ m/s}^2$$

    (b) From Equation 2.7, the time of flight is

    $$t = \frac{v - v_0}{a} = \frac{2800 \text{ m/s} - (0 \text{ m/s})}{46 \text{ m/s}^2} = 61 \text{ s}$$

    ASSESS An acceleration of 46 m/s² or approximately 5g ($g = 9.8$ m/s²), is typical for rockets during liftoff. This enables the rocket to reach a speed of 2.8 km/s in just about one minute.

33. **INTERPRET** The object of interest is the car that undergoes constant deceleration (via braking) and comes to a complete stop after traveling a certain distance.

    DEVELOP The three quantities, displacement, velocity, and deceleration (negative acceleration), are related by Equation 2.11, $v^2 = v_0^2 + 2a(x - x_0)$. This is the equation we shall use to solve for a. Since the distance to the light is in feet, we can convert the initial speed

    $$v_0 = 50 \text{ mi/h}\left(\frac{5280 \text{ ft}}{1 \text{ mi}}\right)\left(\frac{1 \text{ h}}{3600 \text{ s}}\right) = 73.3 \text{ ft/s}$$

    EVALUATE Since the car stops ($v = 0$) after traveling $x - x_0 = 100$ ft from an initial speed of $v_0 = 73.3$ ft/s, Equation 2.11 gives

    $$a = \frac{v^2 - v_0^2}{2(x - x_0)} = \frac{0 - (73.3 \text{ ft/s})^2}{2(100 \text{ ft})} = -27 \text{ ft/s}^2$$

    The magnitude of the deceleration is the absolute value of a: $|a| = 27$ ft/s².

    ASSESS With this deceleration, it would take about $t = v_0/|a| = (73 \text{ ft/s})(27 \text{ ft/s}^2) = 2.7$ s for the car to come to a complete stop. The value is in accordance with our driving experience.

35. **INTERPRET** The object of interest is a piece of the fragments of the meteor that undergoes constant deceleration. We are asked to calculate its negative acceleration (i.e., deceleration) upon impacting the Earth, given its final velocity and the distance over which it experienced this negative acceleration.

    DEVELOP The three quantities, displacement, velocity, and deceleration (negative acceleration), are related by Equation 2.11, $v^2 = v_0^2 + 2a(x - x_0)$. Use this equation to solve for the initial speed v of the fragment.

EVALUATE   Consider a particular fragment that followed a straight-line path to the bottom, perpendicular to the desert surface. From Equation 2.11, its initial speed is

$$v_0 = \pm\sqrt{v^2 - 2a(x - x_0)} = \sqrt{0 \text{ m/s} - 2(-4\times10^5 \text{ m/s}^2)(180 \text{ m})} = 1\times10^4 \text{ m/s}$$

where we report the result to a single significant figure because we are given the accleration of the fragment to a single significant figure. The two signs indicate velocity toward or away from the Earth, and we choose the positive sign arbitrarily to indicate the velocity toward the Earth.

ASSESS   With this rate of deceleration, it took only about 0.03 s for the fragment to penetrate 180 m deep into the Earth and come to a complete stop (can you confirm this number?). The impact force that created this gigantic hole was enormous!

Section 2.5 The Acceleration of Gravity

37.   **INTERPRET**   This problem involves constant acceleration due to gravity. We are asked to calculate the distance traveled by the rock before it hit the water.

DEVELOP   We chose a coordinate system where the positive-x axis is downward. We are given the rock's constant acceleration (gravity, $g = 9.8$ m/s^2), its initial velocity $v_0 = 0.0$ m/s, and its travel time t = 4.4 s. Insert this data into Equation 2.10 and solve for the displacement $x - x_0$.

EVALUATE   From Equation 2.10, we find

$$x - x_0 = v_0 t + \frac{1}{2}at^2 = v_0 t + \frac{1}{2}gt^2$$

$$= (0.0 \text{ m/s})(4.4 \text{ s}) + \frac{1}{2}(9.8 \text{ m/s}^2)(4.4 \text{ s})^2 = 95 \text{ m}$$

ASSESS   When the travel time of the sound is ignored, the depth of the well is quadratic in t. The depth of the well is about the length of an American football field. If we use the speed of sound $s = 340$ m/s, how will that change our answer?

39.   **INTERPRET**   The problem involves constant acceleration due to gravity. We are asked to find the maximum altitude reached by a model rocket that is launched upward with the given velocity. In addition, we need to find the speed and altitude at three different times, counting from the launch time.

DEVELOP   We choose a coordinate system in which the upward direction corresponds to the positive-x direction. We are given the initial velocity, $v_0 = 49$ m/s, and we know that the velocity at the peak of the rocket's flight is $v = 0$ m/s, the rocket's acceleration is $a = g = -9.8$ m/s^2 (i.e., it accelerates downward toward the Earth), and its initial position is $x_0 = 0$ m. Equation 2.11, $v^2 = v_0^2 + 2a(x - x_0)$, relates these quantities to the rocket's displacement x. For parts (b), (c), and (d), use Equation 2.7, $v = v_0 + at$, to find the rocket's speed at the different times, and then Equation 2.9, $x - x_0 = (v_0 + v)t/2$, to find its displacement (i.e., altitude).

EVALUATE   (a) At the peak of the rocket's flight, Equation 2.11 gives

$$v^2 = v_0^2 + 2a(x - x_0)$$

$$x = \frac{v^2 - v_0^2}{2a} + x_0 = \frac{(0.0 \text{ m/s})^2 - (49 \text{ m/s})^2}{2(-9.8 \text{ m/s}^2)} + 0.0 \text{ m} = 123 \text{ m}$$

(b) At $t = 1$ s, the speed and the altitude are

$$v = v_0 - gt = 49 \text{ m/s} - (9.8 \text{ m/s}^2)(1 \text{ s}) = 39 \text{ m/s}$$

$$x = x_0 + v_0 t - \frac{1}{2}gt^2 = 0.0 \text{ m/s} + (49 \text{ m/s})(1 \text{ s}) - \frac{1}{2}(9.8 \text{ m/s}^2)(1 \text{ s})^2 = 44 \text{ m}$$

The first quantity (39 m/s) is known to two significant figures because we know the intial velocity to this precision, so subtacting a less-precise quantity from it does not change its precision. The second quantity should be rounded to 40 m because both non-zero terms in Equation 2.9 are known to a single significant figure.

(c) At $t = 1$ s, the speed and the altitude are

$$v = v_0 - gt = 49 \text{ m/s} - (9.8 \text{ m/s}^2)(4 \text{ s}) = 9.8 \text{ m/s}$$

$$x = x_0 + v_0 t - \frac{1}{2}gt^2 = 0.0 \text{ m/s} + (49 \text{ m/s})(4 \text{ s}) - \frac{1}{2}(9.8 \text{ m/s}^2)(4 \text{ s})^2 = 118 \text{ m}$$

Again, we need to round the second result to a single significant figure, which gives 100 m as the final answer.
(d) At $t = 7$ s, the speed and the altitude are

$$v = v_0 - gt = 49 \text{ m/s} - (9.8 \text{ m/s}^2)(7 \text{ s}) = -20 \text{ m/s}$$

$$x = x_0 + v_0 t - \frac{1}{2}gt^2 = 0.0 \text{ m/s} + (49 \text{ m/s})(7 \text{ s}) - \frac{1}{2}(9.8 \text{ m/s}^2)(7 \text{ s})^2 = 103 \text{ m}$$

Again, we need to round the second result to a single significant figure, which gives 100 m as the final answer.
ASSESS   As the rocket moves vertically upward, its velocity decreases due to gravitational acceleration, which is oriented downward. Upon reaching its maximum height, the velocity reduces to zero. It then falls back to Earth with a negative velocity. From **(c)** and **(d)**, we see that the velocities have different signs at $t = 4$ s and $t = 7$ s, so we conclude that the rocket reaches its maximum height between 4 and 7 s. Calculating the time it takes to reach its maximum height using Equation 2.7 gives $t = (v - v_0)/a = (0.0 \text{ m/s} - 49 \text{ m/s})/(-9.8 \text{ m/s}^2) = 5.0 \text{ s}$, in agreement with our expectation.

41.   **INTERPRET**   This problem involves one-dimensional motion under the influence of gravity. We are asked to calculate what initial velocity of the rock is needed so that it is traveling at 3 m/s when it reaches the Frisbee.
DEVELOP   Choose a coordinate system in which the positive-x direction is upward. When the rock hits the Frisbee, its velocity and height are $v = 3$ m/s and $x = 6.5$ m, and the rocks initial position is $x_0 = 1.3$ m. These quantities are related by Equation 2.11:

$$v^2 = v_0^2 + 2a(x - x_0)$$

EVALUATE   Solving this equation for the initial velocity, we obtain

$$v^2 = v_0^2 + 2a(x - x_0)$$

$$v_0 = \pm\sqrt{v^2 - 2a(x - x_0)} = \pm\sqrt{(3 \text{ m/s}) - 2(-9.8 \text{ m/s}^2)(6.5 \text{ m} - 1.3 \text{ m})} = 11 \text{ m}$$

where we have chosen the positive square root because the rock must be travelling upward.
ASSESS   The initial velocity v_0 must be positive since the rock is thrown upward. In addition, v_0 must be greater than the final velocity 3 m/s. These conditions are met by our result.

PROBLEMS

43.   **INTERPRET**   This is a one-dimensional problem involving two travel segments. We are asked to calculate the average velocity for the second segment of the trip.
DEVELOP   The trip can be divided into two time intervals, t_1 and t_2 with $t = t_1 + t_2 = 40$ min $= 2/3$ h. The total distance traveled is $x = x_1 + x_2 = 25$ mi, where x_1 and x_2 are the distances covered in each time interval.
EVALUATE   During the first time interval, $t_1 = 15$ min (or 0.25 h), and with an average speed of $\bar{v}_1 = 20$ mi/h, the distance traveled is

$$x_1 = \bar{v}_1 t_1 = (20 \text{ mi/h})(0.25 \text{ h}) = 5 \text{ mi}$$

Therefore, the remaining distance $x_2 = x - x_1 = 25$ mi $- 5$mi $= 20$ mi must be covered in

$$t_2 = t - t_1 = 40 \text{ min} - 15 \text{ min} = 25 \text{ min} = \frac{5}{12} \text{ h}$$

This implies an average speed of

$$\bar{v}_2 = \frac{x_2}{t_2} = \frac{20 \text{ mi}}{5 \text{ h}/12} = 48 \text{ mi/h}$$

ASSESS The overall average speed was pre-determined to be

$$\bar{v} = \frac{x}{t} = \frac{25 \text{ mi}}{2 \text{ h}/3} = 37.5 \text{ mi/h}$$

When you drive slower during the first segment, you make it up by driving faster during the second. In fact, the overall average speed equals the time-weighted average of the average speeds for the two parts of the trip:

$$\bar{v} = \frac{x}{t} = \frac{x_1 + x_2}{t} = \frac{\bar{v}_1 t_1 + \bar{v}_2 t_2}{t} = \left(\frac{t_1}{t}\right)\bar{v}_1 + \left(\frac{t_2}{t}\right)\bar{v}_2 = \left(\frac{15 \text{ min}}{40 \text{ min}}\right)(20 \text{ mi/h}) + \left(\frac{25 \text{ min}}{40 \text{ min}}\right)(48 \text{ mi/h}) = 37.5 \text{ mi/h}$$

45. **INTERPRET** This is a one-dimensional problem involving a number of evenly divided time intervals. The key concept here is the average velocity.

DEVELOP The average velocity in a given time interval can be found by using Equation 2.1, $\bar{v} = \Delta x / \Delta t$. Note that each time interval consists of 2 h of driving at a velocity of 105 km/h and a 30 min (0.5 h) stop. Therefore, the distance traveled for each segment is $\Delta x = (105 \text{ km/h})(2 \text{ h}) = 210 \text{ km},$ and the time interval for each segment is $\Delta t = 2 \text{ h} + 0.5 \text{ h} = 2.5 \text{ h}$ (where we keep the extra significant figure because these are intermediate results, note that we know Δt to a single significant figure). The ratio of these two quantities is the average speed, and the total distance divided by the average speed is the time it takes for the entire trip.

EVALUATE **(a)** Inserting the quantities calculated above into Equation 2.1, the average velocity is

$$\bar{v} = \frac{\Delta x}{\Delta t} = \frac{210 \text{ km}}{2.5 \text{ h}} = 84 \text{ km/h}$$

which we must round down to 80 km/h because the input (2 h) has only a single significant figure. Because each 2.5 h interval covers the same distance of 210 km, we conclude our result is also the average velocity of the entire trip (but see Assess below).

(b) The amount of time required for this coast-to-coast trip is

$$\Delta t = \frac{\Delta x}{\bar{v}} = \frac{4600 \text{ km}}{84 \text{ km/h}} = 54.8 \text{ h}$$

which we must round down to 50 h, as per part (a).

ASSESS This time is a little over 2 days, which means that our driver likely slept some during the trip, which is not taken into account in our calculation. Will this time spent sleeping increase or decrease the average velocity?

47. **INTERPRET** This is a one-dimensional kinematics problem that asks us to calculate the point at which two jetliners will meet given their starting points and average velocities.

DEVELOP Given the average speed, the distance traveled during a time interval can be calculated using Equation 2.1, $\Delta x = \bar{v} \Delta t$. An important point here is to recognize that at the instant the airplanes pass each other, the sum of the total distance traveled by both airplanes is $\Delta x = 4600 \text{ km.}$

EVALUATE Suppose that the two planes pass each other after a time Δt from take-off. We then have

$$\Delta x = \Delta x_1 + \Delta x_2 = \bar{v}_1 \Delta t + \bar{v}_2 \Delta t = (\bar{v}_1 + \bar{v}_2)\Delta t$$

which yields

$$\Delta t = \frac{\Delta x}{\bar{v}_1 + \bar{v}_2} = \frac{4600 \text{ km}}{1100 \text{ km/h} + 700 \text{ km/h}} = 2.56 \text{ h} \approx 2.6 \text{ h}$$

Thus, the encounter occurs at a point about $\Delta x_1 = \bar{v}_1 \Delta t = (1100 \text{ km/h})(2.56 \text{ h}) = 2811 \text{ km} \approx 2800 \text{ km}$ from San Francisco, or $\Delta x_2 = \bar{v}_2 \Delta t = (700 \text{ km/h})(2.56 \text{ h}) = 1789 \text{ km} \approx 2000 \text{ km}$ from New York. The approximate results are those with the correct number of significant figures.

ASSESS The point of encounter is closer to New York than San Francisco. This makes sense because the plane that leaves from New York travels at a lower speed. If we sum the distances covered by the two airplanes when they encounter, we find $\Delta x = 2811 \text{ km} + 1789 \text{ km} = 4600 \text{ km}$, which is the distance from San Francisco to New York, as expected.

49. **INTERPRET** This is a one-dimensional kinematics problem involving finding the instantaneous velocity as a function of time, given the position as a function of time. We must also show that the average velocity from $t = t_1 = 0$ to any arbitrary time $t = t_2$ is one-fourth of the instantaneous velocity at t_2.

DEVELOP The instantaneous velocity $v(t)$ can be obtained by taking the derivative of $x(t)$. The derivative of a function of the form bt^n can be obtained by using Equation 2.3. The average velocity for any arbitrary time interval $\Delta t = t_2 - t_1$ may be calculated by using Equation 2.1, $\bar{v} = \Delta x/\Delta t$, where Δx is determined by evaluating $x = bt^4$ for the two times t_1 and t_2.

EVALUATE The instantaneous velocity is $v(t) = dx/dt = d/dt\left(bt^4\right) = 4bt^3$. The average velocity over the time interval from $t = 0$ to any time $t > 0$ is

$$\bar{v} = \frac{\Delta x}{\Delta t} = \frac{x(t) - x(0)}{t - 0} = \frac{bt^4}{t} = bt^3$$

which is just ¼ of $v(t)$ from above.

ASSESS Note that $\bar{v}$ is not equal to the average of $v(0)$ and $v(t)$, as stated in Equation 2.8. That is applicable only when acceleration is constant, which is clearly not the case here.

51. **INTERPRET** This is a one-dimensional kinematics problem involving finding the instantaneous velocity, instantaneous acceleration, average velocity, and average acceleration as a function of time, given position as a function of time.

    DEVELOP The instantaneous velocity $v(t)$ can be obtained by taking the derivative of $x(t)$ and by differentiating $v(t)$ with respect to t, we obtain the instantaneous acceleration $a(t)$. The derivative of a function of the form bt^n can be obtained by using Equation 2.3. The average velocity and acceleration may be found using Equations 2.1 and 2.4, respectively.

    EVALUATE (a) Differentiating $x(t)$, we find the instantaneous velocity is

    $$v(t) = \frac{dx}{dt} = \frac{d}{dt}\left(bt^3\right) = 3bt^2 \quad \Rightarrow \quad v(2.5\text{ s}) = 3\left(1.5\text{ m/s}^3\right)\left(2.5\text{ s}\right)^2 = 28.1\text{ m/s}$$

    (b) Differentiating $v(t)$ [or taking the second derivative of $x(t)$], we find the instantaneous acceleration is

    $$a(t) = \frac{dv}{dt} = \frac{d}{dt}\left(3bt^2\right) = 6bt \quad \Rightarrow \quad a(2.5\text{ s}) = 6\left(1.5\text{ m/s}^3\right)\left(2.5\text{ s}\right) = 22.5\text{ m/s}^2$$

    (c) The average velocity during the first 2.5 s is

    $$\bar{v} = \frac{\Delta x}{\Delta t} = \frac{x(2.5\text{ s}) - x(0)}{2.5\text{ s}} = \frac{\left(1.5\text{ m/s}^3\right)\left(2.5\text{ s}\right)^3 - 0}{2.5\text{ s}} = 9.38\text{ m/s}$$

    (d) Similarly, the average acceleration during the first 2.5 s is

    $$\bar{a} = \frac{\Delta v}{\Delta t} = \frac{v(2.5\text{ s}) - v(0)}{2.5\text{ s}} = \frac{28.1\text{ m/s} - 0\text{ m/s}}{2.5\text{ s}} = 11.3\text{ m/s}^2$$

    ASSESS The acceleration in this problem is not constant in time, but varies linearly in t. In a situation where the acceleration a is a constant, the displacement is a quadratic function of time.

53. **INTERPRET** This as a one-dimensional problem involving a car subjected to constant deceleration. We need to relate the car's stopping distance to its stopping time.

    DEVELOP For motion with constant acceleration, the stopping distance and the stopping time are related by Equation 2.9, $x - x_0 = \left(v_0 + v\right)t/2$.

    EVALUATE Let v_0 be the initial velocity and $v = 0$ be the final velocity. Equation 2.9 can then be rewritten as

    $$x - x_0 = \frac{1}{2}\left(v_0 + v\right)t = \frac{1}{2}v_0 t$$

    Thus, we see that the stopping distance, $x - x_0$, is proportional to the stopping time, t, so both are reduced by the same amount (55%).

    ASSESS Anti-lock brakes optimize the deceleration by controlling the wheels so that they roll just at the point of skidding.

55.   **INTERPRET** We interpret this as a one-dimensional kinematics problem with the hockey puck being the object of interest.

DEVELOP We are told that the hockey puck undergoes constant deceleration while moving through the snow. Equation 2.9, $x = x_0 + \frac{1}{2}(v_0 + v)t$, provides the connection between the initial velocity $v_0 = 32$ m/s, the final velocity $v = 18$ m/s, the travel time t, and the distance traveled $x = 0.35$ m. For part **(b)**, we use Equation 11, $v^2 = v_0^2 + 2a(x - x_0)$ to find the acceleration, and then use the same equation again to find the minimum thickness of the snow, x_{min}, needed to stop the hockey puck entirely (v=0).

EVALUATE (a) Solving for the time

$$t = \frac{(x - x_0)}{\frac{1}{2}(v_0 + v)} = \frac{(0.35 \text{ m} - 0)}{\frac{1}{2}(32 \text{ m/s} + 18 \text{ m/s})} = 0.014 \text{ s}$$

(b) First we solve for the acceleration

$$a = \frac{(v^2 - v_0^2)}{2(x - x_0)} = \frac{\left((18 \text{ m/s})^2 - (32 \text{ m/s})^2\right)}{2(0.35 \text{ m} - 0)} = -1000 \text{ m/s}^2$$

Then we plug this back in to the same equation to find the minimum snow thickness for stopping the puck

$$x_{min} = \frac{(v^2 - v_0^2)}{2a} = \frac{\left(0 - (32 \text{ m/s})^2\right)}{2(-1000 \text{ m/s}^2)} = 0.51 \text{ m} = 51 \text{ cm}$$

ASSESS We find the minimum thickness to be proportional to v_0^2 and inversely proportional to the deceleration $-a$. This agrees with our intuition: The greater the speed of the puck, the thicker the snow needed to bring it to a stop; similarly, less snow would be needed with increasing deceleration.

57.   **INTERPRET** This is a one-dimensional kinematics problem. We assume the jetliner slows down on the runway with constant deceleration.

DEVELOP Equation 2.9, $x = x_0 + \frac{1}{2}(v_0 + v)t$, relates distance, initial velocity, and final velocity. The equation can be used to solve for the shortest runway.

EVALUATE With $t = 29$ s $= (29/3600)$ h, and the final velocity v set to zero, Equation 2.9 gives

$$x - x_0 = \frac{1}{2}(v_0 + v)t = \frac{1}{2}(220 \text{ km/h})(29/3600 \text{ h}) = 0.89 \text{ km}$$

ASSESS The length is a bit short compared to the typical minimum landing runway length of about 1.5 km for full-size jetliners.

59.   **INTERPRET** This is a one-dimensional kinematics problem in which we are asked to find the initial velocity of a racing car given its initial velocity, it acceleration, the distance covered, and the time interval.

DEVELOP We chose a coordinate system in which the positive-x direction is in the direction of the car's velocity. We are told that the car undergoes constant acceleration, so we can use the equations from Table 2.1. For part (a), we are given the distance, time, and final velocity, so we can use Equation 2.9, $x - x_0 = (v_0 + v)t/2$, to find the initial velocity. For part (b), find the acceleration of the car and use the result in Equation 2.11, $v^2 = v_0^2 + 2a(x - x_0)$, to solve for the distance travelled.

EVALUATE (a) The distance covered $x - x_0 = 140$ m, the time interval is $t = 3.6$ s, and the final velocity is $v = 53$ m/s. Inserting these quantities into Equation 2.9 and solving for the intial velociy v_0 gives

$$x - x_0 = \frac{1}{2}(v_0 + v)t$$

$$v_0 = \frac{2(x - x_0)}{t} - v = \frac{2(140 \text{ m})}{3.6 \text{ s}} - 53 \text{ m/s} = 24.8 \text{ m/s} = 25 \text{ m/s}$$

(b) From Equation 2.7, we find the acceleration to be

$$a = \frac{v - v_0}{t} = \frac{53 \text{ m/s} - 24.8 \text{ m/s}}{3.6 \text{ s}} = 7.84 \text{ m/s}^2$$

Upon substituting the result into Equation 2.11, the distance traveled starting from rest ($v_0 = 0$) to a velocity $v = 53$ m/s is

$$x - x_0 = \frac{v^2 - v_0^2}{2a} = \frac{(53 \text{ m/s})^2 - 0}{2(7.84 \text{ m/s}^2)} = 180 \text{ m}$$

to two significant figures.

ASSESS   Comparing parts (**a**) and (**b**), the car travels a distance of 179 m *from rest* to the end of the 140-m distance. Using Equation 2.11, we can show that the additional 39 m (=179 m − 140 m) is the distance traveled to bring the car from rest to an initial speed of $v_0 = 24.8$ m/s:

$$x - x_0 = \frac{v_0^2}{2a} = \frac{(24.8 \text{ m/s})^2}{2(7.84 \text{ m/s}^2)} = 39 \text{ m}$$

61.   **INTERPRET**   We interpret this as *two* problems involving one-dimensional kinematics with constant acceleration. We are asked to find the acceleration needed so that the two runners arrive at the finish line simultaneously.

DEVELOP   Calculate the speed of the runner B (the leader) from the distance she's already covered. This gives

$$v^B = \frac{\Delta x}{\Delta t} = \frac{(9 \text{ km} + 0.1 \text{ km})(10^3 \text{ m/km})}{(35 \text{ min})(60 \text{ s/min})} = 4.33 \text{ m/s}$$

The remaining 900 m will take her $t = (900 \text{ m})/4.33 \text{ m/s} = 207.7 \text{ s}$ to cover. The initial speed of the trailing runner A is

$$v_0^A = \frac{\Delta x}{\Delta t} = \frac{9000 \text{ m}}{(35 \text{ min})(60 \text{ s/min})} = 4.29 \text{ m/s}$$

Use these results in Equation 2.10 to find the acceleration needed so that both runners finish at the same time.

EVALUATE   The acceleration needed so that both runners finish simultaneously can be found by inserting the time into Equation 2.10, and solving for the acceleration, which gives

$$x_A = v_0^A t + \frac{1}{2}at^2$$

$$a = \frac{2(x_A - v_0^A t)}{t^2} = \frac{2[1000 \text{ m} - (4.29 \text{ m/s})(207.7 \text{ s})]}{(207.7 \text{ s})^2} = 0.0051 \text{ m/s}^2$$

ASSESS   For runner A to catch up to runner B, he must run faster than the speed at which he was initially running, so his acceleration is positive. When runner A crosses the finish line, his speed is
$v^A = v_0^A + at = 4.29 \text{ m/s} + (0.0051 \text{ m/s}^2)(207.7 \text{ s}) = 5.34 \text{ m/s}$, or an increase of about 25% with respect to his initial speed.

63.   **INTERPRET**   This as a one-dimensional kinematics problem in which we are asked to find the initial velocity of an object given its acceleration due to gravity (on Mars) and its maximum height.

DEVELOP   Choose a coordinate system in which x indicates the upward direction from the surface of Mars, with the origin at the surface (i.e., $x_0 = 0$). Use Equation 2.11, $v^2 = v_0^2 + 2a(x - x_0)$, to describe the vertical motion of the Mars rover Spirit. Because the impact speed is the same as the rebound speed, both are given by v_0 (note that the impact velocity is opposite in sign to the rebound velocity). The spacecraft attains a maximum height of $x = 15$ m when $v = 0$. Note that the gravitational acceleration of Mars is $g_{\text{Mars}} = 3.74 \text{ m/s}^2$ (Appendix E).

EVALUATE   Solving Equation 2.11 with $a = -g_{\text{Mars}} = -3.74 \text{ m/s}^2$, the impact speed is

$$v_0 = \sqrt{v^2 - 2a(x - x_0)} = \sqrt{2g_{\text{Mars}}(x - x_0)} = \sqrt{2(3.74 \text{ m/s}^2)(15 \text{ m})} = 10.59 \text{ m/s} = 11 \text{ m/s}$$

where we have retained two significant figures in the answer.

ASSESS   We find the impact speed to be proportional to $\sqrt{x - x_0}$, which is the squareroot of the rebound height. This agrees with our expectation that the greater the impact speed, the higher the rover will rebound.

67.   **INTERPRET**   This is a one-dimensional kinematics problem that involves finding the vertical distance of an object as a function of time.

DEVELOP   Choose a coordinate system in which the positive-x direction is upward. Equation 2.10, $x(t) = x_0 + v_0 t + at^2/2$, describes the vertical position $x(t)$ of an object falling from x_0 as a function of time. Because the object was dropped from a stationary position, $v_0 = 0$ so $x(t) = x_0 + at^2/2$. Furthermore, we are free to choose the origin of the x axis where we like, so we let $x_0 = 0$, which gives $x(t) = at^2/2$.

Finally, the acceleration is $a = -g = -9.8$ m/s^2, which points downward, so our Equation 2.10 takes the form $x(t) = -gt^2/2$. The problem states that $x(t) - x(t-1) = x(t)/4$, from which we can solve for t, which we can insert into $x(t)$ to find x (i.e., the height from which it was dropped). Notice that x will be negative because the object's final position is below its initial position.

EVALUATE

$$x(t) - x(t-1) = \frac{x(t)}{4}$$

$$-\frac{1}{2}gt^2 - \left[-\frac{1}{2}g(t-1)^2 \right] = -\frac{1}{8}gt^2$$

$$\frac{1}{2}g(1-2t) = -\frac{1}{8}gt^2$$

$$t^2 - 8t + 4 = 0$$

$$t = 4 \pm 2\sqrt{3} \text{ m/s}$$

(We discarded the negative square root because $t > 1$ s.) Inserting this result into $x(t)$ gives

$$x(t) = -\frac{1}{2}gt^2 = -\frac{1}{2}\left(9.8 \text{ m/s}^2\right)\left[\left(4 + 2\sqrt{3}\right)\text{s} \right]^2 = -270 \text{ m}$$

to two significant figures. Thus, the object must be dropped from a height of 273 m.

ASSESS   During a free fall, the vertical distance traveled is proportional to t^2. Therefore, we expect the object to travel a greater distance during the latter time interval. In general, we must also take into consideration air resistance.

67.   **INTERPRET**   This is a gravitational acceleration problem where two balls are dropped at the same time, but they have different initial positions and velocities.

DEVELOP   The first ball starts at a height of $y_{10} = h/2$ and velocity of $v_{10} = 0$. The second ball starts at a height of $y_{20} = h$, but we are asked to find its initial velocity. The goal is to have them hit the ground ($y_1 = y_2 = 0$) at the same time. We'll use Equation 2.10, $y = y_0 + v_0 t - \frac{1}{2}gt^2$, for each ball.

EVALUATE   The time it takes the first ball to reach the ground is

$$t = \sqrt{\frac{-y_{10}}{-\frac{1}{2}g}} = \sqrt{\frac{2y_{10}}{g}} = \sqrt{\frac{h}{g}}$$

This is the same time for the second ball, so we can use this to find its initial velocity:

$$v_{20} = \frac{1}{2}gt - y_{20}/t = \frac{1}{2}\sqrt{hg} - \sqrt{hg} = -\frac{1}{2}\sqrt{hg}$$

ASSESS   The velocity is negative since the second ball has to be thrown downwards to catch up with the first ball.

69.   **INTERPRET**   We interpret this as two problems involving one-dimensional kinematics with constant acceleration due to gravity. We are asked to find the final velocity of two divers given their initial speed, and to find which diver hits the water first and by how much time.

DEVELOP   We choose a coordinate system in which the positive-x direction is upward. Let A be the diver who jumps upward at 1.80 m/s, and B be the one who steps off the platform. The velocity of diver A as he passes B on his way down is $v = -1.80$ m/s, which can be found by inserting $x = x_0$ in Equation 2.11, $v^2 = v_0^2 + 2a(x - x_0)$ with $v_0 = 1.80$ m/s. Thus, the initial velocity of diver A for the remainder of his trajectory is $v_{0,A} = -1.80$ m/s. The initial velocity of diver B is $v_{0,B} = 0.00$ m/s. Applying Equation 2.11 to both divers gives

$$v_A^2 = v_{0,A}^2 - 2g(x - x_0)$$

$$v_B^2 = v_{0,B}^2 - 2g(x - x_0) = -2g(x - x_0)$$

which we can solve to find the speeds at the water. Note that the acceleration is $a = -g$, which points downward. For part (b), use Equation 2.10, $x = x_0 + v_0 t + at^2/2$, to express the vertical position of the divers as a function of time.

EVALUATE   (a) At the water's surface, $x = 0$, and the speeds of the divers are

$$v_A = \pm\sqrt{v_0^2 - 2g(x - x_0)} = \pm\sqrt{(-1.80 \text{ m/s})^2 - 2(9.82 \text{ m/s}^2)(0.00 \text{ m} - 3.00 \text{ m})} = -7.88 \text{ m/s}$$

$$v_B = \pm\sqrt{-2g(x - x_0)} = \pm\sqrt{-2(9.82 \text{ m/s}^2)(0.00 \text{ m} - 3.00 \text{ m})} = -7.67 \text{ m/s}$$

Note that we have chosen the negative square roots for v_A and v_B because the divers are moving downward.

(b) From Equation 2.10, the vertical position of the divers as a function of time is

$$x_A(t) = x_0 + v_0 t + \frac{1}{2}at^2 = (3.00 \text{ m}) + (-1.80 \text{ m/s})t - \frac{1}{2}(9.82 \text{ m/s}^2)t^2$$

$$x_B(t) = x_0 + \frac{1}{2}at^2 = (3.00 \text{ m}) - \frac{1}{2}(9.82 \text{ m/s}^2)t^2$$

The divers hit the water when $x(t) = 0$. Solving the equations above, we find $t_A = 1.61 \text{ s}$ and $t_B = 0.782 \text{ s}$. Note that we've chosen the negative square roots for v_A and v_B because the divers are moving downward; however, we're asked to report their speeds, so our answers shouldn't include the minus signs.

ASSESS   We expect diver A to hit the water first because he has a non-zero initial velocity for the trajectory from the platform to the water.

71.   INTERPRET   This is a one-dimensional kinematics problem involving a spacecraft that undergoes free fall under the influence of the gravitational acceleration of the Moon. We are asked to find the spacecraft's impact speed and the time of its fall given the height from which it falls.

DEVELOP   We choose a coordinate system in which the positive-x direction is downward. Using Equation 2.10, $x = x_0 + v_0 t + at^2/2$, the vertical position of the spacecraft falling from x_0 as a function of time is

$$x(t) = x_0 + v_0 t + \frac{1}{2}at^2 = x_0 + \frac{1}{2}g_{\text{Moon}}t^2$$

because $v_0 = 0$ (the spacecraft falls from a stationary position), and the gravitational acceleration of the Moon is g_{Moon} is downward. Note that when the spacecraft impacts the Moon, it will have fallen $x - x_0 = 12 \text{ m}$. From Appendix E, we find that $g_{\text{Moon}} = 1.62 \text{ m/s}^2$.

EVALUATE   Solving this equation for the time t, we find that the amount of time it takes the spacecraft to drop 12 m from rest is

$$t = \sqrt{\frac{2(x - x_0)}{g_{\text{Moon}}}} = \sqrt{\frac{2(12 \text{ m})}{1.62 \text{ m/s}^2}} = 3.85 \text{ s} \approx 3.9 \text{ m/s}$$

From Equation 2.7, the velocity at impact is $v = v_0 + g_{\text{Moon}}t = 0.00 \text{ m/s} + (1.62 \text{ m/s}^2)(3.85 \text{ s}) = 6.2 \text{ m/s}$.

ASSESS   Our result indicates that t is proportional to $g^{-1/2}$. Therefore, the greater the gravitational acceleration, the less time it takes for the free fall and the higher the velocity at impact. The same fall on the Earth would result in a velocity at impact of $v = (9.8 \text{ m/s}^2)\left[2(12 \text{ m})/(9.8 \text{ m/s}^2)\right]^{1/2} = 15 \text{ m/s}$.

73.   INTERPRET   We're asked to find the relative speed between the two subway trains when they collide. We can interpret this as *two* problems involving one-dimensional kinematics with constant acceleration. The two objects of interest are the two trains.

DEVELOP   Let the fast train be A and the slow train be B. While B maintains a constant speed, A tries to slow down to avoid collision with a constant deceleration. We take the origin $x = 0$ and $t = 0$ at the point where A begins decelerating, with positive x in the direction of motion. Position as a function of time is given by Equation 2.10, $x = x_0 + v_0 t + at^2/2$. We write two versions of this equation, one for x_A and one for x_B. The condition that both trains collide may be expressed as $x_A = x_B$.

EVALUATE   We first rewrite the initial speeds of the trains as

$$v_{0A} = 80 \text{ km/h} = \left(80 \frac{\text{km}}{\text{h}}\right)\left(\frac{1000 \text{ m}}{1 \text{ km}}\right)\left(\frac{1 \text{ h}}{3600 \text{ s}}\right) = 22.22 \text{ m/s}$$

$$v_{0B} = 25 \text{ km/h} = \left(25 \frac{\text{km}}{\text{h}}\right)\left(\frac{1000 \text{ m}}{1 \text{ km}}\right)\left(\frac{1 \text{ h}}{3600 \text{ s}}\right) = 6.94 \text{ m/s}$$

We can express the positions of trains A and B as

$$x_A = v_{0A}t + \frac{1}{2}at^2 = (22.22 \text{ m/s})t + \frac{1}{2}(-2.1 \text{ m/s}^2)t^2$$

$$x_B = x_{B0} + v_{0B}t = 50 \text{ m} + (6.94 \text{ m/s})t$$

When the trains collide, $x_A = x_B$. The above equations then give

$$\frac{1}{2}at^2 + (v_{0A} - v_{0B})t - x_{B0} = 0 \rightarrow (-1.05 \text{ m/s}^2)t^2 + (15.28 \text{ m/s})t - (50 \text{ m}) = 0$$

Using the quadratic formula to solve for the smaller root, we find

$$t = \frac{(15.28 \text{ m/s}) - \sqrt{(15.28 \text{ m/s})^2 - 4(1.05 \text{ m/s}^2)(50 \text{ m})}}{2(1.05 \text{ m/s}^2)} = 4.97 \text{ s}$$

The velocity of train A at the time of the collision is

$$v_A = v_{A0} + a_1 t = (22.22 \text{ m/s}) - (2.1 \text{ m/s}^2)(4.97 \text{ s}) = 11.78 \text{ m/s}$$

Therefore, their relative speed at the collision is

$$v_{rel} = v_A - v_{B0} = 11.78 \text{ m/s} - 6.94 \text{ m/s} = 4.84 \text{ m/s}$$

or 17.4 km/h.

ASSESS   The initial relative speed is $v_{rel,0} = v_{A0} - v_{B0} = 22.22 \text{ m/s} - 6.94 \text{ m/s} = 15.28 \text{ m/s}$. Braking reduces the speed of train A, and the relative speed between A and B, but the deceleration $a = -2.1 \text{ m/s}^2$ is not enough to prevent collision.

75.   **INTERPRET**   This is a one-dimensional kinematics problem involving two travel segments. The key concept here is the average speed.

DEVELOP   The average speed is the total distance divided by the total time, or $\bar{v} = \Delta x / \Delta t$. For both cases, we shall find the total distance traveled and the time taken.

EVALUATE   (a) Let the distances traveled during the two time intervals be L_1 and L_2. The total distance is the sum of the distances covered at each speed:

$$L = L_1 + L_2 = v_1 \left(\frac{t}{2}\right) + v_2 \left(\frac{t}{2}\right) = \frac{1}{2}(v_1 + v_2)t$$

so

$$\bar{v} = \frac{L}{t} = \frac{1}{2}(v_1 + v_2)$$

(b) In this case, let t_1 and t_2 be the two time intervals. The total time is the sum of the times traveled at each speed:

$$t = t_1 + t_2 = \frac{L/2}{v_1} + \frac{L/2}{v_1} = \frac{L}{2}\left(\frac{v_1 + v_2}{v_1 v_2}\right)$$

Therefore, the average speed is

$$\bar{v} = \frac{L}{t} = \frac{2v_1 v_2}{v_1 + v_2}$$

ASSESS   The average speed $\bar{v}$ is the time-weighted average of the separate speeds: $\bar{v} = (t_1/t)v_1 + (t_2/t)v_2$. With this in mind, the result in part (a) may be rewritten as

$$\bar{v} = (1/2)v_1 + (1/2)v_2$$

and for part (b),

$$\bar{v} = \left(\frac{t_1}{t}\right)v_1 + \left(\frac{t_1}{t}\right)v_2 = \left(\frac{v_2}{v_1 + v_2}\right)v_1 + \left(\frac{v_1}{v_1 + v_2}\right)v_2 = \frac{2v_1 v_2}{v_1 + v_2}$$

77.   **INTERPRET**   This as a one-dimensional kinematics problem that involves finding the vertical position of a leaping person as a function of time.

DEVELOP   We choose a coordinate system in which the positive-x direction is upward. Using Equation 2.10, the vertical position of a person as a function of time may be written as (setting $x_0 = 0$)

$$x(t) = x_0 + v_0 t + \frac{1}{2}at^2$$

$$\frac{1}{2}gt^2 - v_0 t + x = 0$$

Note that the acceleration is $a = -g$, which points downward. The quadratic formula gives two times when the leaper passes a particular height:

$$t_{\pm} = \frac{v_0 \pm \sqrt{v_0^2 - 2gx}}{g}$$

The smaller value, t_-, corresponds to the time for going up and the larger value, t_+, corresponds to the time for coming down. Therefore, the time spent above that height is

$$\Delta t(x) = t_+ - t_- = \frac{v_0 + \sqrt{v_0^2 - 2gx}}{g} - \frac{v_0 - \sqrt{v_0^2 - 2gx}}{g} = \frac{2\sqrt{v_0^2 - 2gx}}{g}$$

Using Equation 2.11, $v^2 = v_0^2 + 2a(x - x_0)$, we find that in order to reach a maximum height h, the initial velocity must be $v_0 = \sqrt{2gh}$. Therefore, the above expression for $\Delta t(x)$ may be simplified as

$$\Delta t(x) = \frac{2\sqrt{2g(h - x)}}{g}$$

EVALUATE   The total time spent in the air is the time spent above the ground. Setting $x = 0$, we have

$$\Delta t(0) = \frac{2\sqrt{2gh}}{g} = 2\sqrt{\frac{2h}{g}}$$

Similarly, the time spent in the upper half, above $x = h/2$, is

$$\Delta t(h/2) = \frac{2\sqrt{2g(h/2)}}{g} = 2\sqrt{\frac{h}{g}}$$

Therefore,

$$\frac{\Delta t(h/2)}{\Delta t(0)} = \frac{2\sqrt{h/g}}{2\sqrt{2h/g}} = \frac{1}{\sqrt{2}} = 0.707$$

or 70.7%.

ASSESS   Our result indicates that while in the air, a person spends 70.7% of the time on the upper half of the height. Such a large fraction of time is what gives the illusion of "hanging" almost motionless near the top of the leap.

79.   **INTERPRET**   This is a one-dimensional kinematics problem involving constant deceleration. We are asked to calculate an acceleration given the distance and the initial and final velocities.

DEVELOP   Equation 2.11, $v^2 = v_0^2 + 2a(x - x_0)$ relates the distance traveled to the initial speed, the final speed, and the acceleration. We shall use this equation to solve for the acceleration.

EVALUATE   The motorist has to reduce his speed within $x - x_0 = 0.9$ km from $v_0 = 110$ km/h to $v = 70$ km/h. This requires a constant acceleration of

$$a = \frac{v^2 - v_0^2}{2(x - x_0)} = \frac{(70 \text{ km/h})^2 - (110 \text{ km/h})^2}{2(0.9 \text{ km})}$$
$$= -4000 \text{ km/h}^2 = -1.11 \text{ km} \cdot \text{h}^{-1} \cdot \text{s}^{-1} = -0.3 \text{ m/s}^2$$

ASSESS   The result means that the speed must be decreased by 1.11 km/h in each second. So, in 36 seconds, the speed is decreased from 110 km/h − (1.11 km·h^{-1}·s^{-1})(36 s) = 70 km/h.

81.   **INTERPRET**   This problem considers a car falling through a camera's field of view in a given time duration.

DEVELOP   Let's assume the car starts at rest at the position y_0. Let's also define the top of the field of view as y_1 and the bottom as y_2. As the car falls, it reaches y_1 at time t_1 with velocity v_1, and similarly for y_2. By definition, $y_1 - y_2 = h$ and $t_2 - t_1 = \Delta t$. We are looking for the height the car is released above the top of the field of view, $y_0 - y_1 = H$. We can solve for H using the equations in Table 2.1.

EVALUATE   From Equation 2.11, we have

$$v_1^2 = -2g(y_1 - y_0) \;\rightarrow\; H = \frac{v_1^2}{2g}$$

We need to relate v_1 to the variables we were given: h and Δt. We can do that with Equation 2.10:

$$y_2 = y_1 + v_1(t_2 - t_1) - \tfrac{1}{2}g(t_2 - t_1)^2$$
$$\Rightarrow v_1 = \frac{h}{\Delta t}\left(\frac{g\Delta t^2}{2h} - 1\right)$$

Plugging this into the above equation for H gives us

$$H = \frac{h}{4}\left(\frac{2h}{g\Delta t^2}\right)\left(\frac{g\Delta t^2}{2h} - 1\right)^2$$

ASSESS This problem is actually the same as Problem 2.78, with the car and camera view replacing the balloon and window view. If you substitute the values from that problem ($h = 1.3$ m and $\Delta t = 0.22$ s) into the expression for H, you find the answer comes out right ($H = 1.2$ m).

83. **INTERPRET** This problem involves one-dimensional kinematics under constant acceleration. We are asked to find the frequency with which drops of water hit the sink given the initial conditions.

    DEVELOP There are exactly three drops falling at any time: two partway down and one either hitting the sink or just leaving the faucet. Find the time it takes one drop to fall and divide that by three to get the time between drops. Use Equation 2.10, $x = x_0 + v_0 t + \frac{1}{2}at^2$ with $x = 0$, $x_0 = 19.6$ cm $= 0.196$ m, $v_0 = 0$, and $a = -g = -9.8$ m/s^2. The question asks for drops per second, so convert seconds per drop to drops per second for the final answer.

    EVALUATE From Equation 2.10, the time it takes one drop to fall is

    $$x = x_0 + v_0 t + \frac{1}{2}at^2$$

    $$0 = x_0 - \frac{1}{2}gt^2$$

    $$t = \sqrt{\frac{2x_0}{g}} = \sqrt{\frac{2(0.196\text{ m})}{9.8\text{ m/s}^2}} = 0.20\text{ s}$$

    There are three drops that hit the sink in this time interval, so the time between drops is $(0.20\text{ s})/(3\text{ drops}) = 0.067$ s/drop. Thus, the frequency with which the drops hit the sink is $1/(0.067$ s/drop$) = 15$ drops/s.

    ASSESS This is pretty fast for a leaky faucet, but the time looks about right for the distance involved.

85. **INTERPRET** You are asked to integrate Equation 2.7 in order to derive Equation 2.10.

    DEVELOP Recall the general formula for integrating a polynomial

    $$\int t^n dt = \frac{1}{n+1}t^{n+1} + \text{constant}$$

    EVALUATE Let's integrate Equation 2.7 over the time variable, t', from $t' = 0$ to $t' = t$

    $$\int_0 v\,dt = \int_0 (v_0 + at)\,dt$$

    By definition, the time integral of $v(t)$ is $x(t)$, so the equation transforms to

    $$x(t) - x(0) = \left.\left(v_0 t' + \tfrac{1}{2}at'^2\right)\right|_0^t = v_0 t + \tfrac{1}{2}at^2$$

    Since $x(0) = x_0$ by definition, this is Equation 2.10.

    ASSESS For those who want a challenge, it's also possible to derive Equation 2.11 by integrating $v = dx/dt$ over velocity.

87. **INTERPRET** We're asked to interpret the graph of a tiger's velocity.

    DEVELOP The tiger has zero acceleration when the velocity is not changing, i.e., when the curve is flat.

    EVALUATE The acceleration is zero at points C and F.

    The answer is (c).

    ASSESS The tiger first accelerates to the right, but then at point C it starts to slow down and comes to a stop at point E. She then immediately begins to accelerate to the left, but then at point F it starts to slow down and comes to a stop at point H.

89.   **INTERPRET**   We're asked to interpret the graph of a tiger's velocity.

DEVELOP   The tiger has greatest acceleration at the point in the graph where the velocity is changing the fastest, i.e., where the slope is greatest.

EVALUATE   The slope appears to be the greatest at point D.

The answer is (c).

ASSESS   At point C, the tiger is moving quickly to the right, but it suddenly slows down at point D and comes to a stop at point E.

MOTION IN TWO AND THREE DIMENSIONS

EXERCISES

Section 3.1 Vectors

11. **INTERPRET** This problem involves finding the magnitude and direction of a vector in two dimensions.

DEVELOP In two dimensions, a vector can be written as $\Delta \vec{r} = \Delta r_x \hat{i} + \Delta r_y \hat{j}$, where Δr_x and Δr_y are the x- and y-components of the displacement, respectively, and $\hat{i}$ and $\hat{j}$ are unit vectors in the x- and y-directions, respectively. For this problem, $\Delta r_x = -220$ m, $\Delta r_y = 150$ m, $\hat{i}$ indicates east, and $\hat{j}$ indicates north. The magnitude of $\Delta \vec{r}$ is $\Delta r = \sqrt{(\Delta r_x)^2 + (\Delta r_y)^2}$, and the angle $\Delta \vec{r}$ makes with the $+x$ axis is

$$\theta = \mathrm{atan}\left(\frac{\Delta r_y}{\Delta r_x}\right)$$

EVALUATE The magnitude of the vector is therefore

$$|\Delta \vec{r}| = \sqrt{(-220\text{ m})^2 + (150\text{ m})^2} = 266\text{ m/s} \approx 270\text{ m}$$

to two significant figures. The direction of the vector is

$$\theta = \mathrm{atan}\left(\frac{150\text{ m}}{-220\text{ m}}\right) = 146° \approx 150°$$

with respect to the positive-x axis (and to two significant figures).

ASSESS With the coordinate system we have chosen, this vector lies in the second quadrant.

13. **INTERPRET** This problem involves the addition of two displacement vectors in two dimensions and finding the magnitude and direction of the resultant vector.

DEVELOP Using Equation 3.1, we see that in two dimensions, a vector $\vec{A}$ can be written in unit vector notation as

$$\vec{A} = A_x \hat{i} + A_y \hat{j} = A\left[\cos(\theta_A)\hat{i} + \sin(\theta_A)\hat{j}\right]$$

where $A = \sqrt{A_x^2 + A_y^2}$ and $\theta_A = \mathrm{atan}\left(A_y/A_x\right)$. Similarly, we express a second vector $\vec{B}$ as $\vec{B} = B_x \hat{i} + B_y \hat{j} = B\left[\cos(\theta_B)\hat{i} + \sin(\theta_B)\hat{j}\right]$. The resultant vector $\vec{C}$ is

$$\vec{C} = \vec{A} + \vec{B} = (A_x + B_x)\hat{i} + (A_y + B_y)\hat{j} = \left[A\cos(\theta_A) + B\cos(\theta_B)\right]\hat{i} + \left[A\sin(\theta_A) + B\sin(\theta_B)\right]\hat{j} = C_x \hat{i} + C_y \hat{j}$$

EVALUATE From the problem statement, $A = 360$ km and $\theta_A = 135°$ (see figure below). The first segment of travel can thus be written as

$$\vec{A} = A\left(\cos\theta_A \hat{i} + \sin\theta_A \hat{j}\right) = (360\text{ km})\left[\cos(135°)\hat{i} + \sin(135°)\hat{j}\right] = (-254.6\text{ km})\hat{i} + (254.6\text{ km})\hat{j}$$

Similarly, the second segment of the travel can be expressed as (with $B = 400$ km and $\theta_B = 90°$)

$$\vec{B} = B\left[\cos(\theta_B)\hat{i} + \sin(\theta_B)\hat{j}\right] = (400\text{ km})\hat{j}$$

Thus, the resultant displacement vector is

$$\vec{C} = \vec{A} + \vec{B} = C_x \hat{i} + C_y \hat{j} = (-254.6\text{ km})\hat{i} + (254.6\text{ km} + 400\text{ km})\hat{j} = (-254.6\text{ km})\hat{i} + (654.6\text{ km})\hat{j}$$

The magnitude of $\vec{C}$ is

$$C = \sqrt{C_x^2 + C_y^2} = \sqrt{(-254.6 \text{ km})^2 + (654.6 \text{ km})^2} = 702.4 \text{ km} \approx 700 \text{ km}$$

to two significant figures. Its direction is

$$\theta = \tan^{-1}\left(\frac{C_y}{C_x}\right) = \text{atan}\left(\frac{654.6 \text{ km}}{-254.6 \text{ km}}\right) = -68.75°, \text{ or } 111°$$

We choose the latter solution (110° to two significant figures) because the vector (with $C_x < 0$ and $C_y > 0$) lies in the second quadrant.

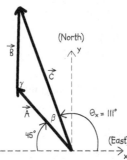

ASSESS   As depicted in the figure, the resultant displacement vector $\vec{C}$ lies in the second quadrant. The direction of $\vec{C}$ can be specified as 111° CCW from the x-axis (east), or $45° + 23.7° = 68.7°$N of W.

15. INTERPRET   We are asked to express a displacement with the unit vectors $\hat{i}$ and $\hat{j}$.

DEVELOP   We're given the magnitude of the vector, $A = 120$ km, and the angle, $\theta = 29°$. The x and y components are given in Equation 3.2:

$$A_x = A\cos\theta = (120 \text{ km})\cos 29° = 105 \text{ km}$$
$$A_y = A\sin\theta = (120 \text{ km})\sin 29° = 58 \text{ km}$$

Since the vector points above the x-axis and to the right of the y-axis, both the x and y components should be positive.

EVALUATE   In terms of unit vectors, the displacement is

$$\vec{A} = A_x\hat{i} + A_y\hat{j} = 105\hat{i} + 58\hat{j} \text{ km}$$

ASSESS   We can verify that indeed this is right by checking that $A = \sqrt{A_x^2 + A_y^2}$ and $\tan\theta = A_y / A_x$ from Equation 3.1.

17. INTERPRET This problem involves finding the magnitude and direction of a given vector.

DEVELOP   Choose a coordinate system where $\hat{i}$ corresponds to the positive-x axis. Use Equations 3.1 to find the magnitude and direction of the given vector $\vec{A}$, whose components are $A_i = 1$ and $A_j = 1$.

EVALUATE   The magnitude of $\vec{A}$ is $A = \sqrt{1^2 + 1^2} = \sqrt{2}$. The direction of $\vec{A}$ with respect to the x axis is $\theta =$ atan(1/1) = 45°.

ASSESS   This vector lies in the first quadrant because both components are positive.

Section 3.2 Velocity and Acceleration Vectors

19. INTERPRET   This problem asks us to find the components of a vector given its magnitude and direction.

DEVELOP   Draw a diagram of the vector (see figure below). Use Equations 3.2, with the magnitude of the vetor being $v = 18$ m/s and the angle $\theta = 220°$.

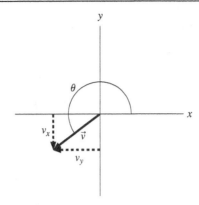

EVALUATE The x component of the vector v is $v_x = v\cos(\theta) = (18 \text{ m/s})\cos(220°) = -14$ m/s and the y component is $v_y = v\sin(\theta) = (18 \text{ m/s})\sin(220°) = -12$ m/s .

ASSESS Note that both components are negative, as expected from the plot.

21. **INTERPRET** We're asked to find the acceleration given the velocity as a function of time.

   DEVELOP Equation 3.6 relates the instantaneous velocity to the instantaneous acceleration: $\vec{a} = d\vec{v}/dt$.

   EVALUATE Taking the derivate of the velocity expression gives

$$\vec{a} = \frac{d}{dt}\left[ct^3\hat{i} + d\hat{j}\right] = 3ct^2\hat{i}$$

   ASSESS The object has a constant velocity in the y-direction, so there is no acceleration in that direction.

23. **INTERPRET** This problem involves calculating the average velocity and average acceleration given the initial and final positions of an object.

   DEVELOP Draw a diagram of the situation (see figure below) to display graphically the difference $\Delta\vec{r} = \vec{r_2} - \vec{r_1}$, where r_1 is the initial position of the tip of the clock hand and r_2 is its final position. Use Equation 3.3 $\bar{\vec{v}} = \Delta\vec{r}/\Delta t$ to find the average velocity and Equation 3.5 $\bar{\vec{a}} = \Delta\vec{v}/\Delta t$ to find the average acceleration. The magnitude of $\vec{r}$ is 2.4 cm, so $\vec{r_1} = (2.4 \text{ cm})\hat{j}$ and $\vec{r_2} = (-2.4 \text{ cm})\hat{j}$. For part (b), note that the tip of the clock moves at a constant speed, which is $v = 2\pi r/(12 \text{ h})$. The direction of the velocity vector at any moment is perpendicular to the clock hand, so $\vec{v_1} = \left[2\pi r/(12 \text{ h})\right]\hat{i}$ and $\vec{v_2} = \left[-2\pi r/(12 \text{ h})\right]\hat{i}$.

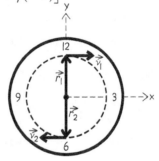

EVALUATE (a) The change in position is $\Delta r = \vec{r_2} - \vec{r_1} = (-2.4 \text{ cm})\hat{j} + (-2.4 \text{ cm})\hat{j} = (-4.8 \text{ cm})\hat{j}$, so the average velocity is $\bar{\vec{v}} = \Delta\vec{r}/\Delta t = (-4.8 \text{ cm})\hat{j}/(6 \text{ h}) = (-0.8 \text{ cm/h})\hat{j} = (-2.2\times10^{-6} \text{ m/s})\hat{j}$. (b) From Equation 3.5, the average acceleration is

$$\bar{\vec{a}} = \frac{\Delta\vec{v}}{\Delta t} = \frac{\vec{v_2} - \vec{v_1}}{\Delta t} = \frac{-4\pi(2.4 \text{ cm})/(12 \text{ h})}{6 \text{ h}}\hat{i} = (-0.42 \text{ cm/h}^2)\hat{i} = (-3.2\times10^{-10} \text{ m/s}^2)\hat{i}$$

ASSESS Notice that the average acceleration is perpendicular to the average velocity, as expected for circular motion (see Figure 3.9).

25. **INTERPRET**   This problem asks us to find the final velocity given the initial velocity, acceleration, and the time interval for the acceleration.

    DEVELOP   Use Equation 3.5, $\vec{a} = \Delta\vec{v}/\Delta t = \left(\vec{v}_2 - \vec{v}_1\right)/\Delta t$ to find the final velocity.

    EVALUATE   Solving Equation 3.5 for $\vec{v}_2$ and inserting the given quantities gives

    $$\vec{v}_2 = \vec{v}_1 + \vec{a}\Delta t = (1.3 \text{ m/s})\hat{i} + \left(0.52 \text{ m/s}^2\right)(4.4 \text{ s})\hat{j} = (1.3 \text{ m/s})\hat{i} + (2.3 \text{ m/s})\hat{j}$$

    The magnitude of the final velocity is $\vec{v}_2 = \sqrt{(1.3 \text{ m/s})^2 + (2.3 \text{ m/s})^2} = 2.6 \text{ m/s}$, and its angle with respect to the positive x axis is $\theta = \text{atan}(2.3 \text{ m/s}/1.3 \text{ m/s}) = 60.4°$.

    ASSESS   Given the components of each vector, we simply added the components together to find the resulting vector.

Section 3.3 Relative Motion

27. **INTERPRET**   This problem involves relative motion. You are asked to find the direction of the velocity with respect to the water so that a boat traverses the current perpendicularly with respect to the shore. You also need to find the time it takes to cross the river.

    DEVELOP   Choose a coordinate system where $\hat{i}$ is the direction perpendicular to the water current, and $\hat{j}$ is the direction of the water current (see figure below). The velocity of the current $\vec{V}$ relative to the ground is $\vec{V} = (0.57 \text{ m/s})\hat{j}$, and the magnitude of the velocity $\vec{v}'$ of the boat relative to the water is $v' = 1.3 \text{ m/s}$. We also known that the velocity $\vec{v}$ of the boat relative to the shore is in the $\hat{i}$ direction, so $\vec{v} = v\hat{i}$, and that the river is $d = 63$ m wide. These three vectors satisfy Equation 3.7, $\vec{v} = \vec{v}' + \vec{V}$, as drawn in the figure below. This allows us to find the direction of $\vec{v}'$ for part (a) and the time it will take to cross the river for part (b).

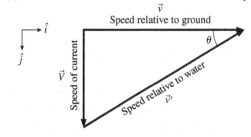

    EVALUATE   **(a)** From the figure above, we see that

    $$\sin(\theta) = V/v'$$
    $$\theta = \text{asin}(V/v') = \text{asin}(0.57 \text{ m/s}/1.3 \text{ m/s}) = 26°$$

    so your heading upstream is 26° above the x axis.

    (b) To find the time to traverse the river, we calculate the speed v with respect to the shore. From the figure above, we see that $v = v'\cos(\theta) = (1.3 \text{ m/s})\cos 26° = 1.17 \text{ m/s}$, so the crossing time t is $t = d/v = (63 \text{ m})/(1.17 \text{ m/s}) = 53.9 \text{ s} = 54 \text{ s}$ to two significant figures.

    ASSESS   In this time, you will have rowed a distance d' relative to the water of $d' = v't = (1.3 \text{ m/s})(53.9 \text{ s}) = 70 \text{ m}$.

29. **INTERPRET**   This problem involves relative velocities. We are asked to calculate the direction at which geese should fly to travel due south given the wind velocity and the bird's air speed.

    DEVELOP   The Equation 3.7, $\vec{v} = \vec{v}' + \vec{V}$ is shown graphically in the figure below. We are given that the magnitude of $\vec{v}'$ is 7.5 m/s, and that the wind velocity is $\vec{V} = (5.1 \text{ m/s})\hat{i}$. This information allows us to calculate the direction θ at which the birds should fly.

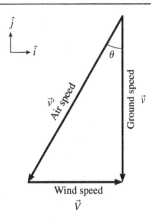

EVALUATE   From trigonometry, we see that $\theta = a\sin(V/v') = a\sin(5.1\ \text{m/s}/7.5\ \text{m/s}) = 43°$, so the geese should fly 43° west of south to travel due south.

ASSESS   The ground speed of the geese is $v = v'\cos(\theta) = (7.5\ \text{m/s})\cos(43°) = 5.5\ \text{m/s}$.

Section 3.4 Constant Acceleration

31.   **INTERPRET**   This problem asks you to determine how far your sailboard goes during a gust of wind that results in a constant acceleration.

DEVELOP   We'll assume the initial velocity is in the positive x direction ($\vec{v}_0 = 6.5\hat{i}$ m/s). The acceleration can be broken up into x and y components as follows:

$$a_x = a\cos\theta = (0.48\ \text{m/s}^2)\cos 35° = 0.393\ \text{m/s}^2$$

$$a_y = a\sin\theta = (0.48\ \text{m/s}^2)\sin 35° = 0.275\ \text{m/s}^2$$

To find the displacement, we can use Equation 2.10 for both the x and y directions.

EVALUATE   The displacement in the x direction is

$$\Delta x = v_0 t + \tfrac{1}{2} a_x t^2 = (6.5\ \text{m/s})(6.3\ \text{s}) + \tfrac{1}{2}(0.393\ \text{m/s}^2)(6.3\ \text{s})^2 = 48.7\ \text{m}$$

The displacement in the y direction is

$$\Delta y = \tfrac{1}{2} a_y t^2 = \tfrac{1}{2}(0.275\ \text{m/s}^2)(6.3\ \text{s})^2 = 5.46\ \text{m}$$

The magnitude and direction of the displacement are

$$\Delta r = \sqrt{\Delta x^2 + \Delta y^2} = \sqrt{(48.7\ \text{m})^2 + (5.46\ \text{m})^2} = 49\ \text{m}$$

$$\theta = \tan^{-1}\left(\frac{\Delta y}{\Delta x}\right) = \tan^{-1}\left(\frac{5.46\ \text{m}}{48.7\ \text{m}}\right) = 6.4°$$

ASSESS   The angle of the displacement is less than that of the acceleration. That makes sense because the initial velocity was along the x axis, and therefore there should be a greater displacement in that direction.

Section 3.5 Projectile Motion

33.   **INTERPRET**   This problem involves an object moving under the influence of gravity near the Earth's surface, so we are dealing with projectile motion. We are given the initial velocity and height of the object, and are asked to find the time of flight and the horizontal distance traveled by the object before it hits the ground.

DEVELOP   Draw a diagram of the situation (see figure below) to define the coordinate system. The difference $y - y_0 = 0\ \text{m} - 8.8\ \text{m} = -8.8\ \text{m}$, and the initial vertical velocity is $v_{y0} = 0$ m/s. Using this information, we can solve Equation 3.13 for t, which we can then insert into Equation 3.12 to find $x - x_0$, given that $v_{x0} = 11$ m/s.

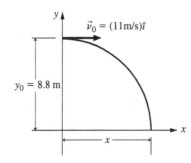

EVALUATE (a) The shingle reaches the ground when

$$y - y_0 = \overbrace{v_{y0}}^{=0} t - \frac{1}{2} g t^2$$

$$t = \sqrt{\frac{2y_0}{g}} = \sqrt{\frac{2(8.8 \text{ m})}{9.8 \text{ m/s}^2}} = 1.34 \text{ s} = 1.3 \text{ s}$$

in which we have retained two significant figures.

(b) The horizontal displacement is $x - x_0 = v_0 t = (11 \text{ m/s})(1.34 \text{ s}) = 15 \text{ m}$ to two significant figures.

ASSESS   The height of 8.8 m indicates that the building in question is likely a two story building, taking into consideration the height of the worker who throws the tiles. The tiles cover a distance of 15.5 m, which is the length of roughly 5 pick-up trucks.

35. **INTERPRET**   This problem involves ink drops moving under the influence of gravity near the surface of the Earth, so it is projectile motion. For this problem, we are given the horizontal distance traveled and the initial horizontal and vertical velocities.

DEVELOP   The horizontal and vertical motions of the drops are independent of each other, so we can consider them separately. From Equation 3.12 we can find the time of flight t, which we can insert into Equation 3.13 to find the vertical displacement (i.e., the distance fallen during the time interval t). This initial conditions are

$x - x_0 = 1.0 \text{ mm} = 1.0 \times 10^{-3} \text{ m}$, $v_{0x} = 12 \text{ m/s}$, $v_{0y} = 0.0 \text{ m/s}$, and $y = 0.0 \text{ m}$.

EVALUATE   From Equation 3.12, the total flight time for an ink drop is

$$t = \frac{x - x_0}{v_{0x}} = \frac{1.0 \times 10^{-3} \text{ m}}{12 \text{ m/s}} = 8.33 \times 10^{-5} \text{ s}$$

Substituting this result into Equation 3.13, we find the distance y_0 that an ink drop falls is

$$y_0 = y + \frac{1}{2} g t^2 = 0.0 \text{ m} + \frac{1}{2}(9.8 \text{ m/s}^2)(8.33 \times 10^{-5} \text{ s})^2 = 3.4 \times 10^{-8} \text{ m} = 34 \text{ nm}$$

to two significant figures.

ASSESS   This distance is an order of magnitude less than the wavelength of visible light, so it is insignificant on the scale of printed, visual matter.

37. **INTERPRET**   This is a problem in projectile motion that asks us to find the horizontal range of a golf ball on the Moon, given the range on Earth.

DEVELOP   Equation 3.15 gives us the horizontal range, which is how far a projectile travels over level ground (i.e., $y = y_0 = 0$):

$$x = \frac{v_0^2}{g} \sin 2\theta_0$$

We assume that the initial velocity and angle are the same on Earth and Moon. The Moon's gravity is 1.62 m/s^2 (Appendix E).

EVALUATE   The range is inversely proportional to the gravity, so

$$x_M = x_E \frac{g_E}{g_M} = (180 \text{ m}) \frac{(9.81 \text{ m/s}^2)}{(1.62 \text{ m/s}^2)} = 1090 \text{ m}$$

ASSESS In 1971 as part of the Apollo 14 mission, the astronaut Alan Shepard hit a golf ball on the Moon, but it didn't travel a kilometer (more like 200-300 m apparently). To Shepard's credit, he was wearing a bulky spacesuit and could only swing with one arm.

Section 3.6 Uniform Circular Motion

39. **INTERPRET** This problem asks us to estimate the acceleration of the Moon given its orbital radius and its orbital period. Because the Moon's orbit is nearly circular, we can use the formulas for uniform circular motion.

    DEVELOP For uniform circular motion, the centripetal (i.e., center-seeking) acceleration is given by Equation 3.16, $a = v^2/r$, where v is the orbital speed and r is the orbital radius. The problem states that $r = 3.85 \times 10^5$ km and that the orbital period T is $T = 27$ days $= 648$ h. The orbital speed is the distance covered in one period divided by the period, or $v = 2\pi r/T$.

    EVALUATE Inserting the given quantities into Equation 3.16, we find

    $$a = \frac{v^2}{r} = \frac{4\pi^2 r}{T^2} = \frac{4\pi^2\left(3.85\times 10^5 \text{ km}\right)}{\left(648 \text{ h}\right)^2} = \left(36 \text{ km/h}^2\right)\left(\frac{10^6 \text{ mm}}{\text{km}}\right)\left(\frac{1 \text{ h}}{3600 \text{ s}}\right)^2 = 2.8 \text{ mm/s}^2$$

    ASSESS The direction of the acceleration is always towards the center of the Earth.

PROBLEMS

41. **INTERPRET** This problem is an exercise in vector addition.

    DEVELOP That the vectors have the same magnitude A and are perpendicular to each other may be expressed mathematically as $\vec{A} = A\hat{i}$ and $\vec{B} = A\hat{j}$. Use Equation 3.1 to find the magnitude of the vector sum.

    EVALUATE (a) $\vec{A} + 2\vec{B} = A\hat{i} + 2A\hat{j}$, so the magnitude is $\left|\vec{A} + 2\vec{B}\right| = \sqrt{A^2 + \left(2A\right)^2} = A\sqrt{5}$. (b) $3\vec{A} - \vec{B} = 3A\hat{i} - A\hat{j}$

    so the magnitude is $\left|3\vec{A} - \vec{B}\right| = \sqrt{\left(3A\right)^2 + \left(-A\right)^2} = A\sqrt{10}$.

    ASSESS The formula for the vector magnitude is simply the Pythagorean Theorem applied to the orthogonal (i.e., perpendicular) components.

43. **INTERPRET** This problem is an exercise in vector addition.

    DEVELOP The vectors are given in component form, so we can perform the indicated algebraic operations on each component individually. The components of $\vec{C}$ are the opposite of the sum of the components of $\vec{A}$ and $\vec{B}$:
    $\vec{C} = -\left(\vec{A} + \vec{B}\right)$.

    EVALUATE Using $\vec{A} = 15\hat{i} - 40\hat{j}$ and $\vec{B} = 31\hat{j} + 18\hat{k}$, and writing the above vector equation in component form gives the components of $\vec{C}$ as

    $$C_i = -15\hat{i}$$
    $$C_j = -\left(-40\hat{j}\right) - 31\hat{j} = 9\hat{j}$$
    $$C_k = -18\hat{k}$$

    so the complete vector is $\vec{C} = -15\hat{i} + 9\hat{j} - 18\hat{k}$.

    ASSESS The magnitude of this vector is $C = \sqrt{\left(-15\right)^2 + \left(9\right)^2 + \left(-18\right)^2} = 25$. This vector is three dimensional, so the angle that its projection onto the x-y plane makes with the positive-x axis is $\theta = \operatorname{atan}(9/-15) = 149°$. The angle that $\vec{C}$ makes with the positive-z axis is

    $$\phi = 180° - \operatorname{atan}\left(\frac{\sqrt{\left(-15\right)^2 + \left(9\right)^2}}{18}\right) = 136°$$

45. **INTERPRET**   We are asked to find when the particle is moving in a particular direction. For this, we will need to derive an expression for the particle's velocity.

DEVELOP   The velocity can be found from Equation 3.4: $\vec{v} = d\vec{r}/dt$. The particle will be moving in the x direction when $v_y = 0$, and it will be moving in the y direction when $v_x = 0$.

EVALUATE   First taking the derivative of the position vector with respect to time:

$$\vec{v} = \frac{d}{dt}\left[\left(ct^2 - 2dt^3\right)\hat{i} + \left(2ct^2 - dt^3\right)\hat{j}\right] = \left(2ct - 6dt^2\right)\hat{i} + \left(4ct - 3dt^2\right)\hat{j}$$

(a) To move in the x direction,

$$v_y = 4ct - 3dt^2 = 0 \ \rightarrow \ t = 4c/3d$$

Note that $t = 0$ is also a solution, but at that time $v_x = 0$, so the particle is not moving at all then.

(b) To move in the y direction,

$$v_x = 2ct - 6dt^2 = 0 \ \rightarrow \ t = c/3d$$

ASSESS   We can plug these times back into the equation for the velocity vector:

$$\vec{v}(c/3d) = \left[4c\left(\frac{c}{3d}\right) - 3d\left(\frac{c}{3d}\right)^2\right]\hat{j} = \frac{c^2}{d}\hat{j}$$

$$\vec{v}(4c/3d) = \left[2c\left(\frac{4c}{3d}\right) - 6d\left(\frac{4c}{3d}\right)^2\right]\hat{i} = -8\frac{c^2}{d}\hat{i}$$

So the particle starts at rest; sometime later it is traveling upwards (+y direction), and then even later it is moving to the left (–x direction). As t gets very large, the particle moves in the direction:

$$\theta = \tan^{-1}\left(\frac{v_y}{v_x}\right) \approx \tan^{-1}\left(\frac{-3dt^2}{-6dt^2}\right) = \tan^{-1}(0.5) = 207°$$

Where we have used the fact that t^2 grows faster than t.

47. **INTERPRET**   This problem involves finding the average acceleration given the initial velocity, the final velocity, and the stopping time.

DEVELOP   Use Equation 3.5, $\vec{a} = \Delta\vec{v}/\Delta t$, to calculate the average acceleration, given that the initial velocity is $\vec{v}_0 = (80 \text{ km}/h)\hat{i} = (22.22 \text{ m/s})\hat{i}$, the final velocity is $\vec{v} = 0$, and the stopping time is $\Delta t = 3.9$ s.

EVALUATE   To reduce the initial velocity to zero in $\Delta t = 3.9$ s, the acceleration must be
$\vec{a} = \Delta\vec{v}/\Delta t = (\vec{v} - \vec{v}_0)/\Delta t = (-22.22 \text{ m/s})\hat{i}/(3.9 \text{ s}) = (-5.7 \text{ m/s}^2)\hat{i}$.

ASSESS   The *average* acceleration only depends on the change of velocity between the starting point and the end point. Note that that although the car is accelerated in the y direction, it is also decelerated by an equal amount in the y direction (since it comes to a stop), so the average acceleration in the y direction is zero.

49. **INTERPRET**   We're asked to find the magnitudes of the average velocity and acceleration of the Ferris wheel over a specified interval. Averages involve differences between final and initial quantities over the interval.

DEVELOP   Let's choose a coordinate system with origin at the center of the Ferris wheel, so that the position vector always has magnitude $r = \frac{1}{2}150 \text{ m} = 75$ m. The speed is the circumference divided by the rotational period: $v = 2\pi \cdot 75 \text{ m}/30 \text{ min} = 0.262$ m/s. We can take the initial position anywhere; although that will affect the vector values of average velocity and acceleration, it won't affect their magnitudes. So let's take the initial position to be at the lowest point, i.e., $\vec{r}_0 = -75\hat{j}$ m, and we'll assume the wheel moves counterclockwise, so $\vec{v}_0 = 0.262\hat{i}$ m/s. After $\Delta t = 5.0$ min, the wheel will have completed 1/6th of its rotation, meaning it will have advanced by 60°. The final position will be –30° from the x direction, while the final velocity will be 60° from the x direction. See the figure below.

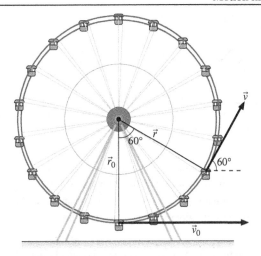

In component form, the final position and velocity are

$$\vec{r} = r\cos(-30°)\hat{i} + r\sin(-30°)\hat{j} = 65.0\hat{i} - 37.5\hat{j} \text{ m}$$

$$\vec{v} = v\cos(60°)\hat{i} + v\sin(60°)\hat{j} = 0.131\hat{i} + 0.227\hat{j} \text{ m/s}$$

EVALUATE (a) The average velocity is change in position divided by the time:

$$\overline{\vec{v}} = \frac{\Delta\vec{r}}{\Delta t} = \frac{(65.0\hat{i} - 37.5\hat{j} \text{ m}) - (-75\hat{j} \text{ m})}{5.0 \text{ min}} = 0.217\hat{i} + 0.125\hat{j} \text{ m/s}$$

We're asked for the magnitude, which is $|\overline{\vec{v}}| = \sqrt{(0.217 \text{ m/s})^2 + (0.125 \text{ m/s})^2} = 0.25 \text{ m/s}$

(b) The average acceleration is change in velocity divided by the time:

$$\overline{\vec{a}} = \frac{\Delta\vec{v}}{\Delta t} = \frac{(0.131\hat{i} + 0.227\hat{j} \text{ m/s}) - (0.262\hat{i} \text{ m/s})}{5.0 \text{ min}} = (-4.37\hat{i} + 7.57\hat{j})\times 10^{-4} \text{ m/s}^2$$

giving magnitude $|\overline{\vec{a}}| = \sqrt{(-4.37\times 10^{-4} \text{ m/s}^2)^2 + (7.57\times 10^{-4} \text{ m/s}^2)^2} = 8.7 \text{ m/s}^2$

ASSESS The magnitude of the average velocity is nearly the same as the instantaneous velocity of 0.26 m/s. The average velocity is smaller because it doesn't take into account the curved path followed by a point on the rim of the wheel. Similarly, the average is close to the instantaneous acceleration $v^2/r = 9.1\times 10^{-4} \text{ m/s}^2$.

51. **INTERPRET** This problem is an exercise in vector addition. We are asked to compare the magnitude of two vectors given that their sum is perpendicular to their difference.

DEVELOP Use Equation 3.1 to express vector $\vec{A}$ in component form: $\vec{A} = A_x\hat{i} + A_y\hat{j}$ where $A = \sqrt{A_x^2 + A_y^2}$ and $\theta_A = \text{atan}(A_y/A_x)$. Similarly, express vector $\vec{B}$ as $\vec{B} = B_x\hat{i} + B_y\hat{j}$. Let $\vec{C}$ be the sum of the two vectors:

$$\vec{C} = \vec{A} + \vec{B} = (A_x + B_x)\hat{i} + (A_y + B_y)\hat{j} = C_x\hat{i} + C_y\hat{j}$$

and $\vec{D}$ be the difference of the two vectors:

$$\vec{D} = \vec{A} - \vec{B} = (A_x - B_x)\hat{i} + (A_y - B_y)\hat{j} = D_x\hat{i} + D_y\hat{j}.$$

If $\vec{C}$ and $\vec{D}$ are to be perpendicular to each other, then we can let them define our coordinate system, so $\vec{C} = C_x\hat{i}$ and $\vec{D} = D_y\hat{j}$.

EVALUATE The conditions set for $\vec{C}$ and $\vec{D}$ imply $C_y = 0$ and $D_x = 0$, or

$$C_y = A_y + B_y = 0 \quad \Rightarrow \quad A_y = -B_y$$
$$D_x = A_x - B_x = 0 \quad \Rightarrow \quad A_x = B_x$$

Using these results, we can show that the magnitudes of $\vec{A}$ and $\vec{B}$ are equal as follows:

$$A = \sqrt{A_x^2 + A_y^2} = \sqrt{B_x^2 + (-B_y)^2} = \sqrt{B_x^2 + B_y^2} = B$$

ASSESS An alternative way to establish the equality between A and B is to note that the vectors $\vec{A} + \vec{B}$ and $\vec{A} - \vec{B}$ form the two diagonals of a parallelogram formed by sides $\vec{A}$ and $\vec{B}$ (see figure below). If the diagonals are perpendicular, the parallelogram is a rhombus, so $A = B$.

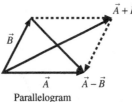 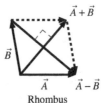

Parallelogram   Rhombus

53. **INTERPRET** This problem involves motion in two dimensions. We are asked to find the acceleration of an object given its initial velocities and the distance it moves in a given time interval.

DEVELOP Note that the displacements in the x and y directions are independent of each other. The x component of the displacement is due to the initial velocity, $\Delta \vec{x} = \vec{v}_0 \Delta t$, and the y component is due to the acceleration, $\Delta \vec{y} = \vec{a} \Delta t^2 / 2$.

EVALUATE The condition that the object moves an equal distance in both directions can be expressed mathematically as $\Delta x = \Delta y$ when $\Delta t = 18 \text{ s}$. This gives

$$v_0 \Delta t = \frac{1}{2} a \Delta t^2$$

$$a = \frac{2v_0}{\Delta t} = \frac{2(4.5 \text{ m/s})}{18 \text{ s}} = 0.50 \text{ m/s}^2$$

ASSESS The answer can be checked by substituting the value of a into Δy:

$$\Delta y = \frac{1}{2} a \Delta t^2 = \frac{1}{2} (0.5 \text{ m/s}^2)(18 \text{ s})^2 = 81 \text{ m}$$

This is equal to $\Delta x = v_0 \Delta t = (4.5 \text{ m/s})(18 \text{ s}) = 81 \text{ m}$.

55. **INTERPRET** This problem involves combined horizontal and vertical motion due to the gravity near the Earth's surface, so it is projectile motion. More specifically, it asks for the initial position of the water, so it involves the trajectory of the water.

DEVELOP We are given the initial height of the water, $y_0 = 1.6$ m, the final height, $y = 0.93$ m, and the range $x = 2.1$ m. In addition, the problem states that the water is fired horizontally, so the initial angle at which the water is fired is $\theta_0 = 0°$. These quantities are related by the trajectory Equation 3.14,

$$y - y_0 = x \tan(\theta_0) - \frac{g}{2v_0^2 \cos^2(\theta_0)} x^2$$

which we can solve to find the initial speed v_0.

EVALUATE With $\theta_0 = 0$, the above equation simplifies to $y - y_0 = -gx^2 / (2v_0^2)$, which gives

$$v_0 = x \sqrt{\frac{-g}{2(y - y_0)}} = (2.1 \text{ m}) \sqrt{\frac{(-9.8 \text{ m/s}^2)}{2(0.93 \text{ m} - 1.6 \text{ m})}} = 5.7 \text{ m/s}$$

ASSESS We can check the answer by solving the problem in a different way. Because the water was fired horizontally $v_{0y} = 0$, so the time it takes to fall from $y_0 = 1.6$ m to $y - 0.93$ m is given by Equation 3.13:

$$t = \sqrt{2(y_0 - y)/g} = \sqrt{2(1.6 \text{ m} - 0.93 \text{ m})/(9.8 \text{ m/s}^2)} = 0.370 \text{ s}$$

Its initial speed, $v_0 = v_0 x$, can be found from Equation 3.12:

$$v_0 = (x - x_0)/t = (2.1 \text{ m})/(0.370 \text{ s}) = 5.7 \text{ m/s}$$

Both approaches lead to the same answer for v_0.

57.   **INTERPRET**   You are asked to calculate the kinematics of two projectiles (the balls) with different starting points.
DEVELOP   Let's assume you are at the origin. Your friend's direction is $45°$ above the horizontal from you, which means your friend is located at the same distance from you in the vertical and horizontal directions, i.e., $y_f = h$ and $x_f = h$. Your ball (ball 1) has an initial velocity of:

$$\vec{v}_0 = v_0 \cos 45° \hat{i} + v_0 \sin 45° \hat{j} = \tfrac{1}{\sqrt{2}} v_0 \left(\hat{i} + \hat{j} \right)$$

This ball therefore follows a trajectory of:

$$x_1 = \tfrac{1}{\sqrt{2}} v_0 t$$

$$y_1 = \tfrac{1}{\sqrt{2}} v_0 t - \tfrac{1}{2} g t^2$$

Your friend's ball (ball 2) has no initial velocity, so it falls straight down from its initial position

$$x_2 = h$$

$$y_2 = h - \tfrac{1}{2} g t^2$$

To verify whether or not the balls collide, we need to first find when ball 1 to reaches the horizontal position of ball 2, i.e. when $x_1 = x_2$. Once we have this time we can compare the vertical position of each ball.
EVALUATE   (a) Solving for the time when $x_1 = x_2$ gives

$$t = \frac{x_1}{\tfrac{1}{\sqrt{2}} v_0} = \frac{\sqrt{2} \cdot h}{v_0}$$

Plugging this time in for the vertical position of each ball gives

$$y_1 = \frac{v_0}{\sqrt{2}} \left(\frac{\sqrt{2}h}{v_0} \right) - \tfrac{1}{2} g \left(\frac{\sqrt{2}h}{v_0} \right)^2 = h - \frac{gh^2}{v_0^2}$$

$$y_2 = h - \tfrac{1}{2} g \left(\frac{\sqrt{2}h}{v_0} \right)^2 = h - \frac{gh^2}{v_0^2}$$

The fact that $y_1 = y_2$ implies a collision. But the assumption is that ball 1 has enough initial velocity to reach the horizontal position of ball 2 before hitting the ground. We will now derive an expression for this minimum velocity.
(b) For the collision to occur, ball 1 has to still be in the air ($y_1 \geq 0$) when it reaches the horizontal position of ball 2:

$$y_1 = h - \frac{gh^2}{v_0^2} \geq 0 \quad \rightarrow \quad v_0 \geq \sqrt{gh}$$

ASSESS   Another way to derive this is by saying that ball 1's horizontal range: $x = v_0^2 \sin 2\theta_0 / g$ (Equation 3.15) has to be large enough to reach the horizontal position of ball 2.

59.   **INTERPRET**   This problem involves motion under the influence of gravity near the Earth's surface, where we are only interested the velocity of an object. Thus, we will use the equations of projectile motion.
DEVELOP   Draw a diagram of the situation (see figure below). Use Equation 3.13, $y = y_0 + v_{y0}t - gt^2/2$, to find the initial velocity in the y direction, with $y - y_0 = 1.5 - 4.2$ m $= -27$ m. This result may be used to find the time of flight to the window sill by inserting it into Equation 3.11, $v_y = v_{y0} - gt$, with $v_y = 0$ m/s because the package presumably attains its maximum height at the window sill. The time t may be inserted into Equation 3.12, $x = x_0 + v_{x0}t$ with $x - x_0 = 3.0$ m.

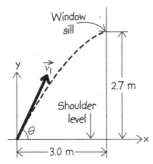

EVALUATE   Solving Equation 3.13 for the initial velocity in the y direction gives

$$v_{0y} = \sqrt{2(9.8 \text{ m/s}^2)(2.7 \text{ m})} = 7.27 \text{ m/s}$$

where we retain more significant figures than warranted because this is an intermediate result. From Equation 3.11, the time of flight to the window sill is

$$t = \frac{v_y - v_{y0}}{g} = \frac{0 \text{ m/s} - 7.27 \text{ m/s}}{9.8 \text{ m/s}^2} = 0.7423 \text{ s}$$

where we have again retained excess significant figures. The initial velocity in the x direction is

$$v_{0x} = \frac{x}{t} = \frac{3.0 \text{ m}}{0.742 \text{ s}} = 4.041 \text{ m/s}$$

From these components, we find the magnitude and direction of $\vec{v}_0$ are

$$v_0 = \sqrt{v_{0x}^2 + v_{0y}^2} = \sqrt{(4.041 \text{ m/s})^2 + (7.275 \text{ m/s})^2} = 8.3 \text{ m/s}$$

$$\theta_0 = \operatorname{atan}\left(\frac{v_{0y}}{v_{0x}}\right) = \operatorname{atan}\left(\frac{7.275}{4.041}\right) = 61°$$

above the x axis. Notice that we have retained only two significant figures in the final result.

ASSESS   We can check the answer by inserting the result for the angle into Equation 3.14. At the window sill, Equation 3.14 gives

$$y = x\tan(\theta_0) - \frac{g}{2v_0^2 \cos^2(\theta_0)}x^2 = (3.0 \text{ m})\tan(60.9°) - \frac{(9.8 \text{ m/s}^2)(3.0 \text{ m})^2}{2(8.32 \text{ m/s})^2 \cos^2(60.9°)} = 2.7 \text{ m}$$

which agrees with the problem statement. Notice that although we evaluated Equation 3.14 for $x = 3.0$ m (i.e., at the window sill), we still used the initial angle θ_0. This is because the initial angle is a constant, whereas the quantities x and y vary along the trajectory.

61.   **INTERPRET**   We are asked to derive a mathematical relation between the time of flight of two projectiles.

DEVELOP   In general, the time of flight over level ground is $t = 2v_{y0}/g$, where we have used Equation 3.13 with $y = y_0$. The initial velocities of the two projectiles have the same magnitude but point at different angles, so the vertical components of the velocities can be written as:

$$v_{y0}^+ = v\sin(45° + \alpha) \; ; \; v_{y0}^- = v\sin(45° - \alpha)$$

To derive the final equation, we will need some of the trigonometric identities from Appendix A.

EVALUATE   Using the vertical components of the initial velocities, the time of flight ratio is

$$\frac{t^+}{t^-} = \frac{v_{y0}^+}{v_{y0}^-} = \frac{\sin(45° + \alpha)}{\sin(45° - \alpha)}$$

Using the equations for $\sin(\alpha - \beta)$ and $\cos(\alpha + \beta)$ from Appendix A and the fact that $\sin 45° = \cos 45°$, we can rewrite the denominator from above as:

$$\sin(45° - \alpha) = \sin 45° \cos\alpha - \cos 45° \sin\alpha$$

$$= \cos 45° \cos\alpha - \sin 45° \sin\alpha$$

$$= \cos(45° + \alpha)$$

Plugging this back into the time of flight ratio equation gives

$$\frac{t^+}{t^-} = \frac{\sin(45° + \alpha)}{\cos(45° + \alpha)} = \tan(45° + \alpha)$$

ASSESS   We could just as easily have derived $\cot(45° - \alpha)$, where the cotangent is equal to 1/tan.

63.   **INTERPRET**   This problem involves projectile motion. We are asked to prove that a projectile launched on level ground reaches its maximum height midway along its trajectory.

DEVELOP   The total flight time can be found by using Equation 3.13, $y = y_0 + v_{0y}t - \frac{1}{2}gt^2$ and setting $y = y_0$ because the problem states that we are on level ground. Similarly, to find the time it takes for the projectile to reach its maximum height y_{max} note that $v_y = 0$ at y_{max} and apply Equation 3.11, $v_y = v_{y0} - gt$.

EVALUATE   From Equation 3.13, the total flight time is

$$y - y_0 = 0 = v_{0y}t_{tot} - \frac{1}{2}gt_{tot}^2$$

$$t_{tot} = \frac{2v_{0y}}{g}$$

Solving Equation 3.11 for the time t' it takes to reach y_{max}, we obtain

$$v_y = 0 = v_{0y} - gt'$$

$$t' = \frac{v_{0y}}{g}$$

Comparing the two expressions, we see that $t_{tot} = 2t'$. Thus, a projectile launched on level ground reaches its maximum height at a time that is one-half the total trajectory time, or at midway along its trajectory.

ASSESS   The result shows that the time of ascent is equal to the time of descent, as expected. An alternative proof may be done by differentiating Equation 3.14 with respect to x:

$$\frac{dy}{dx} = \frac{d}{dx}\left(x\tan(\theta_0) - \frac{g}{2v_0^2\cos^2(\theta_0)}x^2\right) = \tan(\theta_0) - \frac{g}{v_0^2\cos^2(\theta_0)}x$$

Because the maximum height (y_{max}) occurs when this derivative is zero, we find that y_{max} occurs when $x = v_0^2\cos^2(\theta_0)\tan(\theta_0)/g = v_0^2\sin(2\theta_0)/g$. This result is precisely half the x range (see Equation 3.15).

65.   **INTERPRET**   This problem involves projectile motion. You are asked to estimate the initial horizontal speed of the motorcyclist given the range over which he flew.

DEVELOP   Imagine the motorcyclist is traveling at the legal speed, 60 km/h = 16.67 m/s. If we find that his range is less than the 39 m reported, we can conclude that he was probably not speeding. If his range is greater than 39 m, then he was probably speeding. Assume that he is deflected upwards off the car's windshield (which we consider to be a frictionless surface), at 45°, which will maximize his range. We can then use Equation 3.15 to find the range over which he would travel before landing on the road.

EVALUATE   Inserting the initial speed and angle into Equation 3.15 gives

$$x = \frac{v_0^2}{g}\sin(2\theta_0) = \frac{(16.67 \text{ m/s})^2\sin(90°)}{9.8 \text{ m/s}^2} = 28 \text{ m}$$

Because of our assumptions, this would be the motorcyclist's maximum range. The fact that he flew 39 m before landing implies that he was almost certainly speeding.

ASSESS   To estimate the minimum speed at which he was traveling, insert the range of $x = 39$ m into Equation 3.15 and solve for the initial velocity v_0 (again assuming $\theta_0 = 45°$). This gives

$$x = \frac{v_0^2}{g}\sin(2\theta_0)$$

$$v_0 = \pm\sqrt{\frac{xg}{\sin(2\theta_0)}} = \pm\sqrt{\frac{(39 \text{ m})(9.8 \text{ m/s}^2)}{\sin(90°)}} = (19.56 \text{ m/s})\left(\frac{1 \text{ km}}{10^3 \text{ m}}\right)\left(\frac{3600 \text{ s}}{1 \text{ h}}\right) = 70 \text{ km/h}$$

67.   **INTERPRET**   This problem involves projectile motion. We are asked to calculate the angle at which to aim a basketball to score a basket given the horizontal distance to the basket, the launch speed of the basketball, and the initial height difference between the basketball and the basket.

DEVELOP   Draw a diagram of the situation (see figure below). The initial height of the ball is $y_0 = 8.2$ ft, the final height is $y = 10$ fs, the initial speed is $v_0 = 26$ ft/s, and the range is $x = 15$ ft. These quantities are related by the trajectory Equation 3.14

$$y - y_0 = x \tan(\theta_0) - \frac{g}{2v_0^2 \cos^2(\theta_0)} x^2$$

which we shall solve to find the launch angle θ_0. We will also need the acceleration due to gravity in ft/s^2, which we calculate (using the conversion factor from Appendix C) to be $g = (9.8$ m/s$^2)(1$ ft/0.3048 m$) = 32$ ft/s^2.

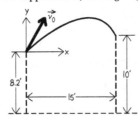

EVALUATE   With origin at the point from which the ball is thrown, the equation of the trajectory, evaluated at the basket, becomes

$$y - y_0 = (10 \text{ ft} - 8.2 \text{ ft}) = (15 \text{ ft}) \tan(\theta_0) - \frac{(32 \text{ ft/s}^2)(15 \text{ ft})^2}{2(26 \text{ ft/s})^2 \cos^2(\theta_0)}$$

$$1.8 \text{ ft} = (15 \text{ ft}) \tan(\theta_0) - \frac{533 \text{ ft}}{\cos^2(\theta_0)}$$

Using the trigonometric identity $1 + \tan^2(\theta_0) = 1 = \cos^2(\theta_0)$, convert this equation into a quadratic in $\tan(\theta_0)$. The result is

$$7.13 - 15 \tan(\theta_0) + 5.33 \tan^2(\theta_0) = 0$$

where we have divided out the units of ft, leaving us with a dimensionless equation. The answers are

$$\theta_0 = \text{atan} \left[\frac{15 \pm \sqrt{15^2 - 4(5.33)(7.13)}}{2(5.33)} \right] = 31.2° \text{ or } 65.7°$$

ASSESS   Like the horizontal range formula for given v_0, there are two launch angles whose trajectories pass through the basket, although in this case they are not symmetric about 45°. Basketball players know that a higher launch angle gives a better chance of scoring a basket. Can you show why this is so?

69.   **INTERPRET**   The plane's 90° turn is part of a circular arc, undertaken at constant speed, so this problem involves uniform circular motion. We are asked to find the minimum-radius turn that the jet can execute without subjecting the pilot to an acceleration that exceeds $5g$.

DEVELOP   The magnitude of the acceleration in circular motion is $a = v^2/r$, which here must not exceed $5g$. The diagram shows that the minimum height for starting the turn is the radius r of the circular turn.

EVALUATE   Setting $a = 5g$ and solving for the radius gives

$$r \geq \frac{v^2}{5g} = \frac{[(1200 \text{ km/h})(1000 \text{ m/km})/(3600 \text{ s/h})]^2}{(5)(9.8 \text{ m/s}^2)} = 2.3 \text{ km}$$

ASSESS To give a margin of safety, the pilot should start the turn at well above this height. Although we ignored gravity, including it would not have changed our 2-significant-figure answer.

71.   **INTERPRET** You want to know what is the farthest you can throw a stone horizontally, given the height you're able to throw it straight up.

DEVELOP When throwing the stone straight up, $v_{y0} = v_0$. Once it reaches its maximum height, h, its velocity will be zero, so using Equation 2.11 we have $v_0^2 = 2gh$. We assume you throw the ball with the same initial speed, but at a different angle. The maximum horizontal range can be found from Equation 3.15: $x = v_0^2 \sin 2\theta_0 / g$.

EVALUATE Using the expression for the initial velocity from the first throw, the range is

$$x = \frac{2gh \sin 2\theta_0}{g} = 2h \sin 2\theta_0$$

Since the maximum that $\sin 2\theta_0$ can be is 1, the maximum distance you can throw the stone is $2h$.

ASSESS If $\sin 2\theta_0 = 1$, then $\theta_0 = 45°$. We will prove that indeed this is the optimum throwing angle in Problem 3.75.

73.   **INTERPRET** This problem involves projectile motion. We are asked to find the horizontal distance at which a trajectory intercepts a 15° slope (see figure below).

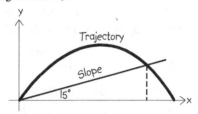

DEVELOP We need to find the intersection of the ball's trajectory (expressed by Equation 3.14) with a 15° line through the same origin, which is mathematically expressed as $y = x \tan(\theta_1)$, where $\theta_1 = 15°$. Equating these two expressions gives

$$y = x \tan(\theta_1) = x \tan(\theta_0) - \frac{g}{2v_0^2 \cos^2(\theta_0)} x^2$$

$$x = 2v_0^2 \cos^2(\theta_0) \left[\frac{\tan(\theta_0) - \tan(\theta_1)}{g} \right]$$

To solve for x, we need to know the initial velocity, which we can obtain from the fact that the range of the ball on level ground is $x_f = 28$ m. Thus

$$y(x_f = 28 \text{ m}) = 0 = x_f \tan(\theta_0) - \frac{g}{2v_0^2 \cos^2(\theta_0)} x_f^2$$

$$v_0 = \pm\sqrt{\frac{gx_f}{2\tan(\theta_0)\cos^2(\theta_0)}}$$

Insert this result into the previous expression for the intersection of the trajectory to find the horizontal distance x that the player can kick the ball up a 15° slope.

EVALUATE   The expression for the x coordinate of the intersection of the trajectory and the 15°C slope gives

$$x = 2\left(\frac{gx_f}{2\tan(\theta_0)\cos^2(\theta_1)}\right)\cos^2(\theta_0)\left[\frac{\tan(\theta_0) - \tan(\theta_1)}{g}\right]$$

$$= x_f \frac{\tan(\theta_0) - \tan(\theta_1)}{\tan(\theta_0)} = (28 \text{ m})\frac{\tan(40°) - \tan(15°)}{\tan(40°)} = 19 \text{ m}.$$

Thus, the player can kick the ball a horizontal distance of 19 m up a 15° slope.

ASSESS   The horizontal distance that the player can kick the ball up a slope is linear in the range he has on level ground, and quadratic in the initial velocity he can impart to the ball.

75.   **INTERPRET**   This problem is an exercise in calculus. We are asked to use calculus to show that the maximum range of a projectile occurs for a launch angle of $\theta_0 = 45°$.

DEVELOP   Equation 3.15, $x = v_0^2 \sin(2\theta_0)/g$, gives the range of a projectile over level ground. Differentiate it with respect to θ_0 and set the result to zero. Solving this expression will give the value(s) of θ_0 where Equation 3.15 is at an extremum (maximum or minimum). Interpret the result to determine if the result is a maximum or minimum.

EVALUATE   Differentiating Equation 3.15 with respect to θ_0 gives

$$\frac{dx}{d\theta_0} = \frac{2v_0^2}{g}\cos(2\theta_0) = 0$$

$$\theta_0 = 45°, 135°, 225°, 315°$$

The angles 225° and 315° are below the horizontal and so are unphysical. The angles 45° and 135° correspond to launching a projectile at 45° either to the left or to the right. Thus, the angle for maximum range is 45°.

ASSESS   Looking at Equation 3.15, we see that at $\theta_0 = 45°$, $\sin(2\theta_0) = 1$, which is the maximum value for the sine function. Thus, the answer appears reasonable.

77.   **INTERPRET**   We are given the problem of showing that the *slope* of an equation for a projectile's trajectory is in the direction of the projectile's velocity. The slope of an equation is the derivative. We also need to show that the components of velocity that we find are the same as the components given for projectile motion.

DEVELOP   The trajectory Equation 3.14 is

$$y = x\tan\theta_0 - \frac{g}{2v_0^2 \cos^2(\theta_0)} x^2$$

Use the derivative with respect to x to find the direction of the slope ($\tan\theta = dy/dx$). If we think of the derivative as a fraction, the numerator corresponds to the y component of the velocity, and the denominator corresponds to the x component. The ratio (i.e., slope) that we find should match Equations 3.10($v_x = v_{x0}$) and 3.11 ($v_y = v_{y0} \cdot gt$).

EVALUATE   Differentiating Equation 3.14 gives

$$\frac{dy}{dx} = \tan(\theta_0) - \frac{g}{v_0^2 \cos^2(\theta_0)} x = \frac{\sin(\theta_0)}{\cos(\theta_0)} - \frac{gx}{v_0^2 \cos^2(\theta_0)} = \frac{v_0^2 \sin(\theta_0)\cos(\theta_0) - gx}{v_0^2 \cos^2(\theta)_0}$$

The initial components of velocity are $v_{x0} = v_0 \cos(\theta_0)$ and $v_{x0} = v_0 \sin(\theta_0)$, which we insert into the above expression for dy/dx to obtain

$$\frac{dy}{dx} = \frac{\left[v_0 \sin(\theta_0)\right]\left[v_0 \cos(\theta_0)\right] - gx}{v_0^2 \cos^2(\theta_0)} = \frac{v_{y0}v_{x0} - gx}{v_{x0}^2} = \frac{v_{y0} - gx/v_{x0}}{v_{x0}}$$

Distance divided by velocity is time, so $x/v_{x0} = t$. Inserting this into the expression above gives

$$\frac{dy}{dx} = \frac{v_{y0} - gt}{v_{x0}}$$

Comparing this result with Equations 3.10 and 3.11, we see that the numerator is v_y and the denominator is v_x. **ASSESS** Note that dy/dx is not the velocity itself, but it is a dimensionless ratio that is the same as the dimensionless ratio $\tan\theta = v_x/v_y$.

79. **INTERPRET** This problem asks you to find the initial angle, θ_0, that gives the maximum range, x, for a projectile launched with speed v_0 from a height h. Recall that the maximum occurs when the derivate, $dx/d\theta_0$, is zero.
DEVELOP We need to find an equation that relates x and θ_0. Let's assume the projectile is launched from the origin, so that it lands at a vertical position of $y = -h$. We can find the range from Equation 3.14,

$$y = -h = x\tan\theta_0 - \frac{g}{2v_0^2 \cos^2\theta_0}x^2$$

Let's rearrange this equation by multiplying through by $\cos^2\theta_0$ and defining $H = v_0^2/2g$ (which is the maximum height of the stone's trajectory using Equation 2.11)

$$x^2 - 4xH\sin\theta_0 \cos\theta_0 - 4hH\cos^2\theta_0 = 0$$

Now using the trigonometric identities: $\sin 2\theta = 2\sin\theta\cos\theta$ and $\cos 2\theta = 2\cos^2\theta - 1$, we have

$$x^2 - 2xH\sin 2\theta_0 - 2hH\left(\cos 2\theta_0 + 1\right) = 0$$

We could solve for x using the quadratic formula, but that will get messy. Instead, we will leave the equation like this and take the derivative with respect to θ_0. We can then set $dx/d\theta_0$ equal to zero and then solve for the angle that gives the maximum range.

EVALUATE In taking the derivative of the above equation, we are careful to apply the chain rule and product rule from Appendix A:

$$2x \cdot \frac{dx}{d\theta_0} - 2H\left[\frac{dx}{d\theta_0} \cdot \sin 2\theta_0 + 2x\cos 2\theta_0\right] - 2hH\left[-2\sin 2\theta_0\right] = 0$$

If we then assume $dx/d\theta_0 = 0$ for the maximum range, we are left with

$$-4Hx_{max}\cos 2\theta_{max} - 4hH\sin 2\theta_{max} = 0 \quad \rightarrow \quad x_{max} = h\tan\theta_{max}$$

where θ_{max} is the angle that gives the maximum range, x_{max}. Notice that $\theta_{max} = 45°$ is undefined except for $h = 0$, which would be the normal case of a trajectory over level ground (see Equation 3.75). To solve for θ_{max} generally, we plug it and the expression for x_{max} into the trajectory equation that we derived above:

$$h^2 \tan^2 2\theta_{max} - 2hH\tan 2\theta_{max}\sin 2\theta_{max} - 2hH\left(\cos 2\theta_{max} + 1\right) = 0$$
$$h\sin^2 2\theta_{max} - 2H\sin^2 2\theta_{max}\cos 2\theta_{max} - 2H\cos^2 2\theta_{max}\left(\cos 2\theta_{max} + 1\right) = 0$$

Using the fact that $\sin^2\alpha = 1 - \cos^2\alpha = \left(1 - \cos\alpha\right)\left(1 + \cos\alpha\right)$, the above equation reduces to:

$$\cos 2\theta_{max} = \frac{1}{1 + 2H/h}$$

Or equivalently

$$\theta_{max} = \tfrac{1}{2}\cos^{-1}\left(\frac{1}{1+v_0^2/gh}\right)$$

ASSESS   If we assume the ground is level ($h = 0$), then the argument in the $\cos^{-1}$ function goes to zero, which means $\theta_{max} = 45°$, as it should when the trajectory is over level ground.

81.   **INTERPRET**   We're asked to interpret a map of three trajectories.

   DEVELOP   The shortest distance between two points is a straight line, so Alice has the shortest path. And Bob takes a shorter circular arc than Carrie.

   EVALUATE   The distances traveled are ordered length C>B>A.

   The answer is (c).

   ASSESS   Carrie's path appears to be a semicircle, which is the largest circular arc without starting in a direction opposite of one's goal.

83.   **INTERPRET**   We're asked to interpret a map of three trajectories.

   DEVELOP   The average speed is the distance traveled divided by the time, which were told is the same for the three walkers.

   EVALUATE   Since Alice had the shortest distance, she must have the slowest average speed. The fastest must have been Carrie, since she had the farthest distance to walk.

   The answer is (c).

   ASSESS   Notice, if the question had asked for the average velocity: $\bar{\vec{v}} = \Delta\vec{r}/\Delta t$ (Equation 3.3), the answer would be they are all equal, since the displacement $\Delta\vec{r}$ is the same for the three students.

4

FORCE AND MOTION

EXERCISES

Section 4.2 Newton's First and Second Laws

13. INTERPRET This problem involves Newton's 2nd law for a locomotive with different loads.

DEVELOP By Equation 4.3, the locomotive accelerates due to the force: $a = F/m$.

EVALUATE (a) The mass in this case is just the locomotive itself

$$a = \frac{(1.2 \times 10^5 \text{ N})}{61 \times 10^3 \text{ kg}} = 2.0 \text{ m/s}^2$$

(b) If the locomotive is pulling a train then the mass is the sum

$$a = \frac{(1.2 \times 10^5 \text{ N})}{\left(61 \times 10^3 \text{ kg}\right) + \left(1.4 \times 10^6 \text{ kg}\right)} = 0.082 \text{ m/s}^2$$

ASSESS These seem like reasonable accelerations. The locomotive by itself could reach 60 mi/h in 13 s, but pulling the train it would take over 5 and a half minutes to reach this speed.

15. INTERPRET This problem involves Newton's second law. The object of interest is the passenger, and we are to calculate the force required to stop the passenger in the given time.

DEVELOP Assume that the seatbelt holds the passenger firmly to the seat, so that the passenger also stops in 0.14 s without incurring any secondary impact. The passenger's average acceleration is $a_{av} = (0 - v_0)/t$ and his mass is 60 kg. Insert these quantities into Newton's second law to find the force.

EVALUATE The average force exerted by the seatbelt on the passenger is

$$F_{av} = ma_{av} = -mv_0/t = -\frac{(60 \text{ kg})}{0.14 \text{ s}}(110 \text{ km/h})\left(\frac{1000 \text{ m}}{\text{km}}\right)\left(\frac{1 \text{ h}}{3600 \text{ s}}\right) = -13 \text{ kN}$$

ASSESS The negative sign indicates that the force is opposite to the direction of the initial velocity.

17. INTERPRET This problem involves Newton's 2nd law for constant mass.

DEVELOP By Equation 4.3, the kinesin force imparts an acceleration on the molecular complex of $a = F/m$.

EVALUATE Recall from Appendix B that the SI prefix pico (p) corresponds to 10^{-12}, so

$$a = \frac{F}{m} = \frac{6.0 \times 10^{-12} \text{ N}}{3.0 \times 10^{-18} \text{ kg}} = 2.0 \times 10^6 \text{ m/s}^2$$

ASSESS This is an extraordinarily large acceleration, but it would only be applied for a fraction of a second, so the final velocity would be reasonable.

19. INTERPRET This problem involves Newton's second law and kinematics. The object of interest is the egg, and we are to calculate the minimum stopping distance so that the egg does not experience a force greater than 1.5 N.

DEVELOP For the average net force on the egg to not exceed the stated limit, the magnitude of the deceleration should satisfy $a_{av} \leq F_{max}/m = 1.5 \text{ N}/0.085 \text{ kg} = 17.6 \text{ m/s}^2$. Insert this acceleration into kinematic Equation 2.11 $v^2 = v_0^2 + 2a(x - x_0)$ to find the minimum stopping distance.

EVALUATE   The minimum stopping distance is

$$x - x_0 \geq \frac{(1.2 \text{ m/s})^2}{35.3 \text{ m/s}^2} = 0.041 \text{ m} = 4.1 \text{ cm}$$

ASSESS   Notice that the units work out to units of distance, as expected.

Section 4.4 The Force of Gravity

21. INTERPRET   This problem involves using Newton's second law to convert the usual units of acceleration (m/s^2) to N/kg. We are also asked to explain why it makes sense to express acceleration in N/kg when speaking of mass and weight.

DEVELOP   Newton's second law relates the units of mass (kg), distance (m), time (s), and force (N). Use this to solve the problem.

EVALUATE   From Newton's second law (for constant mass), $\vec{F}_{net} = m\vec{a}$, we see that force (N) is the same as mass (kg) $\times$ acceleration (m/s^2), which can be expressed mathematically as $\text{N} = \text{kg} \cdot \text{m/s}^2$. This can be rearranged to find $\text{m/s}^2 = \text{N/kg}$. It makes sense to use the units N/kg when speaking of mass and weight because kg is a unit of mass and N is a unit of force (i.e., a weight).

ASSESS   An acceleration is thus a mass per unit force.

23. INTERPRET   This problem asks us to find the mass of an object whose weight on the Moon corresponds to the weight of 35-kg object on the Earth.

DEVELOP   Use Equation 4.5 to find the weight of the block on the Earth. Use the gravitational acceleration g_M from Appendix E to calculate the mass that corresponds to an object of this weight on the Moon.

EVALUATE   To lift a 35-kg block on Earth requires a force at least equivalent to its weight, which is $w = mg = (35 \text{ kg})(9.8 \text{ m/s}^2) = 343 \text{ N}$. The same force on the moon could lift a mass $m = w/g_M = (343 \text{ N})/(1.62 \text{ m/s}^2) = 212 \text{ kg} \approx 210 \text{ kg}$ to two significant figures.

ASSESS   The weight of a 212-kg object on Earth is $w = mg = (212 \text{ kg})(9.8 \text{ m/s}^2) = 2078 \text{ N}$, which is a factor $g/g_M = (9.8 \text{ m/s}^2)/(1.62 \text{ m/s}^2) = 6$ times more than the weight on the Moon. Thus, you can lift 6 times the mass on the Moon than you can on the Earth.

25. INTERPRET   This is an exercise in converting between mass and weight.

DEVELOP   The weight on the US side is 10 tons. From Appendix C, we see that 1 ton is equivalent to the weight of 908 kg. Use this conversion to translate the given weight into a mass in kg.

EVALUATE   If 1 ton = weight of 908 k, 10 = weight of 9080 kg. Thus, you should specify 9000 kg (to a single significant figure) on the Canadian side of the border.

ASSESS   The conversion between mass and weight on Earth is $m = w/g$. Because the English unit of mass (the slug) is rarely used, the direct equivalence between mass in SI units and weight (force) in English units is usually given, as in Appendix C. Thus, 10 tons = 2×10^4 lbs is equivalent to the weight of $(2 \times 10^4 \text{ lb})(0.4536 \text{ kg/lb}) = 9 \times 10^3$ kg.

Section 4.5 Using Newton's Second Law

27. INTERPRET   This problem involves kinematics with constant velocity and Newton's second law. We are asked to find the force on the parachute due to air drag. The other force involved is the gravitational force.

DEVELOP   From kinematics (see, for example, Equation 3.6) we know that a body moving with constant velocity experiences an acceleration of $a = 0$. Inserting this into Newton's second law (for constant mass), $F_{net} = ma$, tells us that the net force is $F_{net} = 0$. From the free-body diagram of the situation (see figure below), we see that a zero net force implies that $F_{drag} = w$, from which we can find the drag force exerted by the air.

EVALUATE   Thus, the drag force is $F_{drag} = w = mg = (50 \text{ kg})(9.8 \text{ m/s}^2) = 490 \text{ N}$.

ASSESS   Because the acceleration gives the change in velocity over time, a body moving with constant velocity experiences zero instantaneous and zero average acceleration (cf. Equations 3.6 and 2.7).

29.   **INTERPRET**   This problem involves Newton's second law. The forces acting on the elevator passenger are the gravitational force and the normal force F_{elev} that the elevator floor applies on her feet (see free-body diagram below). We are asked to find the latter force.

DEVELOP   Because this problem involves forces in only one direction, we can dispense with the vector notation. Apply Newton's second law (for constant mass) $F_{net} = ma$ to find the force applied by the floor of the elevator. The net force is the sum of the forces acting on our passeger, so $F_{net} = F_{elev} - w$ (where $w = mg$), the mass of the passenger is $m = 52 \text{ kg}$, and her acceleration is $a = -2.4 \text{ m/s}^2$.

EVALUATE   Newton's second law gives

$$F_{net} = ma$$
$$F_{elev} - w = ma$$
$$F_{elev} = mg + ma = (52 \text{ kg})(9.8 \text{ m/s}^2 - 2.4 \text{ m/s}^2) = 380 \text{ N}$$

ASSESS   Because the elevator accelerates downward, it does not need to support the entire weight of the person, so the force it applies is slightly less than that necessary to counter the gravitational force on the person. What would happen if the elevator accelerated downward at the 9.8 m/s²? At $a > 9.8 \text{ m/s}^2$?

31.   **INTERPRET**   We assume the rocket's acceleration is constant, so we'll need the equations from Chapter 2, Section 4. Once we know the acceleration, we can find the force from the rocket engines using Newton's 2nd law.

DEVELOP   The rocket has to go from rest to $v = 7200 \text{ km/h}$ in 2 min. We can use Equation 2.7 ($v = v_0 + at$) to find the acceleration. From this we use Equation 4.3 ($F = ma$) to find the force of the rocket and the force on the astronaut.

EVALUATE   The rocket accelerates at

$$a = \frac{v - v_0}{t} = \frac{7200 \text{ km/h}}{2.0 \text{ min}} = 66.7 \text{ m/s}^2$$

To accelerate a load of 630 Mg, the rocket will need a thrust of

$$F = ma = (630 \times 10^3 \text{ kg})(66.7 \text{ m/s}^2) = 4.2 \times 10^7 \text{ N}$$

During launch, a 75-kg astronaut experiences a force of

$$F = ma = (75 \text{ kg})(66.7 \text{ m/s}^2) = 5.0 \times 10^3 \text{ N}$$

ASSESS   This is nearly 7 g of acceleration, but astronauts and modern pilots are often trained to handle up to around 9 g without losing consciousness.

Section 4.6 Newton's Third Law

33.   **INTERPRET**   This problem involves Newton's third law. We are asked to find the third-law force that pairs with the gravitational force from the Earth pulling the elephant toward the Earth.

DEVELOP   As shown in Figure 4.17, the third-law force that pairs with the Earth's gravitational pull is the gravitational force exerted by the elephant on the Earth, pulling the Earth upward. Apply Newton's third law to calculate this force.

EVALUATE   Newton's third law gives $F_{eE} = F_{Ee} = mg = (5600 \text{ kg})(9.8 \text{ m/s}^2) = 55 \text{ kN}$.

ASSESS   Note that the magnitudes of the forces in a third-law force pair are equal, but they are oriented in opposite directions.

35.   **INTERPRET**   This is a one-dimensional problem that involves calculating a force using Hooke's law, and applying Newton's third law to find the force necessary to stretch the spring.

DEVELOP   Choose a coordinate system in which the extension of the spring is in the positive x direction. Hooke's law (Equation 4.9) states that a spring will resist compression or extension with a force proportional to the change in the spring's length, or $F_{sp} = -kx$, where k is the spring constant an x is the extension ($x > 0$) or compression ($x < 0$) of the spring. We are given $k = 270$ N/m and $x = 48$ cm $= 0.48$ m, so we can use Hooke's law to solve the problem.

EVALUATE   Inserting the given quantities into Hooke's law gives $F_{sp} = -(270 \text{ N/m})(0.48 \text{ m}) = -130 \text{ N}$. This means the spring exerts a force in the negative-x direction of 130 N, so by Newton's third law, we must apply a force $F_{app} = -F_{sp} = 130$ N (i.e., in the positive-x direction).

ASSESS   If we stretch the spring too far, it will permanently deform and Hooke's law will no longer apply.

37.   **INTERPRET**   This is a one-dimensional problem that involves Hooke's law and Newton's third law. We are asked to find the distance a spring with a given spring constant is stretched if we apply a given force to it.

DEVELOP   We apply the same reasoning as per Problem 4.36, except that we choose a coordinate system in which the applied force is in the negative-x direction. The problem states that $k = 340$ N/m and the applied force is the gravitational force (Equation 4.5) on the fish: $F_{app} = w = mg = -(6.7 \text{ kg})(9.8 \text{ m/s}^2)$.

EVALUATE   Inserting the given quantities into Hooke's law gives

$$x = -\frac{F_{sp}}{k} = \frac{F_{app}}{k} = \frac{-(6.7 \text{ N})(9.8 \text{ m/s}^2)}{340 \text{ N/m}} = -0.19 \text{ m} = -19 \text{ cm}$$

Thus the spring stretches 19 cm downward.

ASSESS   Notice that the spring is extended in the negative-x direction, as expected if we apply a force in that direction.

PROBLEMS

39.   **INTERPRET**   This is a one-dimensional problem that involves Newton's second law. We are asked to find an acceleration given the forces acting on a body.

DEVELOP   We use a coordinate system where the upward direction corresponds to the positive-x direction. For constant mass, Newton's second law is $\vec{F}_{net} = m\vec{a}$. The forces acting on your body are the gravitational force, $\vec{F}_g = (-mg)\hat{i}$, and the normal force $\vec{n}$ of your seat pushing upward, which is what you feel as your weight ($n = w = mg$, see Equation 4.5). We are told that your weight is 70% of its usual value, so we set $\vec{n} = 0.7\vec{w} = (0.7mg)\hat{i}$. Insert these quatnities into Newton's second law to find the plane's acceleration.

EVALUATE   From Newton's second law, the plane's acceleration is

$$\vec{F}_{net} = (-mg)\hat{i} + (0.7mg)\hat{i} = m\vec{a}$$
$$\vec{a} = (-0.3g)\hat{i} = (-0.3)(9.81 \text{ m/s}^2)\hat{i} = (-2.94 \text{ m/s}^2)\hat{i}$$

ASSESS   The airplane accelerates downward, as expected.

41. **INTERPRET** This is a one-dimensional problem that involves Newton's second law. We are asked to find the acceleration of the dancer given the forces acting on him.

DEVELOP We choose a coordinate system in which the positive -y direction is upward. For constant mass, Newton's second law is $\vec{F}_{net} = m\vec{a}$, where $\vec{F}_{net}$ is the sum of all the forces acting on the dancer. These forces are the gravitational force, which is his weight $\vec{w} = (-mg)\hat{j}$ (see Equation 4.5) pulling him down, and the normal force of the floor, which we are told is $\vec{n} = (1.5mg)\hat{j}$. Sum these to find the net force and his acceleration.

EVALUATE Inserting the known quantities into Newton's second law gives

$$\vec{F}_{net} = m\vec{a}$$
$$(-mg + 1.5mg)\hat{j} = m\vec{a}$$
$$\vec{a} = (0.50g)\hat{j} = (0.5)(9.8 \text{ m/s}^2)\hat{j} = (4.9 \text{ m/s}^2)\hat{j}$$

where we have given the answer to two significant figures to match the precision of the input.

ASSESS Notice that the acceleration is upward, as expected.

43. **INTERPRET** The is a one-dimensional problem that involves Newton's second law and kinematics. We are asked to compute the minimum stopping time for the elevator that allows the passengers to remain on the floor.

DEVELOP We shall take the positive-y axis as upward. To use Newton's second law, $\vec{F}_{net} = m\vec{a}$, we need to know all the forces acting on the passenger. There are two vertical forces on a passenger, the gravitational force $\vec{F}_g = \vec{w} = -mg\hat{j}$ downward (see Equation 4.5), and the upward normal force $\vec{n} = n\hat{j}$ of the floor. The latter is a contact force and always acts in a direction perpendicular to and away from the surface of contact. If the magnitude of this force drops below zero, our passenger will have lost contact with the floor.

EVALUATE Inserting the forces into Newton's second law and demanding that n > 0 gives us a condition whereby the passenger stays in contact with the floor:

$$\vec{F}_{net} = m\vec{a}$$
$$n - mg = ma$$
$$n = ma + mg > 0$$
$$a > -g$$

Note that m is positive. Using Equation 3.8, the time required for the elevator to stop ($v = 0$) from an initial upward velocity ($v_0 = 5.2$ m/s) is $t = (v - v_0)/a = -v_0/a$. Inserting the limiting condition of $a = -g$ gives $t > v_0/g = (5.2 \text{ m/s})/(9.8 \text{ m/s}^2) = 0.53 \text{ s}$.

ASSESS Half a second is a reasonable value. The condition $n = 0$ is the limit for the person and the floor to remain in contact. As long as the passenger is in contact with the floor, his or her vertical acceleration is the same as that of the floor and the elevator.

45. **INTERPRET** This problem deals with interaction between different pairs of objects. The key concepts involved here are Newton's second and third laws.

DEVELOP Let the three masses be denoted, from left to right, as $m_1, m_2,$ and $m_3,$ as shown in the figure below.

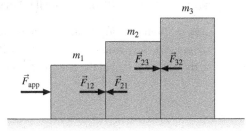

We take the right direction to be +x. We are told that the table is frictionless, so the only horizontal forces are the applied force and the contact forces between the blocks. For example, $\vec{F}_{12}$ denotes the force exerted by block 1 on block 2. Since the blocks are in contact, they all have the same acceleration a, to the right. Newton's second law can be applied to each block separately:

$$\vec{F}_{app} + \vec{F}_{21} = m_1\vec{a}$$

$$\vec{F}_{12} + \vec{F}_{32} = m_2\vec{a}$$

$$\vec{F}_{23} = m_3\vec{a}$$

EVALUATE   Adding all three equations and using Newton's third law ($\vec{F}_{12} + \vec{F}_{21} = 0$, etc.), one finds

$$\vec{a} = \frac{\vec{F}_{app}}{m_1 + m_2 + m_3} = \frac{12\,N}{1.0\,kg + 2.0\,kg + 3.0\,kg} = 2.0\ m/s^2 \quad \text{(to the right)}$$

Thus, the force block 2 exerts on block 3 is

$$F_{23} = m_3 a = (3.0\ kg)(2.0\ m/s^2) = 6.0\ N \quad \text{(to the right)}$$

ASSESS   You might be tempted to assume that the force on block 3 is just the applied force, $\vec{F}_{app}$, but from that you would wrongly conclude that block 3 is accelerating at 4 m/s², which would no longer be the same as for the other 2 blocks.

47.   **INTERPRET**   This is a one-dimensional problem that involves Newton's second and third laws. We are asked to find the force applied by the plane, the tension in the ropes, and the net force on the first glider.

DEVELOP   Make a free-body diagram of the situation (see figure below), on which we have noted all the horizontal forces, the masses of each object, and the coordinate system where the positive-x direction is to the right. Note that we are neglecting the mass of the ropes and any friction forces. From Newton's third law, we know that the third-law force pairs have equal magnitude, but act in opposing directions. Therefore, $\vec{T}_{1,P} = -\vec{T}_{P,1}$ and $\vec{T}_{2,1} = -\vec{T}_{1,2}$. To find the thrust of the propeller, note that the propeller has to accelerate at $\vec{a} = (1.9\ m/s^2)\hat{i}$ a total mass m_T of $m_T = m_P + m_2 + m_1$, which we can insert into Newton's second law to find the thrust. Applying Newton's second law to the airplane, glider 1, and glider 2 will also allow us to find the tension in the two ropes, which will then allow us to find the net force on the first glider.

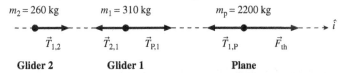

EVALUATE   **(a)** The net force on the three-body object is $\vec{F}_{net} = \vec{F}_{th} + \vec{T}_{1,P} + \vec{T}_{P,1} + \vec{T}_{2,1} + \vec{T}_{1,2} = \vec{F}_{th}$, where the last equality follows from Newton's third law. Inserting this into Newton's second law gives

$$\vec{F}_{net} = m_T\vec{a}$$

$$\vec{F}_{th} = (m_1 + m_2 + m_P)\vec{a} = (2200\,kg + 310\,kg + 260\,kg)(1.9\ m/s^2)\hat{i} = (5.26\times10^3\ N)\hat{i}$$

(b) Applying Newton's second law to the airplane gives the tension $\vec{T}_{1,P}$ in the first rope as

$$\vec{F}_{net} = m_P\vec{a}$$

$$\vec{F}_{th} + \vec{T}_{1,P} = m_P\vec{a}$$

$$\vec{T}_{1,P} = m_P\vec{a} - \vec{F}_{th} = (m_2 + m_1)\vec{a} = -(310\ kg + 260\,kg)(1.9\ m/s^2)\hat{i} = (-1.08\times10^3\ N)\hat{i}$$

where we have used $\vec{F}_{th} = (m_1 + m_2 + m_P)\vec{a}$ from part (a). Because the tension force in the rope acts in both directions ($\pm\hat{i}$), we give only the magnitude of the tension force; $T_1 = 1.1 \times 10^3$ N (to two significant figures).

(c) Applying Newton's second law to the first glider gives the tension $\vec{T}_{2,1}$ in the second rope as

$$\vec{F}_{net} = m_1\vec{a}$$

$$\vec{T}_{P,1} + \vec{T}_{2,1} = m_1\vec{a}$$

$$\vec{T}_{2,1} = m_2\vec{a} - \vec{T}_{P1} = m_1\vec{a} - (m_2 + m_1)\vec{a} = (-260\ kg)(1.9\ m/s^2) = (-494\ N)\hat{i}$$

The tension force in the rope is therefore $T_2 = 490$ N (to two significant figures).

(d) The net force on the first glider is $\vec{F}_{net} = \vec{T}_{2,1} + \vec{T}_{P,1} = (-494\ N + 1080\ N)\hat{i} = (590\ N)\hat{i}$ (to two significant figures), where we have used $\vec{T}_{1,P}$ from part (b) and Newton's third law, which gives $\vec{T}_{P,1} = -\vec{T}_{1,P}$.

ASSESS   The tension in the first rope provides the force to accelerate m_2 and m_3, whereas the tension in the second force accelerates only m_3.

49.   **INTERPRET**   This is a one-dimensional, unidirectional problem that involves Newton's second law, Hooke's law, and kinematics. We are asked to find the distance traveled by the car in 1 min, given the information necessary to find the acceleration of the car.

DEVELOP   We must assume that the rope is taught when the truck and car begin to move. Furthermore, because this is unidirectional problem, we will dispense with vector notation. From Hooke's law (Equation 4.9), we know that the elastic towrope exerts a force on the car of magnitude $F_{sp} = \left|-kx\right|$, where $k = 1300$ N/m and $x = 55$ cm = 0.55 m. Insert this force into Newton's second law to find the acceleration of the car (with mass = 1900 kg). Next, use the kinematic Equation 2.10 for constant acceleration, $x = x_0 + v_0 t + at^2/2$, with $v_0 = 0$ and $t = 1$ min = 60 s to find the distance $x - x_0$ traveled by the car.

EVALUATE   From Newton's second law, the acceleration of the car is

$$F_{net} = F_{sp} = \left|-kx\right| = ma$$
$$a = \frac{\left|-kx\right|}{m} = \frac{(1300 \text{ N/m})(0.55 \text{ m})}{1900 \text{ kg}} = 0.376 \text{ m/s}^2$$

The distance traveled is therefore

$$x - x_0 = \overset{=0}{v_0 t} + \frac{1}{2}at^2 = \frac{(0.376 \text{ m/s}^2)(60 \text{ s})^2}{2} = 680 \text{ m}$$

ASSESS   If the towrope is not taught at $t = 0$ s, then the car will undergo a non-constant acceleration until the rope becomes taught (because the towrope will be supplying a time-varying force as it stretches). Because we do not have information regarding this period of non-constant acceleration, we are obliged to disregard it and assume that the rope is taught at $t = 0$ s.

51.   **INTERPRET**   The problem asks us to determine the crumple zone of a car, in order to keep the stopping force on a passenger below a given value.

DEVELOP   We can think of the crumple zone as the distance, Δx, the car and its passengers continue to travel as they go from the initial speed to zero. We can use Equation 2.11 to relate this distance to the deceleration of the car,

$$v^2 = 0 = v_0^2 - 2a\Delta x$$

Note that we have included a negative sign, so that a is a positive quantity. Using Equation 4.3, we can derive a limit on the crumple zone from the requirement that the force on the passenger must be less than 20 times his/her weight:

$$F \leq 20F_g \quad \rightarrow \quad a \leq 20g$$

EVALUATE   The crumple zone is the distance during the crash over which the car comes to rest, so $\Delta x = v_0^2/2a$. Using the limit on the acceleration, the crumple zone must be at least

$$\Delta x = \frac{v_0^2}{2a} \geq \frac{v_0^2}{2(20g)} = \frac{(70 \text{ km/h})^2}{40(9.8 \text{ m/s}^2)} = 0.96 \text{ m}$$

ASSESS   This says the car would have to crumple by almost a meter. That's quite a bit, but the pictures of cars in high-speed collisions seem to imply that modern cars can compress by this much.

53.   **INTERPRET**   This problem involves applying Hooke's law to a spring and applying Newton's second law to the two-block system that is connected by the spring. We are asked to find the horizontal force applied to the system given the compression of the spring.

DEVELOP   Make a free-body diagram of the situation (see figure below). Because the problem is one-dimensional, we will forego the vector notation until the end. From Hooke's law (Equation 4.9) we know that the magnitude of the force exerted on each block by the spring is $F_{sp} = k\left|x\right|$, where $k = 8.1$ kN = 8100 N and $\left|x\right| = 5.1$ cm = 0.051 m. Apply Newton's second law to both blocks and solve for the applied force.

$m_1 = 640$ kg $m_2 = 490$ kg

$\vec{F}_{sp}^{(1)}$ $\vec{F}_{app}$ $\vec{F}_{sp}^{(2)}$ $\hat{i}$

EVALUATE   Applying Newton's second law to both blocks gives

$$F_{net}^{(1)} = F_{app} - F_{sp} = m_1 a$$

$$F_{net}^{(2)} = F_{sp} = m_2 a$$

Solving this, with the help of Hooke's law, for the applied force gives

$$F_{app} = F_{sp} + m_1\left(\frac{F_{sp}}{m_2}\right) = k|x|\left(\frac{m_1 + m_2}{m_2}\right) = (8100 \text{ N/m})(0.051 \text{ m})\left(\frac{640 \text{ m} + 490 \text{ m}}{490 \text{ m}}\right) = 950 \text{ N}$$

This force is applied in the direction indicated in the free-boy diagram, so $\vec{F}_{app} = (950 \text{ N})\hat{i}$.

ASSESS   Does the result make sense in the limiting situations? Letting $m_2 = m_1$ gives $F_{app} = 2F_{sp}$, which makes sense because the applied force has to accelerate both blocks, whereas the spring only accelerates a single block. If $m_2 \to 0$, then from equations above resulting from Newton's second law, we see that $F_{app} = m_1 a$, as expected. Finally, if $m_2 \gg m_1$, then $F_{app} = F_{sp} = m_2 a$, which is reasonable if m_1 is very small.

55.   **INTERPRET**   This problem involves applying Newton's second law to the spacecraft to find the thrust force required to achieve the various accelerations.

DEVELOP   Draw free-body diagrams of the different situations (see figure below), and apply Newton's second law in each situation to find the requisite thrust. Note that the positive-x direction is upward away from the surface of the Earth. For parts (a) and (b), the weight of the rocket is $w = mg$ (see Equation 4.5). For part (c), the weight of the rocket is $w = 0$ because the rocket is in a zero-gravity environment.

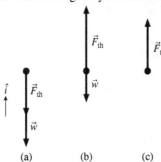

(a)   (b)   (c)

EVALUATE   (a) For the rocket accelerating toward the Earth, Newton's second law gives

$$\vec{F}_{net} = \vec{w} + \vec{F}_{th} = m\vec{a}$$

$$\vec{F}_{th} = m\vec{a} - \vec{w} = m\vec{a} - (-mg)\hat{i} = m(\vec{a} + g\hat{i})$$

for this part, $\vec{a} = (-1.40g)\hat{i}$, so

$$\vec{F}_{th} = m(-1.40g + g)\hat{i} = (-0.40mg)\hat{i}$$

(b) For this part, $\vec{a} = (1.40g)\hat{i}$, so

$$\vec{F}_{th} = m(1.40g + g)\hat{i} = (2.40mg)\hat{i}$$

(c) For this part, $w = 0$ and $\vec{a} = 1.40g\hat{i}$, so

$$\vec{F}_{th} = m\vec{a} = (1.40mg)\hat{i}$$

ASSESS   Notice that for part (c), the direction of the acceleration is in the direction of the force, because there are no other forces (i.e., gravity) to modify the direction of the acceleration. Therefore, the choice of $\hat{i}$ as the direction of the force is arbitrary. To be completely general, we could have written

$$F_{th} = (1.40mg)\frac{\vec{F}_{th}}{F_{th}}$$

where the last factor is simply the unit vector in the direction of the thrust force.

57. **INTERPRET**   We're asked if the thrust from these planes' engines could overcome the weight of the planes. We assume the planes are pointed straight up and then calculate the net force.

DEVELOP   If the planes were trying to fly upwards (like rockets) their wings would not giving them any lift, so the net force would be:

$$F_{net} = F_{thrust} - F_g = ma$$

EVALUATE   Starting with the F-16:

$$F_{net} = 132 \text{ kN} - (12 \times 10^3 \text{ kg})(9.8 \text{ m/s}^2) = 14.4 \text{ kN}$$

As this is positive, the F-16 can climb vertically at an acceleration of

$$a = \frac{F_{net}}{m} = \frac{14.4 \text{ kN}}{12 \times 10^3 \text{ kg}} = 1.2 \text{ m/s}^2$$

Now for the A-380:

$$F_{net} = 1.5 \text{ MN} - (560 \times 10^3 \text{ kg})(9.8 \text{ m/s}^2) = -3.99 \text{ MN}$$

The negative sign here means that A-380 would fall if it didn't have the lift from its wings.

ASSESS   You might have guessed that a fighter can climb straight up, whereas a commercial jet liner cannot.

59. **INTERPRET**   This problem is an exercise in calculus to derive a more general form of Newton's second law for one-dimensional situations.

DEVELOP   From Appendix A, we see that the derivative of a product, $d(ab)/dt = a(db/dt) + b(da/dt)$. Apply this rule to Newton's second law, with $a = m$ and $b = v$.

EVALUATE   Using the product rule, Newton's second law for one-dimensional situations is

$$F_{net} = \frac{d}{dt}(mv) = m\overbrace{\frac{dv}{dt}}^{=a} + v\frac{dm}{dt} = ma + v\frac{dm}{dt}$$

where we have used Equation 2.5, a = dv/dt, which defines the instantaneous acceleration.

ASSESS   The result shows clearly that if the mass is constant in time, $F = ma$, which is the usual form of Newton's second law.

61. **INTERPRET**   This problem involves Newton's second law. We are asked to find the upward acceleration you must have to keep a mass on the other side of the pulley from accelerating. Because the pulley is massless and frictionless, the tension on either side of the pulley is the same.

DEVELOP   First draw free-body diagrams for the hanging mass and for the climber (see figure below). The mass is not accelerating, so the net force on it must be zero. The climber is accelerating, so the upwards tension force on the climber must be greater than the climber's weight. We set the tension on the climber's side equal to the tension on the mass side, and find the resulting acceleration of the climber from Newton's second law (for constant mass and for one dimension), $F = ma$.

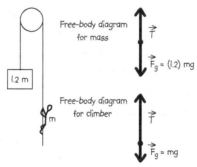

EVALUATE   The net force on the mass is zero, so $\vec{T} = -\vec{F}_g \implies T = 1.2mg$. The net force on the climber is

$\vec{T} + \vec{F}_g = m\vec{a} \implies T - mg = ma \implies T = m(g + a)$. Setting the two tensions equal to each other gives

$$1.2mg = m(g+a)$$
$$1.2g = g+a$$
$$a = 0.2g = (0.2)(9.8 \text{ m/s}^2) = 1.96 \text{ m/s}^2$$

ASSESS   This acceleration is 20% of g. The mass is 20% more than the mass of the climber. Makes sense, no?

63.   **INTERPRET**   We are asked to find the acceleration of your reference frame (the airplane) if objects falling with gravitational acceleration appear in your frame of reference to accelerate upward.

DEVELOP   We choose a coordinate system where the positive-y direction is upward. Because the pretzels are no longer supported by the tray, or anything else, we must conclude that they are accelerating downward at g. In your frame of reference (i.e., the airplane), they are accelerating upward at 2 m/s², so the airplane must be accelerating downward even faster than g. Consider the one-dimensional form of Equation 3.7, $v = v' + V$, where v is the velocity of the pretzel relative to the Earth, v' is the velocity of the pretzel relative to the airplane, and V is the velocity of the airplane relative to the Earth. Differentiating this equation with respect to time gives $a = a' + A$, where $a = -g$ is the acceleration of the pretzel relative to the Earth, $a' = 2.0 \text{ m/s}^2$ is the acceleration of the pretzel relative to the airplane, and A is the acceleration of the airplane relative to the Earth.

EVALUATE   From the equation of relative accelerations, we find that $-g = 2.0 \text{ m/s}^2 + A$, so $A = -g - 2.0 \text{ m/s}^2 = -11.8 \text{ m/s}^2$, or a downward acceleration of 11.8 m/s².

ASSESS   This is a downward acceleration that has a larger magnitude than g. That's what we expected.

65.   **INTERPRET**   We are asked to find the masses, given accelerations and forces. There are two masses, so we'll solve a system of equations. Newton's second law applies.

DEVELOP   Begin with a free-body diagram for each mass, as shown in the figure below. Because this is a one-dimensional problem, we will not use vector notation, but instead use negative values for downward vectors and positive values for upward vectors. Apply Newton's second law to both masses to obtain two equations. From these two equations, solve for the two masses, given that $T = 18 \text{ N}$, $F = 30 \text{ N}$, and $a = 3.2 \text{ m/s}^2$.

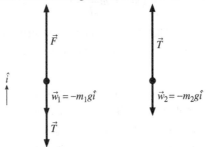

Upper mass   Lower mass

EVALUATE   For mass 1, $F_{net} = ma \Rightarrow F - T - w_1 = m_1 a$, so $F - T = m_1(g+a)$ Solving for m1 gives

$$m_1 = \frac{F-T}{g+a} = \frac{30 \text{ N} - 18 \text{ N}}{9.8 \text{ m/s}^2 + 3.2 \text{ m/s}^2} = 0.92 \text{ kg}$$

Similarly for mass 2, $F_{net} = ma \Rightarrow T - w_2 = m_2 a \Rightarrow T = m_2(g+a)$

$$m_2 = \frac{T}{g+a} = \frac{18 \text{ N}}{9.8 \text{ m/s}^2 + 3.2 \text{ m/s}^2} = 1.4 \text{ kg}$$

ASSESS   One way to check our result is to see what the acceleration of the total mass would be with an upward force of 30 N. Using Newton's second law, we find

$$F - (m_1 + m_2)g = (m_1 + m_2)a$$
$$a = \frac{F}{m_1 + m_2} - g = \frac{30 \text{ N}}{0.92 \text{ kg} + 1.4 \text{ kg}} - 9.8 \text{ m/s}^2 = 3.2 \text{ m/s}^2$$

as expected.

67.   **INTERPRET**   We're asked to calculate the amount of jerk on an amusement ride, where jerk is the rate of change in acceleration.

DEVELOP   The word "rate" implies per time. The jerk is the time derivative of the acceleration. We're given an equation for the force, so the acceleration is just this divided by the mass, M, of the car and passengers.

EVALUATE   The acceleration on the amusement ride is

$$a = \frac{F}{M} = \frac{F_0}{M} \sin \omega t$$

The jerk is the time derivative of this:

$$\frac{da}{dt} = \frac{F_0}{M} \omega \cos \omega t$$

The maximum value of the cosine is 1, so the maximum jerk is equal to $\omega F_0 / M$.

ASSESS   If the maximum jerk is too high, some of the passengers may suffer a whiplash.

69.   INTERPRET   You're asked to analyze data from an accelerometer in your laptop.

DEVELOP   The vertical acceleration is registered in how much the apparent weight diverges from the true weight.

EVALUATE   The apparent weight differs the most from the true weight during interval B.
The answer is (a).

ASSESS   The change in the weigh during interval D appears to be about half that during interval B.

71.   INTERPRET   You're asked to analyze data from an accelerometer in your laptop.

DEVELOP   The acceleration in terms of the apparent (n) and true (mg) weight is:

$$a = g\left(\frac{n}{mg} - 1 \right)$$

EVALUATE   During interval B, the apparent weight appears to be 5.5 lbs, so the acceleration is $0.1g \approx 1 \text{ m/s}^2$.
The answer is (b).

ASSESS   An acceleration of roughly 10% gravity seems reasonable. Compare this to the acceleration experienced by the astronaut in Problem 4.31.

5

USING NEWTON'S LAWS

EXERCISES

Section 5.1 Using Newton's Second Law

13. **INTERPRET** This problem requires an application on Newton's second law in two dimensions. Two forces are exerted on the object of interest (i.e., the 3.1-kg mass) and produce an acceleration. With the mass of the object and one force given, we are asked to find the other force.

    DEVELOP Newton's second law for this mass says $\vec{F}_{net} = \vec{F}_1 + \vec{F}_2 = m\vec{a}$, where we assume no other significant forces are acting. Thus, the second force is given by $\vec{F}_2 = m\vec{a} - \vec{F}_1$.

    EVALUATE Inserting the expressions given in the problem statement for $\vec{F}_1$ and $\vec{a}$, we obtain

    $$\vec{F}_2 = m\vec{a} - \vec{F}_1 = (3.1 \text{ kg})\left[(0.91 \text{ m/s}^2)\hat{i} - (0.27 \text{ m/s}^2)\hat{j}\right] - \left[(-1.2 \text{ N})\hat{i} - (2.5 \text{ N})\hat{j}\right] = (4.0 \text{ N})\hat{i} + (1.7 \text{ N})\hat{j}$$

    ASSESS This force has magnitude $F_2 = \sqrt{(4.02 \text{ N})^2 + (1.66 \text{ N})^2} = 4.3 \text{ N}$ and points in the direction $\theta = \text{atan}(1.66 \text{ N}/4.02 \text{ N}) = 22°$ counterclockwise from the x axis.

15. **INTERPRET** This problem involves Newton's second law applied to a two-dimensional situation to find the acceleration of the skier, then kinematics to find the time it takes him to reach the bottom of the slope.

    DEVELOP Draw a free-body diagram of the situation (see figure below) and apply Newton's second law to find the acceleration. The angle is $\theta = 24°$ and weight of the skier is $w = mg$. Given the acceleration (which is constant), we can use Equation 2.10, $x = x_0 + v_0 t + at^2/2$, with $x - x_0 = 1.3 \text{ km} = 1300 \text{ m}$, to find how long it takes him to reach the bottom.

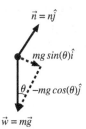

    EVALUATE Applying Newton's second law to the skier gives two equations (one for the x direction and one for the y direction):

    $$\vec{F}_{net} = m\vec{a}$$

    $$n\hat{j} + mg\left[\sin(\theta)\hat{i} - \cos(\theta)\hat{j}\right] = m\vec{a} \quad \Rightarrow \quad \begin{cases} n - mg\cos(\theta) = 0 \\ g\sin(\theta) = a \end{cases}$$

    Solving the second scalar equation for acceleration, we find $a = (9.8 \text{ m/s}^2)\sin(24°) = 3.986 \text{ m/s}^2$, and inserting this result into Equation 2.10 gives

    $$x = x_0 + \overbrace{v_0 t}^{=0} + at^2/2$$

    $$t = \pm\sqrt{\frac{2(x - x_0)}{a}} = \sqrt{\frac{2(1300 \text{ m})}{3.986 \text{ m/s}^2}} = 26 \text{ s}$$

    where we have chosen the positive square root and we have used $v_0 = 0$ because the skier starts at rest.

ASSESS   If the slope becomes vertical, $\theta \rightarrow 90°$ and $a = g$, which is what we expect because the skier would be free-falling.

17.   **INTERPRET**   This is a static problem in which we are looking for the force exerted on the tendon by the two muscles.

DEVELOP   In this case, there are two forces pulling on the tendon as shown below in the figure:

We are told that the horizontal pulls are opposite each other (meaning that the two forces are in the *x-y* plane), and we assume that the net horizontal force is zero: $F_1 \sin 25° - F_2 \sin 25° = 0$, in which case $F_1 = F_2$. The net vertical force pulls up on the tendon with a force equivalent to ten times the gymnast's weight:

$$F_1 \sin 25° + F_2 \sin 25° = 10 \; mg \text{ changes to } F_1 \sin 25° + F_2 \cos 25° = 10 \; mg$$

EVALUATE   Solving for the force in each muscle gives

$$F_1 = F_2 = \frac{10(55 \text{ kg})(9.8 \text{ m/s}^2)}{2 \sin 25°} = 6.4 \text{ kN changes to } F_1 = F_2 = \frac{10(55 \text{ kg})(9.8 \text{ m/s}^2)}{2 \sin 25°} = 3.0 \text{ kN}$$

ASSESS   The Achilles tendon is the thickest and strongest tendon in the body. In simply walking, it has to withstand strains of as much as 4 times the body weight.

Section 5.2 Multiple Objects

19.   **INTERPRET**   In this problem, two masses that rest on slopes at unequal angles are connected by a rope that passes over a pulley. We are asked to find the ratio of the two masses if they both remain at rest.

DEVELOP   Let the masses on the right and on the left be denoted as m_R and m_L, respectively. The free-body diagrams for m_R and m_L are shown in the sketch below, where the forces acing on the masses are the normal force from the slope, the weight, and the tension force. Note also that we use different coordinate systems for the left and right mass. If the masses don't slide, the net force on each must be zero (by Newton's second law), from which we can find the ratio of the masses.

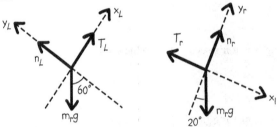

EVALUATE   Applying Newton's second law to the force component that is parallel to each surface gives

$$T_L - m_L g \sin(\theta_L) = 0$$
$$m_R g \sin(\theta_R) - T_R = 0$$

Now, if the masses of the string and pulley are negligible and there is no friction, then the tension must be the same throughout the entire string, so $T_L = T_R$. By adding the two equations, we find

$$m_R g \sin(\theta_R) - m_L g \sin(\theta_L) = 0$$

$$\frac{m_R}{m_L} = \frac{\sin(\theta_L)}{\sin(\theta_R)} = \frac{\sin(60°)}{\sin(20°)} = 2.5$$

ASSESS We conclude that when $\theta_L > \theta_R$, $m_R > m_L$ for the system to remain at rest. This makes sense because when the angle θ_L on the left-hand side increases, there is a greater tendency for m_L to slide down. To counter this, the mass m_R on the right-hand-side must go up. We may also consider the extreme case where $\theta_R = 0$. This situation corresponds to having a mass m_R on a frictionless horizontal surface, with m_L hanging over the edge and connected to m_R by the string. For this system to remain stationary, m_R must be infinitely massive compared to m_L.

21. **INTERPRET** This is a two-dimensional problem that involves applying Newton's second law to two climbers, tied together by a rope and sliding down an icy mountainside. The physical quantities of interest are their acceleration and the force required to bring them to a complete stop.

DEVELOP We choose two coordinate systems where the x axes are parallel to the slopes and y axes are perpendicular to the slopes (see figure below). Assume that the icy surface is frictionless and that the climbers move together as a unit with the same magnitude of down-slope acceleration a. If the rope is not stretching, the tension forces are equal in magnitude, so $T_1 = T_2 \equiv T$. To find the acceleration of the climbers, apply Newton's second law in the direction of the slope. To find the force F_{ax} exerted by the ax, again apply Newton's second law, but this time include F_{ax} and set the acceleration to zero; $a = 0$.

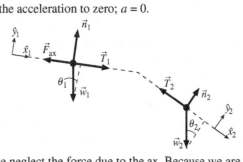

EVALUATE (b) For this part, we neglect the force due to the ax. Because we are now working in one-dimension (the x dimension), we forego vector notation, and insert the sign ($\pm$) according to the direction of the force. Of course, at the end we must interpret the sign of the resulting force as indicating its direction (positive or negative x direction). The magnitude of the net force in the $\hat{x}_1$ and $\hat{x}_2$ directions (downward positive) is thus

$$F_{net} = m_1 g \sin(\theta_1) + \overbrace{T_1 - T}^{=0}{}_2 + m_2 g \sin(\theta_2)$$
$$= (75 \text{ kg})(9.8 \text{ m/s}^2) \sin(12°) + (63 \text{ kg})(9.8 \text{ m/s}^2) \sin(38°) = 533 \text{ N}$$

Thus, the magnitude of the acceleration of the pair is

$$a = \frac{F_{net}}{m_1 + m_2} = \frac{533 \text{ N}}{75 \text{ kg} + 63 \text{ kg}} = 3.9 \text{ m/s}^2$$

so the pair accelerate down the slope at 3.9 m/s².

(b) After they have stopped, we include the force of the ax. Thus, the magnitude of the force due to the ax is

$$F_{net} = -F_{ax} + m_1 g \sin(\theta_1) + \overbrace{T_1 - T}^{=0}{}_2 + m_2 g \sin(\theta_2) = m \overbrace{a}^{=0} = 0$$
$$F_{ax} = m_1 g \sin(\theta_1) + m_2 g \sin(\theta_2) = 530 \text{ N}$$

so two significant figures. That is, the force exerted by the ax must be 530 N up the slope.

ASSESS   If the two climbers were not roped together, then their acceleration would have been

$$a_1 = \frac{F_1}{m_1} = g \sin(\theta_1) = (9.8 \text{ m/s}^2) \sin(12°) = 2.04 \text{ m/s}^2$$

$$a_2 = \frac{F_2}{m_2} = g \sin(\theta_2) = (9.8 \text{ m/s}^2) \sin(38°) = 6.03 \text{ m/s}^2$$

The acceleration of the pair is the mass-weighted average of the individual accelerations:

$$a = \frac{F_{net}}{m_1 + m_2} = \frac{m_1 g \sin(\theta_1) + m_2 g \sin(\theta_2)}{m_1 + m_2} = \left(\frac{m_1}{m_1 + m_2}\right) g \sin(\theta_1) + \left(\frac{m_2}{m_1 + m_2}\right) g \sin(\theta_2)$$

$$= \left(\frac{m_1}{m_1 + m_2}\right) a_1 + \left(\frac{m_2}{m_1 + m_2}\right) a_2 = 3.9 \text{ m/s}^2$$

Section 5.3 Circular Motion

23.   **INTERPRET**   In this problem we are asked to show that the force required to keep a mass m in a circular path of radius r with period T is $4\pi^2 mr/T^2$.

DEVELOP   To derive the formula, we first note that for an object of mass m in uniform circular motion, the magnitude of the net force is given by Equation 5.1: $F = ma = mv^2/r$. Next, we make use of the fact that the period of the motion (i.e., time for one revolution) is the circumference $C = 2\pi r$ divided by the speed v. Thus, $T = 2\pi r/v$.

EVALUATE   Combining the two expressions, the force can be rewritten as

$$F = \frac{mv^2}{r} = \frac{m}{r}\left(\frac{2\pi r}{T}\right)^2 = \frac{4\pi^2 mr}{T^2}$$

ASSESS   Our result indicates that for a fixed radius r, the centripetal force is inversely proportional to T^2. For example, if T is very large (i.e., it takes a very long time for the mass to complete one revolution), then the speed v is very small and the centripetal force F is also very small.

25.   **INTERPRET**   You're asked to find the velocity of the subway while rounding the curve. We'll assume uniform circular motion, in which case all the objects in the subway are accelerating according to $a = v^2/r$.

DEVELOP   The only information you have is that a strap is dangled at 15° to the vertical during the turn. You can assume that the strap hung straight down before the turn and that its displacement was outward with respect to the center of the curve. We can think of the strap as a mass hanging from an attachment (perhaps a chain or belt) that provides a tension, as shown in the figure below.

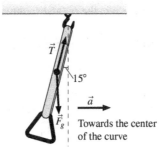

As we said above, the strap and everything else in the subway experience an acceleration, which obeys Newton's 2nd law: $\vec{F}_{net} = \vec{F}_g + \vec{T} = m\vec{a}$. We'll choose our coordinate system such that the acceleration is in the +x-direction. Therefore, in component form, the second law is

$$T_x = T \sin 15° = ma = \frac{mv^2}{r}$$

$$T_y - F_g = T \cos 15° - mg = 0$$

We'll evaluate these equations to find the velocity of the subway.

EVALUATE   We see from above that $T = mg / \cos 15°$, so $a = g \tan 15°$ and

$$v = \sqrt{ar} = \sqrt{(9.8 \text{ m/s}^2)(\tan 15°)(132 \text{ m})} = 18.6 \text{ m/s} = 67.0 \text{ km/h}$$

The train did exceed the 45 km/h speed limit on this curve by 22 km/h, thus provoking the derailment.

ASSESS   As part of the derivation, we found that the acceleration obeys: $a = g \tan \theta$. Does that make sense? If the strap were hanging straight down $\theta = 0$, the acceleration would be zero, as we would expect for a train that is moving straight at a constant speed.

27. **INTERPRET**   This problem involves circular motion and Newton's second law. The object of interest is the plane. By analyzing the force acting on the plane while it travels a circular path, its speed can be determined.

DEVELOP   We follow the strategy outlined in Example 5.6. There are two forces acting on the plane: the gravitational force $\vec{w}$ and the normal force $\vec{n}$. Applying Newton's second law gives $\vec{F}_{net} = \vec{w} + \vec{n} = m\vec{a}$. Breaking this vector equation up into its components gives

$$x: \quad n \sin(\theta) = \frac{mv^2}{r}$$

$$y: \quad n \cos(\theta) = mg$$

EVALUATE   Dividing the two equations allows us to eliminate n and obtain

$$\frac{n \sin(\theta)}{n \cos(\theta)} = \frac{mv^2/r}{mg} \quad \Rightarrow \quad \tan(\theta) = \frac{v^2}{rg}$$

With $\theta = 28°$ and $r = 3.6$ km $= 3600$ m, the speed of the plane is

$$v = \pm\sqrt{gr \tan(\theta)} = \pm\sqrt{(9.8 \text{ m/s}^2)(3600 \text{ m}) \tan(28°)} = \pm140 \text{ m/s} = \pm490 \text{ km/h}$$

ASSESS   The result shows that the speed of the plane is proportional to $\sqrt{\tan \theta}$. If we want to increase the speed v while the radius r is kept fixed, then the banking angle would also need to be increased. This is due to the fact that the horizontal component of the normal force $n \sin(\theta)$ is what keeps the plane in circular motion. The $\pm$ sign indicates that the result is the same whether the plane travels clockwise or counterclockwise.

Section 5.4 Friction

29. **INTERPRET**   This problem involves Newton's second law, friction, and kinematics. The object of interest is the hockey puck. Three forces are involved: gravity, the normal force, and friction, and friction is what causes the acceleration of the puck (in the direction opposing the motion of the puck). The physical quantity to be computed is the coefficient of kinetic friction.

DEVELOP   Draw a free-body diagram of the situation (see figure below). We define the positive-x direction to be the distance traveled by using the kinematic Equation 2.11, $v^2 = v_0^2 + 2a(x - x_0)$. The result is

$$a = -\frac{v_0^2}{2(x - x_0)}$$

where we have used the fact that the final velocity is $v = 0$. Notice that the acceleration is negative, meaning that the puck decelerates. By applying Newton's second law in the vertical direction, we know that the normal force must have the same magnitude as the weight, because the puck does not accelerate vertically. Thus, $n = w$. The force of friction is the only horizontal force acting on the puck, so Newton's second law and Equation 5.3 tell us that $F_{net} = -f_k = -\mu_k mg = ma$, where we have inserted a minus sign because the friction force acts to oppose the motion, which we take to be in the positive-x direction. the direction of the puck's initial velocity $v_0 = 14$ m/s. Given that it travels a distance of $x - x_0 = 56$ m, we can find

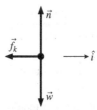

EVALUATE   Inserting the known quantities into the Newton's second law gives

$$-\mu_k mg = ma$$

$$\mu_k = -\frac{a}{g} = -\frac{-v_0^2}{2g(x-x_0)} = \frac{(14 \text{ m/s})^2}{2(9.8 \text{ m/s}^2)(56 \text{ m})} = 0.18$$

ASSESS   The result $a = -\mu_k g$ shows that increasing the coefficient of friction would result in a greater acceleration. This makes sense because friction is what causes the acceleration.

31.   **INTERPRET**   In this problem, the car is moving in uniform circular motion thanks to the friction provided by the road.

DEVELOP   To maintain uniform circular motion through the unbanked turn, the car must be able to accelerate at $a = v^2 / r$. This requires a force of friction: $f_s = ma = mv^2 / r$, directed inward along the curve's radius. As explained in Example 5.9, this friction is static because the car's motion is perpendicular to this force. Using Equation 5.2: $f_s \le \mu_s n$, we will find the minimum value for the coefficient of friction.

EVALUATE   Because the curve is unbanked, the normal is just equal to the weight of the car (see Figure 5.24). Solving for the coefficient of friction

$$\mu_s \ge \frac{v^2}{rg} = \frac{(90 \text{ km/h})^2}{(120 \text{ m})(9.8 \text{ m/s}^2)}\left(\frac{1 \text{ m/s}}{3.6 \text{ km/h}}\right)^2 = 0.53$$

ASSESS   This seems reasonable for the minimum friction coefficient. Rubber tires on a dry concrete road will typically have $\mu_s \approx 1$, but notice that if the concrete is wet, the coefficient drops to about 0.4.

PROBLEMS

33.   **INTERPRET**   This problem involves a block with an initial velocity sliding up a frictionless ramp. The quantity of interest is the distance it travels before coming to a complete stop. We will apply Newton's second law and kinematics to solve this problem.

DEVELOP   Draw a free-body diagram of the situation (see figure below). Applying Newton's second law in the $\hat{\imath}$ direction gives $-w\sin(\theta) = ma$, or $a = -g\sin(\theta)$. The stopping distance can be calculated by solving the kinematic Equation 2.11: $v^2 = v_0^2 + 2a\Delta x$.

EVALUATE   Inserting the acceleration into Equation 2.11, the distance traveled by the block is

$$\Delta x = \frac{v^2 - v_0^2}{2a} = \frac{v^2 - v_0^2}{2\left[-g\sin(\theta)\right]} = \frac{(0 \text{ m/s})^2 - (2.2 \text{ m/s})^2}{-2(9.8 \text{ m/s}^2)\sin(35°)} = 0.43 \text{ m}$$

ASSESS   The result shows that the distance traveled is inversely proportional to $\sin(\theta)$. To see that this makes sense, consider the limit where $\theta \to 0$. This situation would correspond to a frictionless horizontal surface. In this case, we expect the block to travel indefinitely, in agreement with our expression for Δx.

35. **INTERPRET** This problem involves Newton's second law. We are asked to find the tension in a rope needed to support an object of a given mass.

DEVELOP Draw a diagram of the situation (see figure below). Apply Newton's second law in the y direction and solve for the tension of the rope. Note that the tension of the rope is everywhere the same (for a massless rope), so $T_1 = T_2 = T$,

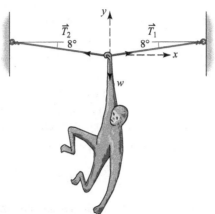

EVALUATE Applied in the y direction, Newton's second law gives

$$T_1 \sin(\theta) + T_2 \sin(\theta) = w$$

$$T = \frac{mg}{2\sin(\theta)} = \frac{(15\text{ kg})(9.8\text{ m/s}^2)}{2\sin(8°)} = 530\text{ N}$$

The monkey's weight is w = mg = (15 kg)(9.8 m/s2) = 150 N (to two significant figures). This is over three times less than the tension force in the rope.

ASSESS Notice that the $T \to \infty$ as $\theta \to 0$ because there is a vanishingly small component of the tension acting in the vertical direction. The majority of the tension simply serves to pull the two support points together.

37. **INTERPRET** The key concepts in this problem are circular motion and Newton's second law. The object of interest is the mass m_1 that travels in a circular path. By analyzing the force acting on m_1 and m_2, the tension in the string and the period of the circular motion of m_1 can be determined.

DEVELOP Apply Newton's second law to each mass (see Figure 5.31). Because the table is frictionless, the only force acting on m_1 in the horizontal plane is the tension. Assuming the massless rope does not encounter any friction when it goes through the hole in the table, the tension T acting on each mass is of the same magnitude. By Newton's second law, and because m_2 does not accelerate, this tension must cancel the force due to gravity acting on m_2. Thus, we have

$$m_1: \quad F_{net} = T = m_1 a_1 = \frac{m_1 v^2}{R} \quad \Rightarrow \quad T = \frac{m_1 v^2}{R}$$

$$m_2: \quad F_{net} = T - m_2 g = m_2 a_2 = 0 \quad \Rightarrow \quad T = m_2 g$$

EVALUATE **(a)** Newton's second law applied to the stationary mass m_2 yields $T = m_2 g$.

(b) The tension in the string also provides the centripetal force for m_1. Let $\tau = 2\pi R/v$ be the period of the circular motion. The above equation for m_1 then gives

$$T = \frac{m_1 v^2}{R} = \frac{m_1}{R}\left(\frac{2\pi R}{\tau}\right)^2 = \frac{4\pi^2 m_1 R}{\tau^2}$$

But from **(a)**, we also have $T = m_2 g$. By combining the two equations, we obtain

$$\frac{4\pi^2 m_1 R}{\tau^2} = m_2 g$$

$$\tau = 2\pi\sqrt{\frac{m_1 R}{m_2 g}}$$

ASSESS   From the expression for τ, we can draw the following conclusions: **(i)** Because $\tau \propto \sqrt{R}$, the larger the radius; the longer the period. **(ii)** With the tension (or m_2) kept fixed, increasing m_1 also leads to a longer period. Note that the derivation above only applies for uniform circular motion. If m_1 is pulled in toward the hole (i.e., R is reduced), then the motion is no longer circular (although the acceleration is still centripetal), and the formula $a = v^2/R$ does not apply.

39.   **INTERPRET**   This problem involves Newton's second and third laws and uniform circular motion. The objects of interest are the roller-coaster seat, the seat belt, and the rider. We are asked to find the force exerted on a rider at the top of the turn by the roller-coaster seat and by the seatbelt, and to determine what would happen should the rider unbuckle his seatbelt.

DEVELOP   The free-body diagram for the rider at the top of the track will be the same as for the roller coaster (see Figure. 5.16). The rider thus has two forces acting on him, the normal force due to the seat and the force due to gravity, both of which are pushing him down. By Newton's second law, the sum of these forces must be proportional to the acceleration. Expressed mathematically, we have

$$F_{net} = n + mg = ma = \frac{mv^2}{R}$$

where $m = 60$ kg, $v = 9.7$ m/s, and $R = 6.3$ m. Solve this equation for n, which is the force exerted by the roller-coaster seat on the rider. To find the force exerted by the seatbelt on the rider, make a free-body diagram of the seatbelt to find the force exerted by the rider on the seatbelt, then use Newton's third law to find the force exerted by the seatbelt on the rider.

EVALUATE   **(a)** From Newton's second law, the seat exerts a force

$$n = \frac{mv^2}{R} - mg = \left(60 \text{ kg}\right) \left[\frac{\left(9.7 \text{ m/s}\right)^2}{6.3 \text{ m}} - \left(9.8 \text{ m/s}^2\right) \right] = 310 \text{ N}$$

on the rider. As indicated in Figure 5.16, the normal force is oriented downward, accelerating the rider toward the Earth.

(b) A free-body diagram of the seatbelt is exactly like that for the rider (see Figure 5.16), but the mass of the seatbelt is different. Therefore, Newton's second law applied to the seatbelt gives

$$F_{net} = n_{R\text{-}SB} + m_{SB}g = m_{SB}a = \frac{m_{SB}v^2}{R}$$

where $n_{R\text{-}SB}$ is now the force exerted by the rider on the seatbelt and m_{SB} is the (unknown) mass of the seatbelt. From Newton's third law, we know that the force exerted by the seatbelt on the rider has the same magnitude as $n_{R\text{-}SB}$, but is oriented in the opposite direction. Therefore,

$$n_{SB\text{-}R} = -\frac{m_{SB}v^2}{R}$$

The negative sign means that this force is oriented upward, as expected. Without knowledge of the mass of the seatbelt, we cannot find the force $n_{SB\text{-}R}$, but we can see that this force is simply the force needed to give the seatbelt the same centripetal acceleration $a = v^2/R$ as the rider.

(c) If the rider unbuckles at this point, nothing would happen because the seatbelt does not enter into Newton's second law in part (a). Thus, the rider would remain in his seat.

ASSESS   Because the seat pushes down on the rider with the normal force, the ride's centripetal acceleration a toward the Earth is greater than that due to gravity; $a > g$. This can be verified by calculating the centripetal acceleration, which gives $a = v^2/R = (9.7 \text{ m/s})^2/(6.3 \text{ m}) = 15 \text{ m/s}^2 > g$. If the normal force were to go to zero, the rider's acceleration due to gravity g would be greater than the centripetal acceleration of the roller coaster, and the rider would fall out of his seat (unless the seat belt were there to restrain him).

41.   **INTERPRET**   This problem involves applying Newton's second law to an object that is executing uniform circular motion. The object of interest is the plane, and we are to find the minimum turning radius it can make given the maximum angle at which it can bank.

DEVELOP   Apply Newton's second law $\vec{F}_{net} = m\vec{a}$ to the plane as sketched in Figure 5.33. In component form, this gives

$$F_w \cos(\theta) - mg = ma_y = 0$$

$$F_w \sin(\theta) = ma_x = \frac{mv^2}{r}$$

where the subscripts x and y indicate the horizontal (to the left) and upward directions, respectively. Note that the acceleration in the horizontal direction is centripetal acceleration given by Equation 5.1, because the plane is executing uniform circular motion. Given the maximum bank angle is $\theta = 40°$ and the speed is $v = 950$ km/h = 263.89 m/s, solve for the turning radius r.

EVALUATE   The minimum turning radius r is

$$F_w \cos(\theta) = mg \quad \Rightarrow \quad F_w = \frac{mg}{\cos\theta}$$

$$F_w \sin(\theta) = \frac{mv^2}{r}$$

$$r = \frac{mv^2 \cos(\theta)}{mg \sin(\theta)} = \frac{v^2}{g \tan(\theta)} = \frac{(263.89 \text{ m/s})^2}{(9.8 \text{ m/s}^2) \tan(40°)} = 8500 \text{ m} = 8.5 \text{ km}$$

where we report the result to two significant figures.

ASSESS   This may seem like a rather wide turn—but the speed of the plane is high and the bank angle is low. These are actually reasonable values for a passenger jet. Will the passengers' coffee spill from their cups during such a turn?

43.   **INTERPRET**   This problem involves Newton's second law. The object of interest is the child and sled, and we are to find the frictional force needed for them to have zero acceleration.

DEVELOP   Draw a free-body diagram of the situation (see figure below). Note that the frictional force is drawn so that opposes the motion (although this is not necessary—were we to draw it in the opposite direction we would find a negative friction force). If the child does not accelerate, Newton's second law gives $\vec{F}_{net} = 0$. In component form, this is

$$n - mg \cos(\theta) = 0$$

$$mg \sin(\theta) - f_k = 0$$

where we have used $w = mg$. From Equation 5.3, we know that $f_k = \mu_k n$, so we can solve this system of equations for the coefficient of kinetic friction μ_k.

EVALUATE   Inserting the expression for f_k and using the first equation to eliminate the normal force gives

$$mg \sin(\theta) - \mu_k mg \cos(\theta) = 0$$

$$\mu_k = \tan(\theta) = \tan(12°) = 0.21$$

ASSESS   If the coefficient of kinetic friction does not depend on the speed of the sled, then the child will slide down the hill at her initial speed. Thus, if she takes a running start, she will continue at the speed at which she runs.

If she sits on her sled with no running start, she will remain stationary. If both situations, she experiences no acceleration.

45.   **INTERPRET**   The interpretation of the problem is the same as given in Example 5.4, except that the rock now has a fourth force acting on it, which is the force due to kinetic friction.

DEVELOP   Because friction always acts to oppose motion, the force f_k due to kinetic friction acting on the rock will be oriented in the negative-x direction. With the addition of the friction force, Newton's second law applied to the rock gives $\vec{T}_r + \vec{F}_{gr} + \vec{n} + \vec{f}_k = m\vec{a}_r$. Writing this vector equation in component form gives

$$x:\quad T_r - \mu_k n = m_r a$$
$$y:\quad n - m_r g = 0$$

where we have inserted a negative sign for the friction force and have used the fact that the rock does not accelerate in the y direction, and that the force due to gravity on the rock is $F_{gr} = m_r g$. Applying Newton's law to the climber gives the same result as in Example 5.4:

$$T - m_c g = -ma$$

Note that the acceleration of the climber and the rock has the same magnitude, although they act in different directions. This is so because the rope is considered to not stretch, so both objects must move at the same rate.

EVALUATE   Solve the 3 equations above for the acceleration, which gives

$$\mu_k m_r g - m_r a - m_c g = m_c a$$

$$a = g\left(\frac{m_c - \mu_k m_r}{m_c + m_r}\right) = \left(9.8 \text{ m/s}^2\right)\left[\frac{70 \text{ kg} - (0.057)(940 \text{ kg})}{70 \text{ kg} + 940 \text{ kg}}\right] = 0.16 \text{ m/s}^2$$

Inserting this result into the kinematic Equation 2.10 gives $t = \sqrt{2(51 \text{ m})/(0.159 \text{ m/s}^2)} = 25$ s.

ASSESS   Notice that our expression for the acceleration reverts to that found in Example 5.4 if we let $m_k \to 0$.

47.   **INTERPRET**   This problem involves Newton's second law and kinematics. The object of interest is the train, and we are asked to find if the train can stop within 150 m if the wheels maintain static contact with the rails (i.e., the wheels do not skid on the rails).

DEVELOP   Considering all the wheels as one point of contact, make a free-body diagram for the train (see figure below). Applying Newton's law to the train wheels gives $\vec{F}_{net} = \vec{f}_s + \vec{n} + \vec{F}_g = m\vec{a}$, and writing this in component form gives

$$x:\quad F_g + n = 0$$
$$y:\quad -f_s = -ma$$

where we have used the fact that there is zero acceleration in the x direction and we have explicitly noted the sign of the friction force and the acceleration to emphasize that they are in the same direction (negative-x direction). The force due to static friction is $f_s \leq \mu_s n$ and the force due to gravity is $F_g = -mg$ (because gravity acts in the downward direction). Insert these values into the above equations to find the maximum acceleration possible without having the wheels slip on the rails, then use the kinematic Equation 2.11 $v^2 = v_0^2 + 2a(x - x_0)$ to find the stopping distance.

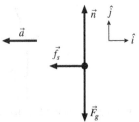

EVALUATE Newton's second law thus gives

$$ma = f_s \le \mu_s n = \mu_s mg$$
$$a \le \mu_s g$$

so the maximum acceleration possible is $\mu_s g$. Inserting this result for the acceleration into Equation 2.11 gives a stopping distance of

$$\overset{=0}{\overbrace{v^2}} = v_0^2 + 2a(x - x_0)$$

$$x - x_0 = \frac{v_0^2}{2a} = \frac{v_0^2}{2\mu_s g} = \frac{(140 \text{ km/h})^2}{2(0.58)(9.8 \text{ m/s}^2)}\left(\frac{10^3 \text{ m}}{\text{km}}\right)^2\left(\frac{\text{h}}{3600 \text{ s}}\right)^2 = 130 \text{ m}$$

so the train will stop before hitting the car.

ASSESS The stopping time for the train is

$$x - x_0 = (v_0 + v)t/2$$

$$t = \frac{2(x - x_0)}{v_0} = \frac{2(133 \text{ m})}{38.9 \text{ m/s}} = 6.8 \text{ s}$$

which should be just enough time for the passengers to get out of the car.

49.　**INTERPRET** This problem involves kinematics, Newton's second law, and frictional forces. We are given information to find the acceleration of the textbook, and are asked to calculate what coefficient of static friction is necessary to keep the paperback book stuck to the textbook during the acceleration, and the maximum coefficient of static friction possible that would still let the paperback book slide off the textbook during deceleration.

DEVELOP Draw a free-body diagram for the paperback book for both the case of acceleration a_1 and deceleration a_2 (see figure below). The accelerations can be calculated using the kinematic Equation 2.7 $v = v_0 + at$. This gives $\vec{a}_1 = \vec{v}/t = (0.96 \text{ m/s})/(0.42 \text{ s})\hat{i} = (2.286 \text{ m/s}^2)\hat{i}$ and $\vec{a}_2 = -\vec{v}_0/t = -(0.96 \text{ m/s})/(0.33 \text{ s}) = -(2.909 \text{ m/s}^2)\hat{i}$. Apply Newton's second law to the paperback book in both situations to find the coeffecient of static friction.

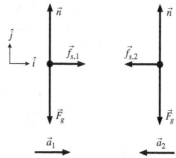

EVALUATE When the paperback book accelerates, Newton's second law (in component form) gives

$$\left.\begin{array}{c} ma_1 = f_{s,1} \le \mu_{s,1}n \\ n - mg = 0 \end{array}\right\} \mu_{s,1} \ge \frac{a_1}{g} = \frac{2.286 \text{ m/s}^2}{9.8 \text{ m/s}^2} = 0.23$$

When the paperback book decelerates, we have

$$\left.\begin{array}{c} ma_2 > f_{s,2}^{max} = \mu_{s,2}n \\ n - mg = 0 \end{array}\right\} \mu_{s,2} = \frac{a_2}{g} = \frac{0.33 \text{ m/s}^2}{9.8 \text{ m/s}^2} = 0.30$$

Thus, the actual coefficient of static friction must lie in the range $0.23 \le \mu_s \le 0.30$.

ASSESS Notice that the force due to friction acts to oppose any potential velocity of the paperback book with respect to the reference book.

51.   **INTERPRET**   This problem involves Newton's second law. We are asked to find the maximum acceleration of a front-wheel-drive car, given that 70% of its mass rests on the front wheels.

DEVELOP   Draw a free-body diagram for the combined front wheels of the car, which will be the same diagram as for the sled on level ground in the preceding problem. Note that the mass resting on the front wheels is $m_f = 0.7m$, so $F_g = 0.7mg$. The mass entering into Newton's second law, however, is m, because the entire car is being accelerated. Applying Newton's second law to the combined front-wheel system gives

$$x: \quad f_s = ma$$
$$y: \quad n - 0.7mg = 0$$

EVALUATE   Using Equation 5.2 $f_s \leq \mu_s n$ in Newton's second law gives

$$a = \frac{f_s}{m} \leq \frac{\mu_s n}{m} = \frac{\mu_s (0.7mg)}{m}$$

$$a \leq 0.7\mu_s g = 0.7(0.61)(9.8 \text{ m/s}^2) = 4.2 \text{ m/s}^2$$

Thus, the car's maximum acceleration is 4.2 m/s^2.

ASSESS   Note that as the car accelerates, the proportion of weight carried by the back wheels increases. To take advantage of this effect, some people drive backwards in front-wheel-drive cars when the road is very slick.

53.   **INTERPRET**   This problem involves Newton's second law, kinematics, and the force due to kinetic friction. The object of interest is a swimmer that slides down the slide, and we are asked to find the coefficient of kinetic friction given the relative time it takes a swimmer to slide down the slide when it's wet and when it's dry.

DEVELOP   The free-body diagram for this problem is the same as that for the sled on the slope in Problem 5.50. Apply Newton's second law to a swimmer sliding down the dry slide ($\mu_k \neq 0$) to find the acceleration, then let $\mu_k \to 0$ to find the acceleration for the wet slide. This gives

$$\left. \begin{array}{l} x: \quad mg\sin(\theta) - f_k = ma \\ y: \quad n - mg\cos(\theta) = 0 \end{array} \right\} \quad a = g\left[\sin(\theta) - \mu_k \cos(\theta)\right]$$

To find the coefficient of kinetic friction, use the kinematic Equation 2.10 $x = x_0 + v_0 t + at^2/2$ to relate the acceleration to the time it takes to travel down the slide with $v_0 = 0$.

EVALUATE   The time it takes to slide down the dry slide is

$$x = x_0 + \overset{=0}{v_0} t + at^2/2$$

$$t_{dry} = \pm\sqrt{\frac{2(x - x_0)}{a}} = \pm\sqrt{\frac{2(x - x_0)}{g\left[\sin(\theta) - \mu_k \cos(\theta)\right]}}$$

Letting $\mu_k \to 0$, we get the time it takes to descend the wet slide:

$$t_{wet} = \pm\sqrt{\frac{2(x - x_0)}{g\sin(\theta)}} = t_{dry}\sqrt{1 - \mu_k \cot(\theta)}$$

We are told that $t_{wet}/t_{dry} = 1/3$, so

$$\frac{t_{wet}}{t_{dry}} = \frac{1}{3} = \sqrt{1 - \mu_k \cot(\theta)}$$

$$\mu_k = \frac{8\tan(35°)}{9} = 0.62$$

ASSESS   Notice that we did not need to evaluate the acceleration, we simply needed to find that the ratio of the accelerations is $1 - \mu_k \cot(\theta)$.

55.   **INTERPRET**   This problem involves Newton's second law and kinematics. The object of interest is the box, and the forces involved are gravity $\vec{F}_g$, the normal force $\vec{n}$, static and kinetic friction $\vec{f}_k$ and $\vec{f}_s$. We are asked to find how far up a slope the box will travel given its initial speed and coefficient of kinetic friction.

DEVELOP   Draw a free-body diagram for the situation (see figure below). Use Newton's second law to find the acceleration of the block as it travels up the slope, then insert the result into the kinematic Equation 2.11 $v^2 = v_0^2 + 2a(x - x_0)$ to find the distance the box travels up the slope.

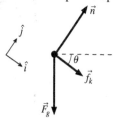

EVALUATE   (a) In component form, Newton's second law gives

$$\left. \begin{array}{l} x: \quad f_k + mg\sin(\theta) = ma \\ y: \quad n - mg\cos(\theta) = 0 \end{array} \right\} a = g\sin(\theta) + \mu_k g\cos(\theta)$$

where we have used $F_g = mg$ and Equation 5.3 $f_k = \mu_k n$ for the force due to kinetic friction. Inserting this result into Equation 2.11 and solving for the distance $x - x_0$ gives

$$\overset{=0}{\cancel{v^2}} = v_0^2 + 2a(x - x_0)$$

$$x_0 - x = \frac{v_0^2}{2a} = \frac{v_0^2}{2\left[g\sin(\theta) + \mu_k g\cos(\theta)\right]} = \frac{(1.4 \text{ m/s})^2}{2(9.8 \text{ m/s}^2)\left[\sin(22°) - (0.70)\cos(22°)\right]} = 0.10 \text{ m}$$

Thus, the block travels 10 cm up the slope.

(b) When the block has stopped, Newton's second law still applies, but with the force due to static friction, which will be oriented up the incline to resist motion down the incline. Thus, Newton's second law gives us

$$\left. \begin{array}{l} x: \quad -f_s + mg\sin(\theta) = ma \\ y: \quad n - mg\cos(\theta) = 0 \end{array} \right\} a \geq g\sin(\theta) - \mu_k g\cos(\theta)$$

where we have used Equation 5.2 $f_s \leq \mu_s n$. The acceleration will be positive if $\mu_s < \tan(\theta) = \tan(22°) = 0.404$. However, this value is less than that for kinetic friction, so it is likely that μ_s exceeds this value, because $\mu_s > \mu_k$ is generally true. Thus, we conclude that the block does not slide back down the slope.

ASSESS   Notice how the direction of the frictional force depends on the circumstances: to resist the motion it was oriented down the incline in part (a) and up the incline in part (b).

57.   **INTERPRET**   Your speed can be determined by assuming uniform circular motion through the turn. Because the curve is banked, the centripetal force is provided by the normal force, but there may also be a contribution from the friction between your tires and the road.

DEVELOP   Example 5.6 describes a curve designed for a certain speed must be banked at an angle given by:

$$\tan\theta = \frac{v_d^2}{gr} = \frac{(80 \text{ km/h})^2}{(9.8 \text{ m/s}^2)(210 \text{ m})} = 0.240$$

In this case, however, we investigate what would have happened if you went faster than the designed speed. To stay in your lane (at the same radius), you would need to turn the steering wheel slightly, causing some friction parallel to the road and perpendicular to the car's motion. The three forces acting on the car are represented in the figure below.

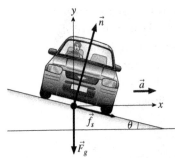

Notice that we have chosen the coordinate axes so that the centripetal acceleration, $\vec{a} = \left(v^2 / r \right) \hat{r}$, points in the $+x$ direction. The sum of the forces for the two components are

$$x: \quad n \sin \theta + f_s \cos \theta = mv^2 / r$$
$$y: \quad n \cos \theta - f_s \sin \theta - F_g = 0$$

The friction has a maximum limit: $f_s \leq \mu_s n$. We will use this to find the maximum limit on your speed.

EVALUATE We don't know the mass of the car, but we can combine the two component equations to eliminate the mass. With some algebra, we arrive at:

$$\frac{f_s}{n} = \frac{\frac{v^2}{gr} \cos \theta - \sin \theta}{\cos \theta + \frac{v^2}{gr} \sin \theta} \leq \mu_s$$

Dividing the numerator and denominator by $\cos \theta$, and using $\tan \theta = v_d^2 / gr$, we get

$$\frac{v^2}{gr} - \frac{v_d^2}{gr} \leq \mu_s \left(1 + \frac{v^2 v_d^2}{\left(gr \right)^2} \right)$$

With a little more algebra and using the value of v_d^2 / gr from above, we obtain

$$v \leq v_d \sqrt{\frac{1 + \mu_s \frac{gr}{v_d^2}}{1 - \mu_s \frac{v_d^2}{gr}}} = \left(80 \text{ km/h} \right) \sqrt{\frac{1 + 0.15 / 0.240}{1 - 0.15 \cdot 0.240}} = \left(80 \text{ km/h} \right) \left(1.30 \right) = 100 \text{ km/h}$$

So you could have been speeding around the curve by as much as 20 km/h over the posted limit.

ASSESS We can check whether our final expression for the maximum speed makes sense. If $\mu_s = 0.9$ (characteristic of a dry road), we get $v \leq 200 \text{ km/h}$. If $\mu_s = 0$, our expression becomes to $v \leq v_d$, which just says that we must drive at the posted speed if the road is frictionless. If the road is not banked, then $v_d \to 0$, and our expression reduces to $v \leq \sqrt{\mu_s gr}$, which agrees with our derivation for the unbanked curve in Problem 3.31.

59.   **INTERPRET** This problem involves Newton's second law and centripetal acceleration. The forces involved are gravity $\vec{F}_g = m\vec{g}$ and the normal force $\vec{n}$ that is everywhere perpendicular to the track. The aim of this problem is to establish the condition under which a car moving too slowly as it goes around a loop-the-loop roller coaster would leave the track.

DEVELOP Assume the roller coaster travels counter-clockwise around the loop-the-loop. Draw a free-body diagram for the roller coaster at an arbitrary point on the track (see figure below), with the angle ϕ of departure from the track indicated. Applying Newton's second law to the roller coaster gives $\vec{F}_g + \vec{n} = m\vec{a}$. For the roller coaster to stay on the track, the radial component of the net force (toward the center of the track) must equate to the mass times the centripetal acceleration,

$$n + mg \cos \left(\phi \right) = \frac{mv^2}{r}$$

The tangential component of the net force is not of interest for this problem.

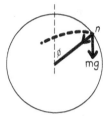

EVALUATE   The car leaves the track when the normal force becomes zero (no more contact):

$$mg \cos \phi = \frac{mv^2}{r} \quad \Rightarrow \quad \cos(\phi) = \frac{v^2}{gr}$$

which is the expression given in the problem statement.

ASSESS   The result implies that the car leaves the track when the speed is too small: $v < \sqrt{gr}$. Otherwise, the car never leaves the track, as in Example 5.7.

61.   **INTERPRET**   This problem involves Newton's second law, uniform circular motion, and frictional forces. The forces involved are the force due to gravity $\vec{F}_g = m\vec{g}$, the normal force $\vec{n}$ that acts perpendicular to the wall, and the force due to static friction $\vec{f}_s$. We are to find the frequency of revolution needed to prevent the book from slipping down the wall.

DEVELOP   Draw a free-body diagram of the situation (see figure below). Because this is uniform circular motion (i.e. the centrifuge rotates at a constant rate), Equation 5.1 $a = v^2/r$ applies for the centripetal acceleration. The direction of this acceleration is toward the center of the circle of rotation, as indicated in the drawing. Applying Newton's second law in the vertical and horizontal directions gives

$$x: \quad n = ma = mv^2/r$$
$$y: \quad f_s - mg = 0$$

Using Equation 5.2 $f_s = \mu_s n$ in Newton's second law leads to $g/\mu_s = v^2/r$. From this expression, we can find the frequency needed to maintain the book in place.

EVALUATE   The frequency of revolution ϕ is related to the tangential velocity by $v = 2\pi r\phi$. Inserting this into the expression above and solving for ϕ gives

$$\frac{g}{\mu_s} = \frac{(2\pi r\phi)^2}{r}$$

$$\phi \geq \left| \pm\sqrt{\frac{g}{4\pi^2 r\mu_s}} \right| = \sqrt{\frac{(9.8 \text{ m/s}^2)}{4\pi^2 (5.1 \text{ m})(0.62)}} = (0.28 \text{ s}^{-1})\left(\frac{60 \text{ s}}{1 \text{ min}}\right) = 17 \text{ min}^{-1}$$

Thus, the centrifuge must rotate at a frequency of 17 min^{-1} or greater.

ASSESS   The two possible signs for the frequency correspond to rotations clockwise and counter clockwise.

63.   **INTERPRET**   This problem involves Newton's second law, uniform circular motion, and frictional forces. The object of interest is the car, and we are to find whether braking in a straight line will stop the car before it hits the truck, or whether it's better to swerve in as tight a circular turn as possible. The forces acting on the car are the force due to gravity $\vec{F}_g = m\vec{g}$ and the force due to kinetic friction $\vec{f}_k$ for the former option and the force due to static friction $\vec{f}_s$ for the latter option.

DEVELOP   For the braking option, Newton's second law applied to the car in the x and y directions gives

$$\left. \begin{array}{l} x: \quad f_k = ma \\ y: \quad n - mg = 0 \end{array} \right\} \mu_s g = a,$$

where we have used Equation 5.3 $f_s = \mu_s n$. For the swerve option, Newton's second law applied in the x and y directions gives

$$\left.\begin{array}{l} x: \quad f_s = ma = mv^2/r \\ y: \quad n - mg = 0 \end{array}\right\} \mu_s g = v^2/r$$

Use the kinematic Equation 2.11 $v^2 = v_0^2 + 2a(x - x_0)$ to find the stopping distance in the braking option, and calculate the turning radius r for the swerve option. Compare these results to decide which option to take.

EVALUATE   For the braking option, the stopping distance is

$$\overset{=0}{\overbrace{v^2}} = v_0^2 + 2a(x - x_0)$$

$$x - x_0 = -\frac{v_0^2}{2a} = -\frac{v_0^2}{2(-\mu_s g)} = \frac{v_0^2}{2\mu_s g}$$

where the acceleration has a negative sign because it is oriented opposite to the velocity. For the swerving option, the turning radius is $r = v^2/\mu_s g = (x - x_0)$. Thus the turning radius is greater then the stopping distance, so you should chose to brake in a straight line rather than swerve.

ASSESS   Note that if the coefficient of static friction decreases from its maximum value of μ_s, the turning radius will get larger, and the linear acceleration will decrease, as expected.

65.   **INTERPRET**   This problem involves Newton's second law, Hooke's law (see Equation 4.9), and uniform circular motion. The object of interest is the $m = 2.1$-kg mass, and we are to find the radius of its circular trajectory given the spring constant and the tangential speed at which it travels. Because the table is frictionless, we only need consider the forces acting horizontally, so the only force of interest is the radial force due to the spring.

DEVELOP   In the horizontal plane, Newton's second law gives $k(r - r_0) = ma = mv^2/r$, where $r_0 = 18$ cm is the unstretched spring length and r is the stretched spring length that we are to find.

EVALUATE   Inserting the given quantities into the above expression gives

$$k(r - r_0) = mv^2/r$$

$$kr^2 - kr_0 r - mv^2 = 0$$

$$r = \frac{kr_0 \pm \sqrt{k^2 r_0^2 + 4kmv^2}}{2k} = \frac{1}{2}\left[0.18 \text{ m} \pm \sqrt{(0.18 \text{ m})^2 + 4(2.1 \text{ kg})(1.4 \text{ m/s})^2/(150 \text{ N/m})}\right] = 0.28 \text{ m}$$

ASSESS   Can you convince yourself that the units under the radical work out to be meters?

67.   **INTERPRET**   We need to find the minimum force necessary to move the trunk in Example 5.11, for an arbitrary value of μ_k. We also need to find the angle at which this minimum occurs. We can do this using calculus to find the minimum of the function.

DEVELOP   For simplicity, we define the reduced tension

$$T' \equiv \frac{T}{mg} = \frac{\mu_k}{\cos(\theta) + \mu_k \sin(\theta)}$$

Then we can find the minimum value of T' by setting $dT'/d\theta = 0$ and solving for the optimal angle θ.

EVALUATE

$$\frac{dT'}{d\theta} = 0 = \mu_k \frac{d}{d\theta}\left[\cos(\theta) + \mu_k \sin(\theta)\right]^{-1} = -\mu_k \frac{-\sin(\theta) + \mu_k \cos(\theta)}{\left[\cos(\theta) + \mu_k \sin(\theta)\right]^2}$$

The numerator must be zero for this equation to be satisfied, so

$$\sin\theta = \mu_k \cos\theta_k$$

$$\theta = \text{atan}(\mu_k)$$

We substitute this value of θ into the expression for T' to obtain

$$T' = \frac{\mu_k}{\cos\left[\mathrm{atan}\left(\mu_k\right)\right] + \mu_k \sin\left[\mathrm{atan}\left(\mu_k\right)\right]}$$

$$= \frac{\mu_k}{\left(\sqrt{1+\mu_k^2}\right)^{-1} + \mu_k^2\left(\sqrt{1+\mu_k^2}\right)^{-1}}$$

$$= \frac{u_k\sqrt{1+\mu_k^2}}{1+\mu_k^2}$$

$$= \frac{u_k}{\sqrt{1+\mu_k^2}}$$

ASSESS We can check our answer by substituting $\mu_k = 0.75$ in our answer and comparing the result with the graphical solution from Problem 5.66. We find $\theta = \mathrm{atan}(0.75) = 36.9°$ and $T' = 0.75\big/\sqrt{1+(0.75)^2} = 0.6$. These answers match previously with what we obtained.

69. **INTERPRET** We need to develop an equation for the trajectory of an object with an initial horizontal velocity. The object has a velocity-dependent drag force like that of Problem 68, and the initial x velocity is equal to the terminal velocity in Problem 68. We also need to find the maximum horizontal distance the object can move.

DEVELOP We find the horizontal and vertical motions separately. We can do this because the force is linear with velocity, as are the individual components. The equation for the vertical component of velocity has been solved in Problem 68: $v_y(t) = \frac{mg}{b}\left(e^{-bt/m} - 1\right)$; we will need to integrate this once more to find vertical position $y(t)$. As for the horizontal motion, the only force is the drag ($F = -bv_x$), so by Newton's second law we have differential equation: $-bv_x = m\frac{dv_x}{dt}$. We find the solution to this, then integrate it to get the equation for horizontal motion as a function of time, $x(t)$, and then take the limit as $t \to \infty$ to find the maximum range. We then invert the equation for horizontal motion to get an equation for $t(x)$, and plug this into our equation for $y(t)$ to find $y(x)$.

EVALUATE First, we find $y(t)$ by separating the y variable from the t variable in the velocity equation:

$$v_y(t) = \frac{dy}{dt} = \frac{mg}{b}\left(e^{-bt/m} - 1\right) \quad \to \quad dy = \frac{mg}{b}\left(e^{-bt/m} - 1\right)dt$$

Integrating both sides of the equation gives

$$\int dy = \int \frac{mg}{b}\left(e^{-bt/m} - 1\right)dt \quad \to \quad y(t) = \frac{mg}{b}\left[-\frac{m}{b}e^{-bt/m} - t + C_1\right]$$

We'll assume the object starts at the origin, so $y(0) = 0 = \frac{mg}{b}[-\frac{m}{b} + C_1]$, which implies $C_1 = \frac{m}{b}$. In final form, the vertical position as a function of time is

$$y(t) = \frac{mg}{b}\left[\frac{m}{b}\left(1 - e^{-bt/m}\right) - t\right]$$

Now, turning to the horizontal direction, the differential equation for the velocity can be solved with a function of the form $v_x(t) = Ae^{-bt/m}$. The initial horizontal velocity is $v_{x0} = \frac{mg}{b}$ so $A = \frac{mg}{b}$ and $v_x(t) = \frac{mg}{b}e^{-bt/m}$. We integrate this like we did $v_y(t)$ to find the horizontal position:

$$v_x(t) = \frac{dx}{dt} = \frac{mg}{b}e^{-bt/m} \quad \to \quad x(t) = \int \frac{mg}{b}e^{-bt/m}dt = -\frac{m^2 g}{b^2}e^{-bt/m} + C_2$$

We have assumed $x(0) = 0$, so it follows that $C_2 = \frac{m^2 g}{b^2}$ and

$$x(t) = \frac{m^2 g}{b^2}\left(1 - e^{-bt/m}\right)$$

Since x gets larger as t gets larger, the maximum value of x is

$$x_{max} = \lim_{t \to \infty}[x(t)] = \frac{m^2 g}{b^2} = \frac{mv_{x0}}{b}$$

We now invert the equation $x(t)$ by rearranging the terms and taking the natural log of both sides:

$$t(x) = -\frac{m}{b} \ln\left(1 - \frac{x}{x_{max}}\right)$$

Plugging this into the $y(t)$ equation

$$y(t) = \frac{mg}{b}\left[\frac{m}{b}\left(1 - \exp\left(\frac{-b}{m}\left\{\frac{-m}{b}\ln\left(1 - \frac{x}{x_{max}}\right)\right\}\right)\right) - \left\{\frac{-m}{b}\ln\left(1 - \frac{x}{x_{max}}\right)\right\}\right]$$

$$= x_{max}\left[\left(1 - \left(1 - \frac{x}{x_{max}}\right)\right) + \ln\left(1 - \frac{x}{x_{max}}\right)\right]$$

$$= x + x_{max}\ln\left(1 - \frac{x}{x_{max}}\right)$$

ASSESS   Does our answer make sense? Plugging in $x = 0$, $y = 0 + x_{max}\ln(1) = 0$, which is exactly what we assumed for the starting position. As $x \to x_{max}$, $y \to -\infty$, which is what we would expect as the object continues to fall straight down through the fluid.

71.   **INTERPRET**   We need to find the tension in the three strings holding up the flower arrangement. If the tension in any one segment is more than 100 N, then we need heavier string. The flowers are in equilibrium, so we use Newton's first law: The components of the forces on each pot will sum to zero.

   DEVELOP   We start with a free-body diagram, as shown in the figure below. After resolving each tension into the horizontal and vertical components, we set the sum of components along each axis, for each pot, equal to zero and solve for the tensions.

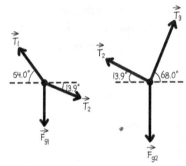

EVALUATE   For the left pot,

$$\left.\begin{aligned} T_{x1} &= -T_1\cos(54.0°) \\ T_{y1} &= T_1\sin(54.0°) \\ T_{x2} &= T_2\cos(13.9°) \\ T_{y2} &= -T_2\sin(13.9°) \\ F_{g1} &= -m_1 g \end{aligned}\right\} \quad \begin{aligned} -T_1\cos(54.0°) + T_2\cos(13.9°) &= 0 \\ T_1\sin(54.0°) - T_2\sin(13.9°) - m_1 g &= 0 \end{aligned}$$

For the right pot,

$$\left.\begin{aligned} T_{x3} &= T_3\cos(68.0°) \\ T_{y3} &= T_3\sin(68.0°) \\ T_{x2} &= -T_2\cos(13.9°) \\ T_{y2} &= T_2\sin(13.9°) \\ F_{g2} &= -m_2 g \end{aligned}\right\} \quad \begin{aligned} T_3\cos(68.0°) - T_2\cos(13.9°) &= 0 \\ T_3\sin(68.0°) + T_2\sin(13.9°) - m_3 g &= 0 \end{aligned}$$

This problem is over-specified: There are three unknowns and four equations. However, we can still estimate if any of the forces are greater than 100 N.

Start with the left pot and solve the top equation for T_1 then substitute that value into the second equation to find T_2:

$$T_1 = T_2 \frac{\cos(13.9°)}{\cos(54.0°)}$$

$$T_2 \frac{\cos(13.9°)}{\cos(54.0°)} \sin(54.0°) - T_2 \sin(13.9°) = m_1 g$$

$$T_2 (1.336 - 0.240) = (3.85 \text{ kg})(9.8 \text{ m/s}^2)$$

$$T_2 = 34.4 \text{ N}$$

Now substitute this back into the equation for T_1 to find that tension: $T_1 = T_2 \cos(13.9°)/\cos(54°) = 56.9$ N. Do the same thing for the right pot.

$$T_3 = T_2 \frac{\cos(13.9°)}{\cos(68.0°)}$$

$$T_2 \frac{\cos(13.9°)}{\cos(68.0°)} \sin(68.0°) + T_2 \sin(13.9°) = m_3 g$$

$$T_2 (2.403 + 0.240) = (9.28 \text{ kg})(9.8 \text{ m/s}^2)$$

$$T_2 = 34.4 \text{ N}$$

Now substitute this back into the equation for T_3 to find $T_3 = T_2 \cos(13.9°)/\cos(68.0°) = 89.2$ N. Thus, the 100 N string will suffice.

ASSESS Although 100 N string will work, we would be better off with a bigger margin of error. As it is, the far-right string will break if anyone waters the plants!

73. **INTERPRET** This problem involves uniform circular motion. We need to find the minimum radius, at a given speed for a vertical circle, if the acceleration is not to exceed six times that of gravity. We find the centripetal acceleration of the plane, and remember that gravity is also a factor.

DEVELOP Converting the speed and the acceleration to SI units gives $1.8 \times (340 \text{ m/s}) = 612$ m/s. Note that gravity provides 1g of acceleration no matter how fast the pilot is flying, so the centripetal acceleration has to provide the remaining 5g. Because we assume uniform circular motion, we can use Equation 5.1 $a = v^2/r$ to calculate the centripetal acceleration.

EVALUATE Inserting $a = 5g = 49$ m/s^2 into Equation 5.1 and solving for the radius r gives

$$r = \frac{v^2}{a} = \frac{(612 \text{ m/s})^2}{5(9.8 \text{ m/s}^2)} = 7.6 \text{ km}$$

ASSESS If he wants to make a smaller loop, he will have to slow down! But notice that the acceleration is a quadratic function of velocity ($a \propto v^2$), so a 50% reduction in speed (while maintaining the same radius) would reduce the total acceleration to $5g/4 + 1g = 2.25g$.

75. **INTERPRET** We are asked to analyze the movement of a figure skater.

DEVELOP The forces acting on the skater are gravity, the normal force, and friction. The friction is the only force pointing in the horizontal direction, and it must account for the centripetal acceleration that points to her left.

EVALUATE To turn left, the net force will have to point left.

The answer is (a).

ASSESS If there were no friction force pointing to the left, the leaning skater's foot would slip out from underneath her and she would fall. In Chapter 12, we'll learn how to analyze a problem such as this by considering the net torque on the skater.

77.   **INTERPRET**   We are asked to analyze the movement of a figure skater.

DEVELOP   The three forces do not all act on the same part of the skater's body. The weight is applied at the center of mass, while the normal and friction forces are applied at the skate. If we imagine a line from the skate to the center of mass, there cannot be any net force perpendicular to this line, otherwise the skater will start to rotate around her center of mass (see Problem 5.40). The sum of the vertical normal force and the horizontal friction force is a force pointing in the direction $\theta = \tan^{-1}\left(f_s / n\right)$ with respect to the vertical. The skater should be leaning at this angle to avoid having a rotating force (or torque).

EVALUATE   The normal balances the downward weight, $n = mg$, and the friction is providing the centripetal force: $f_s = ma_c$. Therefore, if the skater tilt is $\theta = \tan^{-1}\left(0.5\right)$, the centripetal acceleration must be $a_c = 0.5g \approx 5 \text{ m/s}^2$.

The answer is (c).

ASSESS   As mentioned in Problem 5.75, we will later learn specific techniques for how to deal with forces applied away from the center of mass.

6

WORK, ENERGY, AND POWER

EXERCISES

Section 6.1 Work

11. **INTERPRET**   This problem involves the concept of work. You are doing work on the shopping cart by pushing it around.

    DEVELOP   Assume the force is constant and is applied in the horizontal direction, in which case this is a one-dimensional problem and Equation 6.1 applies.

    EVALUATE   Inserting the given quantities into Equation 6.1 gives the work done as

    $$W = F \Delta x = (75 \text{ N})(12 \text{ m}) = 900 \text{ J}$$

    ASSESS   If it takes you 30 seconds to cover this distance, the power expended would be $P = W/\Delta t = (900 \text{ J})/(30 \text{ s}) = 30 \text{ W}$. This gives you some appreciation for the energy needed to power a 60-W light bulb.

13. **INTERPRET**   The problem involves work, which is done by the crane on the beam. We are to find the work done to lift the box vertically 23 m.

    DEVELOP   From the definition of work as the scalar product of force and distance (see Equation 6.3), we see that no work is done when the force applied is perpendicular to the displacement. This is the case for the crane when it swings the beam eastward by 18 m. The crane applies a vertical force (to counter gravity) and the displacement is horizontal (eastward). Thus, we need only concern ourselves with the vertical displacement of the beam. Furthermore, if the beam moves with constant speed, we know that the vertical force applied must be constant and be equal to the weight of the box ($F_{app} = mg$, see previous problem). Thus, we can apply Equation 6.1 to the vertical displacement to find the work done.

    EVALUATE   Inserting the given quantities into Equation 6.1 gives the work done:

    $$W = F_{app} \Delta y = mg \Delta y = (650 \text{ kg})(9.8 \text{ m/s}^2)(23 \text{ m}) = 150 \text{ k J}$$

    to two significant figures.

    ASSESS   We could have used the more general Equation 6.5 to find the work. This gives

    $$W = \vec{F} \cdot \Delta \vec{r} = F_{app} \hat{j} \cdot (18 \text{ m} \hat{i} + 23 \text{ m} \hat{j}) = F_{app} (23 \text{ m}) = 150 \text{ k J}$$

    which agrees with our previous result.

15. **INTERPRET**   This problem involves the average force exerted by the meteorite on the Earth. It is a one-dimensional problem because all forces and displacements are in the same direction (i.e., vertical).

    DEVELOP   Because we are interested in the average force, which is constant during the meteorite's deceleration period, we can use Equation 6.1 $W = F \Delta x$ to find the average force. We are given the W = 140 MJ and Dx = 75 cm = 0.75 m.

EVALUATE   Solving Equation 6.1 for the force and inserting the given quantities gives an average force of

$$W = F\Delta x$$

$$F = \frac{W}{\Delta x} = \frac{140 \text{ MJ}}{0.75 \text{ m}} = 190 \text{ MN}$$

to two significant figures.

ASSESS   Notice that we did not need to convert from MJ to J, we simply retained the prefactor M (= 10^6) in our calculation. Thus, the units of MN are units of force. Using the fact that dynamite carries 7.5 MJ/kg of explosive energy, this meteorite impact delivered the equivalent of (140 MJ)/(7.5 MJ/kg) ≈ 19 kg of dynamite (about 41 lbs).

17.   INTERPRET   This problem is an exercise in vector properties. We are asked to show that the scalar product (or dot product) of two vectors is distributive.

DEVELOP   Use the definition of the scalar product (Equation 6.4) to demonstrate the distributive property of the vector scalar product.

EVALUATE   Using the definition of the vector scalar product, we see that

$$\vec{A} \cdot \left(\vec{B} + \vec{C} \right) = A_x \left(B_x + C_x \right) + A_y \left(B_y + C_y \right) + A_z \left(B_z + C_z \right)$$
$$= A_x B_x + A_y B_y + A_z B_z + A_x C_x + A_y C_y + A_z C_z$$
$$= \vec{A} \cdot \vec{B} + \vec{A} \cdot \vec{C}$$

ASSESS   We could also use Equation 6.3 to demonstrate the distributive property. This gives

$$\vec{A} \cdot \left(\vec{B} + \vec{C} \right) = AB \cos \left(\theta_{AB} \right) + AC \cos \left(\theta_{AC} \right) = \vec{A} \cdot \vec{B} + \vec{A} \cdot \vec{C}$$

19.   INTERPRET   This problem involves the concept of work. We are asked to find the distance a stalled car can be moved by a given amount of work.

DEVELOP   Because the force is directed at 17° to the car's displacement vector, we must use Equation 6.2, W = $F\cos(\theta)\Delta r$.

EVALUATE   Solving Equation 6.2 for Dr, and inserting the given quantitites, we find that the distance the car is moved is

$$\Delta r = \frac{W}{F \cos \theta} = \frac{860 \text{ J}}{\left(470 \text{ N} \right) \cos \left(17° \right)} = 1.9 \text{ m}$$

ASSESS   Only the horizontal component of the force, $F_x = F \cos \theta$, does the work. The vertical part of the force simply modifies the normal force experienced by the car.

Section 6.2 Forces that Vary

21.   INTERPRET   This problem involves the work done to stretch a spring from equilibrium to a given distance, and from that distance to a further distance.

DEVELOP   The problem can be solved by using Equation 6.8, from which Equation 6.10 is derived. [Notice that Equation 6.10 applies to the special case where one of the endpoints is the equilibrium position of the spring, which is not the case for part (b) of the problem.] The force applied to the spring is $F(x) = kx$, so Equation 6.8 gives

$$W = \int_{x_1}^{x_2} F \left(x \right) dx = \int_{x_1}^{x_2} \left(-kx \right) dx = \frac{k}{2} \left(x_1^2 - x_2^2 \right)$$

where x_1 and x_2 are the initial and final displacements from equilibrium, respectively.

EVALUATE   (a) The amount of work done in stretching from $x_1 = 0$ m to $x_2 = 0.1$ m is

$$W = \frac{200 \text{ N/m}}{2} \left[\left(0.1 \text{ m} \right)^2 - \left(0 \text{ m} \right)^2 \right] = 1 \text{ J}$$

(b) Similarly, to stretch from $x_1 = 0.1$ m to $x_2 = 0.2$ m from equilibrium requires

$$W = \frac{200 \text{ N/m}}{2} \left[\left(0.2 \text{ m} \right)^2 - \left(0.1 \text{ m} \right)^2 \right] = 3 \text{ J}$$

ASSESS   Another way to solve part (b) is to note that the work to stretch the spring from 0 to 20 cm is four times the work from part (a), or 4.0 J, so the work in part (b) is 4.0 J − 1.0 J = 3.0 J.

23. **INTERPRET** The problem is about work done to stretch a spring. We want to find out how much the spring can be stretched with a given amount of work.

DEVELOP Because the spring is stretched starting from its equilibrium position, the result of Equation 6.10, $W = kx^2/2$ can be applied. In this expression, x represents the distance from equilibrium that the spring is stretched (or compressed).

EVALUATE Solve Equation 6.10 for x and insert the given quantities. This gives

$$x = \pm\sqrt{\frac{2W}{k}} = \sqrt{\frac{2(8.5 \text{ J})}{190 \text{ N/m}}} = 0.299 \text{ m} = 30 \text{ cm}$$

to two significant figures. We have chosen the positive square root to reflect the fact that the spring is stretched, not compressed.

ASSESS Notice that x is inversely proportional to $\sqrt{k}$. This means that the stiffer the spring (greater k), the less it will be stretched, and vice versa. Also note that the work needed to stretch a spring an amount x is the same as is needed to compress it by this same amount.

Section 6.3 Kinetic Energy

25. **INTERPRET** This problem involves kinetic energy. The object of interest is the airplane, and we are to find its kinetic energy given its mass and velocity.

DEVELOP This a straight-forward application of Equation 6.13, $K = mv^2/2$, where K is the kinetic energy, $m = 2.4 \times 10^5$ kg is the mass, and $v = 900$ km/h is the speed.

EVALUATE The kinetic energy of the airplane is thus

$$K = \frac{1}{2}mv^2 = \frac{(2.4 \times 10^5 \text{ kg})(900 \text{ km/h})^2}{2}\left(\frac{10^3 \text{ m}}{\text{km}}\right)^2\left(\frac{\text{h}}{3600 \text{ s}}\right)^2 = 7.5 \times 10^9 \text{ J} = 7.5 \text{ GJ}$$

ASSESS The units work out to be

$$\frac{\text{kg} \cdot \cancel{\text{km}}^2 \cdot \text{m}^2 \cdot \cancel{\text{h}}^2}{\cancel{\text{h}}^2 \cdot \cancel{\text{km}}^2 \cdot \text{s}^2} = \text{N} \cdot \text{m} = \text{J}$$

as expected.

27. **INTERPRET** This problem involves kinetic energy. We are to find the speed at which the small car must travel so that it has the same kinetic energy as the large truck.

DEVELOP We will use Equation 6.13, K =mv2/2, to find the kinetic energy of each vehicle. By setting their kinetic energies equal, we can solve for the speed of the car.

EVALUATE Let the car's variables carry the subscript c, and the truck's variables carry the subscript T. The kinetic energy of each is $K_c = m_c v_c^2/2$ for the car and $K_T = m_T v_T^2/2$ for the truck. Setting these equal and solving for v_c gives

$$\frac{1}{2}m_c v_c^2 = \frac{1}{2}m_T v_T^2$$

$$v_c = \pm v\sqrt{\frac{m_T}{m_c}} = \pm(20 \text{ km/h})\sqrt{\frac{3.2 \times 10^4 \text{ kg}}{950 \text{ kg}}} = \pm 120 \text{ km/h}$$

ASSESS The plus/minus sign indicates that the car can either travel in the same direction as the truck, or in the opposite direction. Notice that we did not need to convert km/h to m/s for this problem, because the units of kg under the radical cancel.

29. **INTERPRET** This problem involves work and the work-energy theorem. Given a force acting on an object and the distance over which the force acts, we are asked to find the initial velocity of the object.

DEVELOP The work-energy theorem, Equation 6.14 ($W_{net} = \Delta K$) tells us that the net work done on the straw is its change in kinetic energy, which involves the straw's initial speed. Because the stopping force acts in the same direction as the straw's displacement in the tree (i.e., it's a one-dimensional problem), and assuming the stopping force is constant, we can use Equation 6.1, $W = F_x \Delta x$ to find the net work done on the straw by the tree. Because the force of the tree acts to oppose the displacement of the straw, the work is negative : $W = -F_x x$, where $x = 4.5$ cm. Equating this to the change in kinetic energy by the work-energy theorem allows us to find the initial velocity of the straw.

EVALUATE Equating the work done by the tree to the change in the straw's kinetic energy, then solving for the initial speed of the straw gives

$$W_{net} = -Fx = \frac{m}{2}\left(\overset{=0}{v_2^2} - v_1^2 \right)$$

$$v_1 = \pm\sqrt{\frac{2Fx}{m}} = \sqrt{\frac{2(70\ \text{N})(0.045)}{0.5 \times 10^{-3}\ \text{kg}}} = 110\ \text{m/s}$$

to two significant figures. Because the plus/minus sign simply indicates an initial velocity to the left or to the right, we have arbitrarily chosen the positive sign.

ASSESS This speed is reasonable for tornados, which usually have wind speeds between 18 and 140 m/s.

Section 6.4 Power

31. **INTERPRET** This problem is an exercise in converting power from kcal/day to Watts.

    DEVELOP From Appendix C, we find that 1 cal = 4.184 J, and we know that 1 day = (24 h)(3600 s/h) = 86,400 s.

    EVALUATE Performing the conversion gives

    $$\frac{2000\ \text{kcal}}{1\ \text{d}}\overbrace{\left(\frac{1\ \text{d}}{86,400\ \text{s}} \right)}^{=1}\overbrace{\left(\frac{1000\ \text{cal}}{1\ \text{kcal}} \right)}^{=1}\overbrace{\left(\frac{4.184\ \text{J}}{1\ \text{cal}} \right)}^{=1} = 97\ \text{J/s} = 97\ \text{W}$$

    ASSESS This is an *average* power. Human power output is higher during exercise.

33. **INTERPRET** This problem involves calculating the power output of a car battery, or the rate at which energy is drained from the battery.

    DEVELOP According to Equation 6.15, if work ΔW is done in time Δt, then the average power is $\overline{P} = \Delta W/\Delta t$.

    EVALUATE Using Equation 6.15, the power output for each of the three cases is

    (a) $$\overline{P} = \frac{\Delta W}{\Delta t} = \frac{(1\ \text{kW} \cdot \text{h})}{(1/60)\ \text{h}} = 60\ \text{kW}$$

    (b) $$\overline{P} = \frac{\Delta W}{\Delta t} = \frac{(1\ \text{kW} \cdot \text{h})}{1\ \text{h}} = 1\ \text{kW}$$

    (c) $$\overline{P} = \frac{\Delta W}{\Delta t} = \frac{(1\ \text{kW} \cdot \text{h})}{24\ \text{h}}\left(\frac{1000\ \text{W}}{\text{kW}} \right) = 41.7\ \text{W}$$

    ASSESS From Equation 6.15, we see that when the amount of work done is fixed, the average power is inversely proportional to Δt. Thus, the average power output is the greatest in case (a) and smallest in case (c).

35. **INTERPRET** This problem involves calculating the total work done, given average power and time.

    DEVELOP From Equation 6.15, if the average power is $\overline{P}$, then the amount of work done over a period Δt is $\Delta W = \overline{P}\Delta t$. Note that we need to convert hp to SI units, which we can do with the help of Appendix C, where we find 1 hp = 746 W.

    EVALUATE The work done by the lawnmower is

    $$\Delta W = \overline{P}\Delta t = (3.5\ \text{hp})(746\ \text{W/hp})(3600\ \text{s}) = 9.4 \times 10^6\ \text{J}$$

    ASSESS Given a constant average power, the work done is proportional to the time interval Δt. Note that the work done is positive, which means that the lawnmower is doing the work on the grass.

37. **INTERPRET**   In this problem we are asked to estimate the power output or rate of work, while doing deep knee bends at a given rate.

DEVELOP   For a single deep knee bend, our final position is the same as the initial position, so our net displacement is zero. Considering that this is a one-dimensional problem, we can use Equation 6.8 to find the total work done in a single deep knee bend, then divide this by the time required for a single deep knee bend to find the power (Equation 6.15, $\overline{P} = \Delta W / \Delta t$).

EVALUATE   Because the final position is the same as the initial position, we have $x_1 = x_2 \equiv x$ in the limits of the integral in Equation 6.8. Thus the work done for a single deep knee bend is

$$W = \int_x^x F(x)\,dx = 0 \text{ W}$$

Thus, no work is done, so (in theory) no power is expended!

ASSESS   We work up a sweat doing deep-knee bends because our bodies are working against a host of frictional forces. Thus, we are not working against gravity, because gravity gives us as much energy on the way down as it takes on the way up. Instead, we get our exercise from working against friction.

39. **INTERPRET**   This problem involves the concept of average power. We are asked to find the time it takes to melt an ice cube given the energy needed for the task and the average power supplied.

DEVELOP   Use the definition of average power (Equation 6.15), $\overline{P} = \Delta W / \Delta t$, to solve the problem, given that W $= 20$ kJ and $\overline{P} = 900$ W .

EVALUATE   The time required to melt the ice cube is

$$\Delta t = \frac{\Delta W}{P} = \frac{20 \times 10^3 \text{ J}}{900 \text{ W}} = 22 \text{ s}$$

ASSESS   This result seems reasonable given common experience with microwave ovens. Note that the result will depend on the mass of the ice cube (can you deduce the relationship?).

PROBLEMS

41. **INTERPRET**   The problem is about calculating work, given force and displacement. The object of interest is the box, which is being pushed up a ramp. For part (b) of the problem, we consider the work-energy theorem.

DEVELOP   Make a free-body diagram of the box (see figure below). Use Equation 6.5, $W = \vec{F} \cdot \Delta \vec{r}$, to calculate the work done in pushing the box up the ramp.

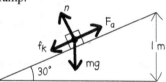

EVALUATE   (a) The box rises $\Delta y = 1$ m vertically. This means that the displacement up the ramp (parallel to the applied force) is

$$\Delta r = \frac{\Delta y}{\sin(\theta)} = \frac{1 \text{ m}}{\sin(30°)} = 2 \text{ m}$$

Therefore, the work done during this process is

$$W_{app} = \vec{F}_{app} \cdot \Delta \vec{r} = (200 \text{ N})(2 \text{ m})\cos(0°) = 400 \text{ J}$$

because the angle between the applied force and the displacement vector is 0°.

(b) To find the mass, we first note that the work done by gravity is

$$W_g = \vec{F}_g \cdot \Delta \vec{r} = \left(-mg\,\hat{j}\right) \cdot \left(\Delta x \hat{i} + \Delta y \hat{j}\right) = -mg\Delta y = -mg\Delta r \sin\theta$$

The work done by friction is

$$W_f = \vec{f}_k \cdot \Delta \vec{r} = -f_k \Delta r = -\mu_k n\, \Delta r = -\mu_k \left(mg\cos\theta\right)\Delta r$$

where in the last step we have used $n - mg\cos(\theta) = 0$, which results from applying Newton's second law to the box in the direction perpendicular to the incline. Because the speed of the box remains unchanged, the work-energy theorem $W = \Delta K$, says the total work must be zero:

$$W_{Tot} = W_{app} + W_g + W_f = 0$$

This implies

$$W_{app} = -W_g - W_f = mg\Delta r \sin\theta + \mu_k (mg\cos\theta)\Delta r = mg\Delta r (\sin\theta + \mu_k \cos\theta)$$

from which the mass is found to be

$$m = \frac{W_a}{g\Delta r (\sin\theta + \mu_k \cos\theta)} = \frac{F_a}{g(\sin\theta + \mu_k \cos\theta)} = \frac{200 \text{ N}}{(9.8 \text{ m/s}^2)[\sin(30°) + (0.18)\cos(30°)]}$$
$$= 31 \text{ kg}$$

ASSESS   The mass could also be found by solving Newton's second law, with zero acceleration:

$$F_{net} = F_{app} - mg(\sin\theta + \mu_k\cos\theta) = ma = 0$$

$$m = \frac{F_a}{g(\sin\theta + \mu_k\cos\theta)}$$

43.   **INTERPRET**   You want to find out how much work you do during a particular exercise.

DEVELOP   You only do work when lifting the weight (gravity does the work to bring the weight back down). The work required to lift the weight the given distance is $W = w\Delta y$ (it's irrelevant at what angle the force from your arms is applied – the net result is that the weight moves up by Δy). We'll need to convert the work to kcal using $1 \text{ kcal} = 4184 \text{ J}$ from Appendix C.

EVALUATE   (a) Each repetition requires you to exert

$$W = w\Delta y = (20 \text{ N})(0.55 \text{ m}) = 11 \text{ J}\left(\frac{1 \text{ kcal}}{4184 \text{ J}}\right) = 2.63 \times 10^{-3} \text{ kcal}$$

To get a 200 kcal workout, the number of reps you'd have to do is

$$N = \frac{200 \text{ kcal}}{2.63 \times 10^{-3} \text{ kcal}} = 76,000$$

(b) If your workout takes 1.0 min, then the power output is just the work divided by the time:

$$P = \frac{W}{\Delta t} = \frac{200 \text{ kcal}}{1.0 \text{ min}}\left(\frac{4184 \text{ J}}{1 \text{ kcal}}\right) = 14 \text{ kW}$$

ASSESS   The answers seem unreasonably large. Typically, lifting weights burns around 300 kcal per hour.

45.   **INTERPRET**   This problem involves calculating the work done as a result of a force acting at a nonzero angle with respect to the displacement. We are asked to find the angle that the rope makes with the horizontal, given the work, force, and distance over which the force acts.

DEVELOP   Because the force is not parallel to the displacement, we must use the more general equation for work; Equation 6.5, $W = \vec{F} \cdot \Delta\vec{r}$. In scalar form, dot product gives $W = F\Delta r\cos(\theta)$, where θ is the angle between the rope and the displacement direction (i.e., horizontal).

EVALUATE   We are given that $W = 2500$ J, $F = 120$ N, $\Delta r = 23$ m, so the angle θ is

$$\theta = \text{ac os}\left(\frac{W}{F\Delta r}\right) = \text{ac os}\left(\frac{2500 \text{ J}}{(120 \text{ N})(23 \text{ m})}\right) = 0.44 \text{ rad} = 25°$$

ASSESS   Notice that the argument of the acos function is dimensionless, as it should be. The angle 25° is a physically reasonable result.

47.   **INTERPRET**   This problem is an exercise in vector multiplication. We are asked to find the scalar product between two vectors of the form $a\hat{i} + b\hat{j}$ and $b\hat{i} - a\hat{j}$, and to find the angle between them, for arbitrary a and b.

DEVELOP   Use Equations 6.3 and 6.4 ($\vec{A} \cdot \vec{B} = AB\cos(\theta)$ and $\vec{A} \cdot \vec{B} = A_x B_x + A_y B_y + A_z B_z$, respectively).

EVALUATE   (a) The scalar product of $a\hat{i} + b\hat{j}$ and $b\hat{i} - a\hat{j}$ is $(a\hat{i} + b\hat{j}) \cdot (b\hat{i} - a\hat{j}) = ab - ab = 0$.

(b) The angle between the two vectors is $\theta = \mathrm{acos}(0) = 90°$.

ASSESS   Thus, for arbitrary a and b, the vectors $a\hat{i} + b\hat{j}$ and $b\hat{i} - a\hat{j}$ are perpendicular.

49.   **INTERPRET**   This problem involves finding the work done by the given force vector that acts through the given displacement.

DEVELOP   Use the general form of the expression for work, Equation 6.5: $W = \vec{F} \cdot \Delta\vec{r}$, with $\vec{F} = 67\hat{i} + 23\hat{j} + 55\hat{k}$ N and

$$\Delta\vec{r} = \vec{r}_2 - \vec{r}_1 = \left(21 - 16\right)\hat{i} + \left(10 - 31\right)\hat{j} + \left(14 - 0\right)\hat{k} \text{ m}$$
$$= 5\hat{i} - 21\hat{j} + 14\hat{k}$$

EVALUATE   Inserting the given force and displacement into Equation 6.5 gives

$$W = \left(67\hat{i} + 23\hat{j} + 55\hat{k} \text{ N}\right) \cdot \left(5\hat{i} - 21\hat{j} + 14\hat{k} \text{ m}\right) \vec{F} \cdot \Delta\vec{r} = \left(335 - 483 + 770\right) \text{ Nm} = 622 \text{ J}$$

ASSESS   Notice that we must keep track of the signs of the individual terms in doing the dot product to be sure to get the correct result.

51.   **INTERPRET**   This problem involves calculating spring constants given the work it takes to deform the springs.

DEVELOP   Use Equation 6.10, $W = kx^2/2$, to express the work W done in terms of the deformation x for each spring. We are given that $2W_A = W_B$ and $x_A = 2x_B$.

EVALUATE   For spring A, $W_A = k_A x_A^2/2$, and for spring B $W_B = k_B x_B^2/2$. Taking the ration of these two equations and using the given relations between springs A and B gives

$$\frac{W_A}{W_B} = \frac{k_A}{k_B}\frac{x_A^2}{x_B^2}$$
$$\frac{1}{2} = 4\frac{k_A}{k_B}$$
$$k_B = 8k_A$$

ASSESS   Note that the spring constant is linear in work, but quadratic in spring deformation.

53.   **INTERPRET**   This is a one-dimensional problem that involves calculating the work done given a non-constant force.

DEVELOP   The force given varies with position, so we need to use the more general expression for work in one dimension; Equation 6.10:

$$W = \int_{x_1}^{x_2} F(x)\,dx$$

with $F(x)$ given in the problem statement. The limit of the integration are from $x_1 = 0$ to $x_2 = x$.

EVALUATE   Evaluating the integral gives

$$W = \int_0^x F_0 \left[\frac{L_0 - x'}{L_0} - \frac{L_0^2}{(L_0 + x')^2} \right] dx'$$
$$= F_0 \left| \frac{1}{L_0}\left(L_0 x' - \frac{x'^2}{2} \right) + \frac{L_0^2}{L_0 + x'} \right|_0^x$$
$$= F_0 \left(x - \frac{x^2}{2L_0} + \frac{L_0^2}{L_0 + x} - L_0 \right)$$

ASSESS   Note that we changed the integration variable from x to x' simply to avoid confusing it with the upper limit x of the integration.

55.   **INTERPRET**   This problem involves calculating the (relative) speed of two particles, given their relative kinetic energy and mass.

DEVELOP   Use Equation 6.13, $K = mv^2/2$ to express the kinetic energy of each particle. Thus, the kinetic energy of particle 1 is $K_1 = mv_1^2/2$ and $K_2 = mv_2^2/2$. The problem states that $K_1 = K_2$, and $m_1 = 4m_2$, so we can find the ratio of the speeds by taking the ratio of the equations.

EVALUATE   Taking ratio K_1/K_2 gives

$$\frac{K_1}{K_2} = \frac{m_1 v_1^2}{m_2 v_2^2}$$

$$1 = 4\frac{v_1^2}{v_2^2}$$

$$v_2 = \pm 2v_1$$

ASSESS   The positive/negative sign indicates that the orientation of the speeds does not matter, only the magnitude matters. In other words, it does not matter if both particles move in the same direction, or if they move in opposite directions.

57.   **INTERPRET**   This is a one-dimensional problem that involves calculating the work done given a non-constant force.

DEVELOP   The force given varies with position, so we need to use the more general expression for work in one dimension; Equation 6.10:

$$W = \int_{x_1}^{x_2} F(x)\,dx$$

with $F(x)$ given in the problem statement. The limits of the integration are from $x_1 = 0$ to $x_2 = x$.

EVALUATE   Evaluating the integral gives

$$W = \int_0^{x_0}\left(\frac{F_0}{x_0}\right)x\,dx = \left(\frac{F_0}{x_0}\right)\frac{x^2}{2}\Big|_0^{x_0} = \left(\frac{F_0}{x_0}\right)\frac{x_0^2}{2} = \frac{1}{2}F_0 x_0$$

ASSESS   Thus, if the force varies linearly with position, the work varies quadratically.

59.   **INTERPRET**   This problem is an exercise in vector multiplication. We are given two vectors of equal magnitude and the relationship between their scalar product. With this information, we are to find the angle between the vectors.

DEVELOP   We are told that $A = B$ and that $\vec{A}\cdot\vec{B} = A^2/3$. Use Equation 6.3 to find the angle θ between the vectors.

EVALUATE   Evaluating the scalar product using Equation 6.3 gives

$$\vec{A}\cdot\vec{B} = AB\cos(\theta) = A^2\cos(\theta) = A^2/3$$

$$\theta = a\cos(1/3) = 70.5°$$

ASSESS   Note that an equivalent condition is $\vec{A}\cdot\vec{B} = B^2/3$ because $A = B$.

61.   **INTERPRET**   This problem involves converting power from W to gallons per day.

DEVELOP   From Appendix C we find that the energy content of oil is 39 kW·h/gal. Let the units guide you in converting from GW to gallons/day.

EVALUATE   The import rate is

$$\left(800\ \cancel{G}\,\cancel{W}\right)\overbrace{\left(\frac{1\ \text{gal}}{39\ \cancel{k}\,\cancel{W}\cdot\cancel{h}}\right)}^{=1}\overbrace{\left(\frac{10^6\ \cancel{k}}{\cancel{G}}\right)}^{=1}\overbrace{\left(\frac{24\ \cancel{h}}{\text{day}}\right)}^{=1} = 490\times10^{12}\ \text{gal/day}$$

ASSESS   This may also be express as 490 Tgal/day.

63. **INTERPRET** You have the mass and power of a car, and need to find the highest rate at which it can climb a given slope. You'll need to use work and energy techniques.

DEVELOP Assume the car is moving at constant speed, such that the net force on the car is zero. That means the force from the engine propelling the car forward along the road, F_c, must balance the component of the gravitational force that is parallel to the ground and points back down the slope. In other words, $F_c = F_g \sin \theta$. This force is related to the car's power through Equation 6.19: $P = \vec{F}_c \cdot \vec{v}$. As we have defined it, the force is in the same direction as the velocity of the car, so $P = F_c v$.

EVALUATE Using all its available power, the car can climb the slope at a speed of

$$v = \frac{P}{F_c} = \frac{P}{mg \sin \theta} = \frac{35 \text{ kW}}{(1750 \text{ kg})(9.8 \text{ m/s}^2) \sin 4.5°} = 26 \text{ m/s}$$

ASSESS This speed (58 mph) seems reasonable for the grade involved. The actual maximum speed will be lower due to air resistance, which is not negligible at this speed. Note, as well, that you can derive the same result by arguing that the car must work against gravity to climb the slope. Therefore, the component of its force pointing straight up must equal mg. The angle between this upward force and the velocity of the car is $90° - 4.5° = 85.5°$, so the power provided by the car is $P = mgv \cos(85.5°)$, which gives the same answer as the above equation.

65. **INTERPRET** This problem involves the concept of work and Newton's second law (for constant mass), $\vec{F}_{net} = m\vec{a}$. The object of interest is the box, and we are asked to find the work done to push it up an inclined slope a given distance.

DEVELOP Draw a free-body diagram of the situation (see figure below). To express the forces in terms of known quantities, apply Newton's second law to the box. This gives

$$\left. \begin{matrix} -f_k + F_{app} - mg \sin(\theta) = 0 \\ n - mg \cos(\theta) = 0 \end{matrix} \right\} -\mu_k mg \cos(\theta) + F_{app} - mg \sin(\theta) = 0$$

which we can solve for μ_k given that we know the work done by you pushing the box up the slope is $F_{app}d = 2.2$ kJ (see Equation 6.1) because the force you apply is in the same direction as the displacement of the box.

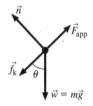

EVALUATE Inserting the known quantities into the expression above and solving for μ_k gives

$$\mu_k = \frac{F_{app} - mg \sin(\theta)}{mg \cos(\theta)} = \frac{(2200 \text{ J})/(3.1 \text{ m}) - (78 \text{ kg})(9.8 \text{ m/s}^2) \sin(22°)}{(78 \text{ kg})(9.8 \text{ m/s}^2) \cos(22°)} = 0.60$$

67. **INTERPRET** The object of interest here is the chest. The physical quantity we are asked to find is the power needed to push the chest against friction. This problem involves the concept of work and power, and we will have to use Newton's second law.

DEVELOP If you push parallel to a level floor, the applied force equals the frictional force (from Newton's second law, $F_{net} = ma$, where the acceleration is zero), so $F_a = f_k$. Because (again by Newton's second law) the normal force equals the weight of the box ($n = mg$) the applied force is

$$F_a = \mu_k n = \mu_k mg$$

Use Equation 6.19, $P = \vec{F} \cdot \vec{v}$, to find the power needed. Because we are applying a force in the same direction as the displacement of the box, we can use Equation 6.1, $W = F\Delta x$, to find the work done.

EVALUATE **(a)** The power required is

$$P_a = F_a v = \mu_k mgv = (0.78)(95 \text{ kg})(9.8 \text{ m/s}^2)(0.62 \text{ m/s}) = 450 \text{ W}$$

which is about 0.6 hp.

(b) The work done by the applied force acting over a displacement $\Delta x = 11$ m is

$$W_a = F_a \Delta x = \mu_k mg \Delta x = (0.78)(95 \text{ kg})(9.8 \text{ m/s}^2)(11 \text{ m}) = 8.0 \text{ kJ}$$

ASSESS   An alternative way to calculate the power is to note that the time required to push the chest 11 m is $\Delta t = \Delta x/v = (11 \text{ m})/(0.62 \text{ m/s}) = 17.74 \text{ s}$. Using Equation 6.17, we have

$$W_a = P_a \Delta t = (450 \text{ W})(17.74 \text{ s}) = 8.0 \text{ kJ}$$

69.   **INTERPRET**   This problem is about the total work done, given the power and time. The object of interest is the machine whose power output is given, and we are to find the total work is done over the given time interval.
   DEVELOP   The power given in this problem is time-varying. Therefore, use Equation 6.18: $W = \int_{t_1}^{t_2} P(t)\,dt$ to find the total work done, with $t_1 = 10$ s, $t_2 = 20$ s, and $P = ct^2$.
   EVALUATE   Inserting the given quantities into Equation 6.18, we obtain

$$W_{t_1 \to t_2} = \int_{t_1}^{t_2} ct^2\,dt = \frac{1}{3}ct^3 \Big|_{t_1}^{t_2} = \frac{c}{3}(t_2^3 - t_1^3) = \frac{1}{3}(18 \text{ W/s}^2)\left[(20 \text{ s})^3 - (10 \text{ s})^3\right] = 42 \text{ kJ}$$

   ASSESS   Because the power increases quadratically with t (i.e., as t^2), as time progresses, more work is done by the machine over the same interval of time. For example, the work done in a 10-s interval from $t_1 = 20$ s to $t_2 = 30$ s ($W_{20\,s \to 30\,s} = 114$ kJ) is greater than the work in a 10-s interval from $t_1 = 10$ s to $t_2 = 20$ s ($W_{10\,s \to 20\,s} = 42$ kJ).

71.   **INTERPRET**   This problem involves the concepts of power and work (or energy). Over a given period of time, the refrigerators will consume different amounts of energy, which we can calculate given their power consumption. We are to find when the cost difference for the energy consumed equals the difference in the price of the refrigerators.
   DEVELOP   To find the energy consumed, use Equation 6.17, $W = P\Delta t$. Thus, the work done (i.e., energy consumed) by the standard refrigerator is $W_s = P_s \Delta t_s$, where $P_s = 425$ W and $\Delta t_s = 0.20\Delta t$. The work done by the energy-efficient refrigerator in the same time interval is $W_{ee} = P_{ee}\Delta t_{ee}$, where $P_{ee} = 225$ W and $\Delta t_{ee} = 0.11\Delta t$. The cost difference Δc for the energy consumed is $\Delta c = p(W_s - W_{ee})$, where $p = 9.5$ ¢/kW·h is the price. We need to find the time interval for which the cost difference is equal to the difference in the price of the refrigerators.
   EVALUATE   The difference in the original price of the refrigerators is $\Delta p = \$1150 - \$850 = \$300$. The time interval to recuperate this difference is

$$\Delta p = \Delta c = p\left(P_s \Delta t_s - P_{ee}\Delta t_{ee}\right) = p\Delta t\left[(0.20)P_s - (0.11)P_{ee}\right]$$

$$\Delta t = \frac{\Delta p}{p\left[(0.20)P_s - (0.11)P_{ee}\right]} = \frac{\cancel{\$}300}{\left(\cancel{\$}0.095 \text{ kW}^{-1}\cdot\text{h}^{-1}\right)\left[(0.20)(0.425 \text{ kW}) - (0.11)(0.225 \text{ kW})\right]} = 5.24 \times 10^4 \text{ h} = 6.0 \text{ y}$$

   ASSESS   Notice that we converted the units so that all quantities were expressed in the same units. The answer is expressed to two significant figures because that is the least number of significant figures in the data.

73.   **INTERPRET**   This problem is about the total work done, given the power and time. The object of interest is the machine, and we are to show that the total work done is finite, even though the machine runs forever.
   DEVELOP   The power given in this problem is time-varying. Therefore, to find the work done in a given time interval, we need to use Equation 6.18, $W = \int_{t_1}^{t_2} P(t)\,dt$.
   EVALUATE   With $P(t) = P_0 t_0^2/(t + t_0)^2$, we obtain

$$W = \int_0^\infty \frac{P_0 t_0^2}{(t+t_0)^2}\,dt = P_0 t_0^2 \int_0^\infty \frac{dt}{(t+t_0)^2} = \frac{P_0 t_0^2}{(t+t_0)}\Big|_0^\infty = P_0 t_0$$

   ASSESS   The result shows that even though the machine operates forever, the total amount of work done is finite. This is not surprising because the power output decreases quadratically with time.

75.   **INTERPRET**   In this one-dimensional problem we are asked to find the work done by a non-constant force that varies with position. We want to show that although the force becomes arbitrarily large as x approaches zero, the work done remains finite.

DEVELOP Because we are dealing with a one-dimensional non-constant force $F(x)$ use Equation 6.8, $W = \int_{x_1}^{x_2} F(x)\, dx$, to find the work done. Let x_1 approach zero to find the limiting expression for the work.

EVALUATE With $F(x) = bx^{-1/2}$ we obtain

$$W_{x_1 \to x_2} = \int_{x_1}^{x_2} bx^{-1/2}\, dx = 2bx^{1/2}\Big|_{x_1}^{x_2} = 2b\left(\sqrt{x_2} - \sqrt{x_1}\right)$$

Thus, we see that $W_{x_1 \to x_2}$ is finite as $x_1 \to 0$. In fact, $W \to 2b\sqrt{x_2}$, for $x_1 \to 0$.

ASSESS The result demonstrates that even though a function $F(x)$ may diverge at some value $x = x_0$, the integral $\int F(x)\,dx$ can be finite at $x = x_0$.

77. **INTERPRET** Your task is to find the work needed to stretch a bungee-jump cord to double its unstretched length. The force exerted by the cord is similar to that of a spring, but with extra terms.

DEVELOP The applied force is equal and opposite to the cord's restorative force, applied to the cord, $\vec{F}_{app} = -\vec{F}$. To find the work required to double the length of the cord, we integrate the applied force from $x = 0$ to $x = L_0$.

EVALUATE (a) Integrating the force equation gives

$$W = \int_0^{L_0} kx + bx^2 + cx^3 + dx^5\, dx = \tfrac{1}{2}kL_0^2 + \tfrac{1}{3}bL_0^3 + \tfrac{1}{4}cL_0^4 + \tfrac{1}{5}dL_0^5$$

(b) With the given values the work becomes

$$W = \tfrac{1}{2}\left(420\tfrac{N}{m}\right)(10\text{m})^2 + \tfrac{1}{3}\left(-86\tfrac{N}{m^2}\right)(10\text{m})^3 + \tfrac{1}{4}\left(12\tfrac{N}{m^3}\right)(10\text{m})^4 + \tfrac{1}{5}\left(-0.50\tfrac{N}{m^4}\right)(10\text{m})^5 = 12\text{ kJ}$$

ASSESS Unlike for a spring, the work formula for the cord is not symmetric around $x = 0$. This is because the cord is easier to stretch than to squeeze. For example, the work needed to squeeze the cord to half its length $\left(x = -\tfrac{1}{2}L_0\right)$ is 11 kJ, which is practically the same as the work to double it.

79. **INTERPRET** In this two-dimensional problem, we need to calculate the work done against a given vector force, along a vector path. We will use the most general integral equation for work to find the work done.

DEVELOP Calculate the work using Equation 6.11, $W = \int_{s_1}^{s_2} \vec{F} \cdot d\vec{r}$. The path taken follows $y = ax^2 - bx$, where $a = 2$ m^{-1} and $b = 4$, so $\tfrac{dy}{dx} = 2ax - b$ and $d\vec{r} = dx\hat{i} + (2ax - b)\,dx\hat{j}$. The force is $\vec{F} = cxy\hat{i} + d\hat{j}$, where $c = 10$ N/m^2 and $d = 15$ N. The position x goes from $x = 0$ to $x = 3$ m.

EVALUATE Inserting the expression for the force and the differential $d\vec{r}$ into

$$W = \int_{x=0}^{x=3\text{m}} \left(cxy\hat{i} + d\hat{j}\right) \cdot \left[\hat{i} + (2ax - b)\,\hat{j}\right]dx = \int_0^3 \left[cxy + d(2ax - b)\right]dx$$

$$W = \int_0^3 \left[cx(ax^2 - bx) + d(2ax - b)\right]dx = \int_0^3 \left(cax^3 - cbx^2 + 2adx - bd\right)dx$$

$$W = \left[\tfrac{1}{4}cax^4 - \tfrac{1}{3}cbx^3 + adx^2 - bdx\right]_0^3 = 405\text{ J} - 360\text{ J} + 270\text{ J} - 180\text{ J} = 135\text{ J}$$

ASSESS Because it is not obvious to what physical situation this problem relates, it's not possible to compare the result with an estimate or a limit gained from our understanding of physics. Notice, however, that the units work out as expected.

81. **INTERPRET** A mass falls a given distance, and we are asked to find the force necessary to stop the mass within a another given distance. From the work-energy theorem (Equation 6.14, $\Delta K = W_{net}$), we see that the work done by gravity on the way down is equal in magnitude to the work done by the stopping force, because there is no change in kinetic energy between the initial (leg on bed) and final (leg on floor) state.

DEVELOP The height dropped is $h = 0.7$ m and the stopping distance is $s = 0.02$ m. The mass of the leg is m = 8 kg. From the work-energy theorem, we know that $|W_{down}| = |W_{stop}|$. The work done by gravity is $W_{down} = mgh$, and the absolute value of the work done by the stopping force is $|W_{stop}| = F_s s$, where F_s is the stopping force.

EVALUATE From the work-energy theorem, we have

$$\left| W_{\text{down}} \right| = \left| W_{\text{stop}} \right|$$

$$mgh = F_{\text{s}} s$$

$$F_{\text{s}} = mg \frac{h}{s}$$

The value $h/s = (0.7 \text{ m})/(0.02 \text{ m}) = 35$, so the average stopping force is 35 times the weight of the leg.

ASSESS   The shorter the distance over which something is stopped, the greater the force required. This is why cars are built to "crumple" on impact: The increased distance traveled by the passengers during the crash means a lower average force on their bodies.

83.   **INTERPRET**   We're asked to analyze a graph of the power a bat imparts on a ball as a function of time.

DEVELOP   As argued in the previous problem, the speed continues to increase as long as the power is non-zero.

EVALUATE   The speed will reach its maximum at the end of the hit, which occurs around 0.185 s on the graph. The answer is (c).

ASSESS   If we neglect wind resistance during the hit, the only horizontal force on the ball is the force from the bat. Consequently, there is nothing to slow the ball down while the bat and ball are in contact. It would be illogical, therefore, for the maximum speed to occur before the bat's force was finished acting on the ball.

85.   **INTERPRET**   We're asked to analyze a graph of the power a bat imparts on a ball as a function of time.

DEVELOP   We can assume that the force provided by the bat and the velocity of the ball are parallel. Therefore, the bat force is given by: $F = P/v$. The power is maximum at the peak in the graph, P_{pk}, whereas the velocity constantly increases while the ball and bat are in contact (recall Problem 6.83).

EVALUATE   We can rule out answer (a), since the power is zero there, which implies the force is too. Near the peak, the power is not changing much (the derivative with respect to time is zero at the maximum). Therefore, at a point slightly before the peak, the power is essentially the same, but the velocity is smaller by some amount we will call Δv. The force at a point before the peak can be approximated as:

$$F_{\text{before}} = \frac{P_{\text{before}}}{v_{\text{before}}} \approx \frac{P_{\text{pk}}}{v_{\text{pk}} - \Delta v} \approx \frac{P_{\text{pk}}}{v_{\text{pk}}} \left(1 + \frac{\Delta v}{v_{\text{pk}}} \right) > F_{\text{pk}}$$

where we have used the binomial approximation from Appendix A: $(1-x)^{-1} \approx 1+x$ for $x \ll 1$. By a similar argument, $F_{\text{after}} < F_{\text{pk}}$, so the force is greatest just before the peak. The answer is (c).

ASSESS   One might question the reasoning above. If the velocity were changing more slowly than the power near the peak, then the force would be maximum at the peak, not before. However, we can show that this leads to a contradiction. The derivative of the force with respect to time is zero when the force is maximum:

$$\frac{dF}{dt} = \frac{d}{dt} \left[\frac{P}{v} \right] = \frac{1}{v} \frac{dP}{dt} - \frac{P}{v^2} \frac{dv}{dt} = 0$$

Assuming the maximum force occurs at the peak, then the derivative of the power would also be zero $(dP/dt = 0)$, since the peak is a maximum of the power as well. The equation above reduces to $dv/dt = 0$, which implies zero acceleration, zero force. But that contradicts the assumption that the peak is a maximum of the force. In conclusion, the maximum force has to occur before the peak.

7 CONSERVATION OF ENERGY

EXERCISES

Section 7.1 Conservative and Nonconservative Forces

11. **INTERPRET** This problem involves calculating the work done by a conservative force (gravity) and comparing the result obtained for the work done over two different paths.

    DEVELOP Take the origin at point 1 in Fig. 7.15 with the x axis horizontal to the right and the y axis vertically upward. Use the same equation for work as we did in Problem 7. 10 (Equation 6.11), but this time the force involved is the force of gravity: $\vec{F}_g = -mg\hat{j}$. For path (a), we use Cartesian coordinates, so $d\vec{r} = dx\hat{i} + dy\hat{j}$. Inserting $\vec{F}_g$ into Equation 6.11 for path (a) thus gives

$$W_a = -\int_{x_1}^{x_2} \overbrace{\left(mg\hat{j}\right) \cdot dx\hat{i}}^{=0} - \int_{y_1}^{y_2} \left(mg\hat{j}\right) \cdot dy\hat{j} = -mg\left(y_2 - y_1\right)$$

    For path (b), we will use radial coordinates, so Equation 6.11 takes the form

$$W_b = -\int_{r_1=0}^{r_2=\sqrt{2}L} mg\hat{j} \cdot d\vec{r} = -\int_{r_1=0}^{r_2=\sqrt{2}L} mg\cos\left(45°\right) dr = -\frac{1}{\sqrt{2}}\int_{r_1=0}^{r_2=\sqrt{2}L} mgdr$$

    EVALUATE Inserting the initial and final positions into the expression for path (a) gives $W_a = -mgL$. For path (b), we find

$$W_b = -\frac{mg}{\sqrt{2}}\left(r_2 - r_1\right) = -\frac{mg}{\sqrt{2}}\left(\sqrt{2}L - 0\right) = -mgL$$

    ASSESS The work done by gravity is the same for both paths, because gravity is a conservative force.

Section 7.2 Potential Energy

13. **INTERPRET** This problem involves finding the potential energy difference between sea level and locations at different heights above sea level.

    DEVELOP The zero of potential energy is at sea level. Use Equation 7.3, $\Delta U = mg\Delta y$, to find the potential energy difference at the other locations.

    EVALUATE (a) Atop Mount Washington, the potential energy difference is $\Delta U = (70 \text{ kg})(9.8 \text{ m/s}^2)(1900 \text{ m}) = 1.3$ MJ.

    (b) In Death Valley, $\Delta y = -86$ m, so the potential energy difference is $\Delta U = (70 \text{ kg})(9.8 \text{ m/s}^2)(-86 \text{ m}) = -59$ kJ.

    ASSESS Notice that the potential energy difference is negative at Death Valley compared to sea level, because Death Valley is below sea level.

15. **INTERPRET** The problem is about the change in gravitational potential energy as the hiker ascends. Given the position of zero potential energy, we are interested in her altitude.

    DEVELOP The change in potential energy with a change in the vertical distance Δy is given by Equation 7.3, $\Delta U = mg\Delta y = mg(y - y_0)$. Knowing ΔU and y_0 allows us to determine y, see the figure below.

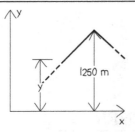

EVALUATE   Equation 7.3 gives

$$\Delta U = U(y) - U(y_0) = mg(y - y_0)$$

From the above expression, we find the altitude of the hiker to be

$$y = y_0 + \frac{\Delta U}{mg} = 1250 \text{ m} + \frac{-240 \text{ kJ}}{(60 \text{ kg})(9.8 \text{ m/s}^2)} = 840 \text{ m}$$

ASSESS   In this problem, the point of zero potential energy is taken to be the top of the mountain with $y_0 = 1250$ m. Since the hiker's potential energy is negative, we expect the hiker's altitude to be lower than y_0.

17.   **INTERPRET**   This problem is similar to Problem 7.16. It is about the potential energy stored in a spring. We'd like to know how much the spring has to be stretched in order to store a given amount of energy.

DEVELOP   The amount of energy stored in a spring is given by Equation 7.4, $U = kx^2/2$, where x is the distance stretched (or compressed) from its natural length.

EVALUATE   Assume one starts stretching from the unstretched position ($x = 0$). Solving Equation 7.4 for x gives

$$x = \pm\sqrt{\frac{2U}{k}} = \pm\sqrt{\frac{2(210 \text{ J})}{1400 \text{ N/m}}} = \pm 55 \text{ cm}$$

ASSESS   The positive and negative signs indicate that you can store the same amount of energy by either compressing the spring or by stretching the spring.

Section 7.3 Conservation of Mechanical Energy

19.   **INTERPRET**   This problem involves potential and kinetic energy. Because the slope is frictionless, the total mechanical energy is conserved, so $K + U$ = constant. We are interested in finding the speed of the skier after he descends each section of the slope.

DEVELOP   We define the zero of potential energy at the top of the hill. Also, because the skier's speed there is zero, his initial kinetic energy is zero. Thus, his initial total mechanical energy is zero. Use Equation 7.3, $U = mg\Delta y$, to express his potential energy at the bottom of each slope, and Equation 6.13, $K = mv^2/2$, to express his kinetic energy at each location. Applying conservation of total mechanical energy to find the speed gives

$$0 = K + U = \frac{1}{2}mv^2 + mg\Delta y$$
$$v = \pm\sqrt{-2g\Delta y}$$

EVALUATE   After the first slope, $\Delta y = -25$ m, so we have

$$v = \pm\sqrt{-2(9.8 \text{ m/s}^2)(-25 \text{ m})} = \pm 22 \text{ m/s}$$

After the second slope, we have

$$v = \pm\sqrt{-2(9.8 \text{ m/s}^2)(-25 \text{ m} - 38 \text{ m})} = \pm 35 \text{ m/s}$$

ASSESS   The plus/minus sign indicates that the result is independent of the direction in which he is skiing. It is the same whether he skis to the left or to the right on the level sections.

21. **INTERPRET** This problem involves the conservative forces of gravity and the elastic force, so we can apply the conservation of mechanical energy to this problem. We are interested in finding the height to which the arrow rises, given its initial elastic potential energy.

    DEVELOP We will take the initial position of the arrow to be the zero of potential energy. The initial total mechanical energy of the arrow is then just the elastic potential energy of the arrow, $U_e = kx^2/2$, with $x = 0.71$ m and $k = 430$ N/m. The final total mechanical energy of the arrow is simply the gravitational potential energy, $U_g = mg\Delta y$, because the arrow has zero speed at the peak of its trajectory, so its kinetic energy there is zero.

    EVALUATE By conservation of total mechanical energy, we equate the initial and final total mechanical energies to find the height Δy to which the arrow rises. The result is

    $$U_e = U_g$$

    $$\frac{1}{2}kx^2 = mg\Delta y$$

    $$\Delta y = \frac{kx^2}{2mg} = \frac{(430 \text{ N/m})(0.71 \text{ m})^2}{2(0.12 \text{ kg})(9.8 \text{ m/s}^2)} = 92 \text{ m}$$

    ASSESS Notice that the height is measured from the arrow's position when the bow is taught, because that is the position at which the arrow has the elastic potential energy.

23. **INTERPRET** We are to find the spring constant needed to launch a toy rocket to a given height. We use the conservation of total mechanical energy: The initial energy is the elastic potential energy of the spring, and the final energy is gravitational potential energy.

    DEVELOP Conservation of total mechanical energy says that $U_i + K_i = U_f + K_f$. For this problem, the initial and final kinetic energies are zero. From Equation 7.4, we know that the initial elastic potential energy of the spring is $U_i = kx^2/2$, and the final gravitational energy is $U_f = mgh$. The spring compression is $x = -0.14$ m, the rocket's mass is $m = 65$ g $= 0.065$ kg, and the desired height is $h = 35$ m.

    EVALUATE Applying the conservation of total mechanical energy and solving for the spring constant k gives

    $$U_i + \overbrace{K_i}^{=0} = U_f + \overbrace{K_f}^{=0}$$

    $$\frac{1}{2}kx^2 = mgh$$

    $$k = \frac{2mgh}{x^2} = \frac{2(0.065 \text{ kg})(9.8 \text{ m/s}^2)(35 \text{ m})}{(0.14 \text{ m})^2} = 2.3 \text{ kN/m}$$

    ASSESS This spring is probably a bit stiff for a kid's toy. It will take a force of 320 N to completely compress the spring, which is about 70 lbs.

Section 7.4 Potential-Energy Curves

25. **INTERPRET** This problem involves conservative forces and conservation of total mechanical energy. At the maximum height of the particle, we know its kinetic energy is zero and its potential energy is maximum. We will define the zero of potential energy as the particle's lowest position, where its kinetic energy will be maximum. We are asked to find the turning point of the particle, which is the x position where the particle stops rising and begins to fall again.

    DEVELOP The particle's trajectory is given by the formula $y = ax^2$, with $a = 0.92$ m^{-1}. The particles potential energy is $U = mgy$, and its kinetic energy is $K = mv^2/2$. Conservation of total mechanical energy tells us that the sum of these two quantities is conserved, and we can find that constant because we are told that the maximum speed (i.e., maximum kinetic energy) is 8.5 m/s, which must occur at the point where the potential energy is minimum (i.e., $y = 0$). Thus, we have

    $$\frac{1}{2}mv_{max}^2 = K_{max}$$

where K_{max} is constant and is the total mechanical energy, which is conserved. We can insert this into the general expression for total mechanical energy to find the turning point of the particle, because we know that the particle kinetic energy will be zero at the turning point.

EVALUATE   The total mechanical energy is

$$K_{max} = U + K = mgy + \frac{1}{2}mv^2 = mg(ax^2) + \frac{1}{2}mv^2$$

At the turning point, v = 0, so we have

$$K_{max} = \frac{1}{2}mv_{max}^2 = mg\left(ax_{turn}^2\right)$$

$$x_{turn} = \pm v_{max}\sqrt{\frac{1}{2ga}} = \pm(8.5 \text{ m/s})\sqrt{\frac{1}{2(9.8 \text{ m/s}^2)(0.92 \text{ m}^{-1})}} = \pm 2.0 \text{ m}$$

ASSESS   The positive/negative sign means that there are two turning points: one at positive 2.0 m from the origin (i.e., to the right) and one at –2.0 m from the origin (i.e., to the left).

27.   **INTERPRET**   For this one-dimensional problem, we are given the potential energy of a particle as a function of the particle's position and are asked to find the force on the particle.

DEVELOP   Because this problem is one-dimensional, we will use Equation 7.8, $F(x) = -dU/dx$, to find the force on the particle. The potential energy of the particle is $U(x) = 16x^2 - b$, with $b = 4.0$ J, so the force is $F(x) = -dU/dx = -32x$.

EVALUATE   (a) At $x = 2.1$ m, the force is $F(x = 2.1 \text{ m}) = -32(2.1 \text{ m}) = -67$ N (the minus sign signifies that the force is in the $-x$ direction).

(b) At $x = 0$ m, the force is $F(x = 0 \text{ m}) = -32(0 \text{ m}) = 0$ N.

(c) At $x = -1.4$ m, the force is $F(x = -1.4 \text{ m}) = -32(-1.4 \text{ m}) = +45$ N (the plus sign signifies that the force is in the $+x$ direction).

ASSESS   Notice that the results are given to two significant figures because the data is given to two significant figures.

PROBLEMS

29.   **INTERPRET**   This problem asks us to calculate the work done around a square path given two different force fields. If the forces are conservative, the work should be zero.

DEVELOP   The work done by the given forces as an object moves around the box can be divided into 4 segments. Starting from the bottom left-hand corner, $d\vec{r} = \hat{i}dx$, $y = 0$ and x goes from 0 to a. Then, $d\vec{r} = \hat{j}dy$, $x = a$ and y goes from 0 to a. After, $d\vec{r} = \hat{i}dx$, $y = a$ and x goes from a to 0. And finally, $d\vec{r} = \hat{j}dy$, $x = 0$ and y goes from a to 0. Mathematically, we can write this as:

$$W = \oint \vec{F} \cdot d\vec{r} = \int_0^a \vec{F} \cdot \hat{i}dx\Big|_{y=0} + \int_0^a \vec{F} \cdot \hat{j}dy\Big|_{x=a} + \int_a^0 \vec{F} \cdot \hat{i}dx\Big|_{y=a} + \int_a^0 \vec{F} \cdot \hat{j}dy\Big|_{x=0}$$

In both cases, the force points in y-direction, so $\vec{F} \cdot \hat{i} = 0$. Recall that we can reverse the endpoints of a definite integral by adding a minus sign: $\int_a^0 dx = -\int_0^a dx$, so the work equation reduces to:

$$W = \oint \vec{F} \cdot d\vec{r} = \int_0^a \vec{F} \cdot \hat{j}dy\Big|_{x=a} - \int_0^a \vec{F} \cdot \hat{j}dy\Big|_{x=0}$$

EVALUATE   For the constant force, $\vec{F}_a = F_0\hat{j}$, in Figure 7.14a, the work is

$$W_a = F_0\int_0^a dy - F_0\int_0^a dy = F_0a - F_0a = 0$$

For the varying force, $\vec{F}_b = F_0(x/a)\hat{j}$, in Figure 7.14b, the work is

$$W_b = F_0\int_0^a (a/a)\, dy - F_0\int_0^a (0/a)\, dy = F_0a - 0 = F_0a$$

ASSESS   The force $\vec{F}_a$ is conservative, but the force $\vec{F}_b$ is not. As far as the work done is concerned, it does matter what path an object takes when moving through the force field in Figure 7.14b.

31. **INTERPRET** This problem involves finding the gravitational potential energy of a brick that is placed in various different positions. We can treat the brick as if all its mass were located at the center of the brick. We are instructed to take the zero of the gravitational potential energy to be at the center of the brick when the brick is lying on its longest side.

DEVELOP Draw a diagram of the brick when it is lying on its longest side, and when it is in the positions for parts (a) and (b) of the problem (see figure below). The gravitational potential energy of the brick is given by Equation 7.3:

$$\Delta U = mg\Delta y = mg(y - y_0)$$

where Δy is the vertical distance of the center of the brick above the point of zero of potential energy, which is y_0. From the figure below, we see that $y_0 = 2.75$ cm.

In position (a) the change in height of the center of the brick is

$$\Delta y_a = y_a - y_0 = 10.0 \text{ cm} - 2.75 \text{ cm} = 7.25 \text{ cm}$$

In (b), the change in height of the brick is

$$\Delta y_b = \tfrac{1}{2}\sqrt{(20.0 \text{ cm})^2 + (5.50 \text{ cm})^2} - 2.75 \text{ cm} = 10.4 \text{ cm} - 2.75 \text{ cm} = 7.62 \text{ cm} .$$

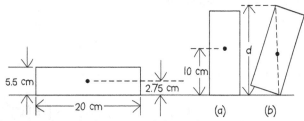

EVALUATE From Equation 7.3, the gravitational potential energies at positions **(a)** and **(b)** are

$$U_a = mg\Delta y_a = (1.50 \text{ kg})(9.82 \text{ m/s}^2)(0.0725 \text{ m}) = 1.07 \text{ J}$$

$$U_b = (1.50 \text{ kg})(9.82 \text{ m/s}^2)(0.0762 \text{ m}) = 1.12 \text{ J}$$

ASSESS The potential energy is larger for part (b) because the center of the brick is higher than it is for part (a). Notice that we had to use the acceleration of gravity to three significant figures for this problem. Had we used $g = 9.8$ m/s2, our results would be $U_a = U_b = 1.1$ J.

33. **INTERPRET** This problem deals with the elastic potential energy stored in a rope. We need to find the elastic potential energy stored in a rope stretched the given amount and compare the result with that of Example 7.3, which uses a different expression for the force exerted by the rope.

DEVELOP The force exerted by the rope is $F(x) = -kx + bx^2 - cx^3$, where x is the length the rope is stretched from its equilibrium position. Use Equation 7.2a, $\Delta U = -\int_{x_1}^{x_2} F(x)\, dx$, to find the elastic potential energy of the rope.

EVALUATE Performing the required integration and inserting the known quantities, we obtain

$$U = -\int_0^x F(x')\, dx' = -\int_0^x \left(-kx' + bx'^2 - cx'^3\right) dx' = \left(\frac{1}{2}kx'^2 - \frac{1}{3}bx'^3 + \frac{1}{4}cx'^4\right)_{x'=0}^{x'=2.62 \text{ cm}}$$

$$= \frac{1}{2}(223 \text{ N/m})(2.62 \text{ m})^2 - \frac{1}{3}(4.10 \text{ N/m}^2)(2.62 \text{ m})^3 + \frac{1}{4}(3.1 \text{ N/m}^3)(2.62 \text{ m})^4$$

$$= 778 \text{ J}$$

In Example 7.3, the energy stored is $U' = 741$ J . Therefore, the percent difference is

$$(100\%)\frac{U - U'}{U'} = (100\%)\frac{778 \text{ J} - 741 \text{ J}}{741 \text{ J}} = 4.90\%$$

ASSESS Adding the term $-cx^3$ increases the potential energy of the system. The negative sign increases the restoring force, and thus the work needed to stretch the spring.

35.   **INTERPRET**   We want to find the energy stored in an unusual spring when it's compressed a certain distance. **DEVELOP**   The energy stored in the spring is just the potential energy. For a one-dimensional force and displacement, we can use Equation 7.2a:

$$\Delta U = -\int_0^x F(x')dx' = -\int_0^x \left(-kx' - cx'^3\right)dx' = \left[\tfrac{1}{2}kx'^2 + \tfrac{1}{4}cx'^4\right]\Big|_0^x = \tfrac{1}{2}kx^2 + \tfrac{1}{4}cx^4$$

EVALUATE   In this case the spring is compressed, which we will define as being in the negative x-direction. The stored energy is therefore:

$$\Delta U = \tfrac{1}{2}(220 \text{ N/m})(-0.15 \text{ m})^2 + \tfrac{1}{4}(3.1 \text{ N/m}^3)(-0.15 \text{ m})^4 = 2.5 \text{ J}$$

ASSESS   The units come out the same (i.e., $\text{N} \cdot \text{m}$) for both terms in the sum, as they should. Notice that for the given force equation, it makes no difference whether the spring is compressed or stretched – the stored energy is the same for the same displacement.

37.   **INTERPRET**   This problem is similar to the preceding problem. We are given a force as a function of position, and we are to find the change in potential energy. For this problem, we are given the zero of the potential energy, so we will find the potential energy with respect to this zero. **DEVELOP**   Use Equation 7.2(a),

$$\Delta U = -\int_{x_1}^{x_2} F(x)\, dx$$

to find the force. Because $U(x = 0) = 0$, this expression reduces to

$$\Delta U = U(x_2) - \overbrace{U(x_1 = 0)}^{=0} \equiv U(x) - \int_0^x F(x')\, dx'$$

EVALUATE   Performing the integration gives

$$U(x) = -\int_0^x F(x')\, dx' = -\int_0^x \left(ax'^2 + b\right) dx' = -\frac{1}{3}ax^3 - bx$$

ASSESS   The potential energy decreases with x, meaning that the force does ever increasing work on the particle as it moves.

39.   **INTERPRET**   This problem involves conservative forces, that due to the spring and that due to gravity. Therefore, assuming the slope is frictionless, conservation of mechanical energy applies so we know that the total mechanical energy at all points on the trajectory of the block is a constant. Without loss of generality, we can take the zero of the gravitational potential energy to be at the position of the block before the spring is released. **DEVELOP**   Initially, before the spring is released, the total mechanical energy is the elastic potential energy of the spring, $U_{\text{Tot}} = kx^2/2$. At the highest point reached by the block, its speed is instantaneously zero, so the final total mechanical energy is its gravitational potential energy, and $U_{\text{Tot}} = mg\Delta y$, where Δy is the height above the starting point. By trigonometry, this height is $\Delta y = r\sin(\theta)$, where r is the distance the block travels up the incline. **EVALUATE**   By conservation of total mechanical energy, we equate the two expressions above for total mechanical energy. Inserting the expression Δy and solving for r gives

$$\frac{kx^2}{2} = mg\Delta y = mgr\sin(\theta)$$

$$r = \frac{kx^2}{2mg\sin(\theta)}$$

ASSESS   The distance traveled up the incline is quadratic in x (the spring compression), so if we compress the spring twice as much, the mass will travel 4 times the distance.

41.   **INTERPRET**   Ignoring air resistance, the only force acting on the object once its release is gravity, which is a conservative force. Thus, this problem involves conservation of total mechanical energy. We will take the final position of the object to be the zero of the gravitational potential energy. Our goal is to derive Equation 2.11 by applying conservation of total mechanical energy.

DEVELOP The instant the ball is released, its total mechanical energy is $K_0 + U_0 = m v_0^2/2 + mgh$, where h is the height above the zero of the potential energy. Without loss of generality, we take the final position to be the zero of the potential energy, so the final mechanical energy is then is $K + U = mv^2/2$. By conservation of total mechanical energy, these two expressions give the same result, so we can equate them and solve for the final speed v.

EVALUATE The final speed of the object is

$$\frac{mv_0^2}{2} + mgy_0 = \frac{mv^2}{2}$$

$$\frac{v^2}{2} = \frac{v_0^2}{2} + gh$$

$$v = \pm\sqrt{v_0^2 + 2gh} = -\sqrt{v_0^2 + 2gh}$$

where we have taken the negative square root because the object is traveling in the negative direction (i.e., from large h to small h).

ASSESS Notice that the formula given in the problem statement should have a $\pm$ sign because the square root is involved.

43. **INTERPRET** The two forces acting on the block are those applied by the springs, so they are conservative forces. In the absence of friction and air resistance, we can apply conservation of total mechanical energy.

DEVELOP When the left hand spring is at its maximum compression, the block is instantaneously motionless, so the total mechanical energy of the block/springs system is just the elastic potential energy of the left-hand spring, so $U_L^{\text{Tot}} = k_L x_L^2/2$. At the opposite end, the total mechanical energy is $U_R^{\text{Tot}} = k_R x_R^2/2$. Between the springs the total energy is just the kinetic energy, so $U_K^{\text{Tot}} = mv^2/2$. By conservation of total mechanical energy, we can equate all three energies to find the compression x_R of the right-hand spring and the speed v of the block between the springs.

EVALUATE (a) At the right-hand end, the spring compresses a distance

$$k_L x_L^2 = k_R x_R^2$$

$$x_R = \pm x_L \sqrt{\frac{k_L}{k_R}} = -(0.16 \text{ m})\sqrt{\frac{130 \text{ N/m}}{280 \text{ N/m}}} = -11 \text{ cm}$$

where we have chosen the negative sign because the right-hand spring compresses.

(b) The speed of the block between the springs is

$$\frac{k_L x_L^2}{2} = \frac{mv^2}{2}$$

$$v = \pm x_L \sqrt{\frac{k_L}{m}} = \pm(0.16 \text{ m})\sqrt{\frac{130 \text{ N/m}}{0.2 \text{ kg}}} = \pm 4 \text{ m/s}$$

where we have kept both signs because the block can move either left-to-right or right-to-left, and we have retained only a single significant figure because we only know the block's weight to a single significant figure.

ASSESS For part (b), the units of the right-hand side are

$$\text{m}\sqrt{\frac{\text{m}}{\text{kg}}} = \text{m}\sqrt{\frac{\text{kg} \cdot \text{m} \cdot \text{s}^{-2}/\text{m}}{\text{kg}}} = \text{m/s}$$

as expected.

45. **INTERPRET** Because the track is frictionless (and we ignore air resistance), the only force acting on the block is gravity, which is a conservative force. Therefore, we can apply the conservation of total mechanical energy to this problem. We will choose the zero of gravitational potential energy to be the base of the loop.

DEVELOP Apply conservation of total mechanical energy, Equation 7.7 ($U_0 + K_0 = U + K$). The initial total mechanical energy is just the gravitational potential energy because the speed (i.e., kinetic energy) is zero at the start. Therefore, $U_0 = mgh$. The energy at the top of the loop is $U + K = 2mgR + mv^2/2$, where R is the radius of the loop (see figure below). For the block to stay on the track, the centripetal acceleration of the block must exceed the acceleration due to gravity at the top of the loop, or $v^2/R \geq g$ (see, e.g., Problems 5.39 or 5.42).

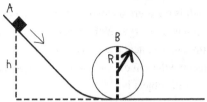

EVALUATE Equating the two expressions for total mechanical energy and using the minimum speed criterion gives

$$mgh = 2mgR + \frac{1}{2}mv^2$$

$$v^2 = 2gh - 4gR \geq gR$$

$$h \geq \frac{5}{2}R$$

ASSESS Because real tracks always have some friction, the actual height needed would be greater than $5R/2$.

47. **INTERPRET** This problem involves the forces of gravity and of an elastic spring, both of which are conservative forces. Therefore, we can apply conservation of total mechanical energy. We take the zero of the gravitational potential energy the height of the spring in equilibrium.

DEVELOP To apply conservation of total mechanical energy (Equation 7.7), we need to express the total mechanical energy for the block before it is dropped and when the spring is maximally compressed by the block (at which point the block is instantaneously motionless, see figure below). For the former, we have $U_0 + K_0 = mgh$. For the latter, we have $U + K = ky^2/2 - mgy$, where y is the distance from equilibrium that the spring is compressed.

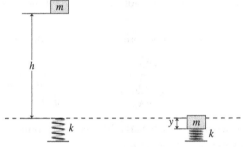

EVALUATE Equating the two expressions above for total mechanical energy and solving for the maximum spring compression y gives

$$mgh = \frac{1}{2}ky^2 - mgy$$

$$\left(\frac{k}{2}\right)y^2 + (-mg)y + (-mgh) = 0$$

$$y = \frac{mg \pm \sqrt{m^2g^2 + 2kmgh}}{k} = \frac{mg}{k}\left(1 + \sqrt{1 + 2kh/mg}\right)$$

ASSESS We have retained the positive sign because the spring would not be compressed if $y < 0$.

49. **INTERPRET** This is a one-dimensional problem in which we are to derive an expression for the potential energy given the force acting on a object as a function of position.

DEVELOP The force is conservative because it is a function of position (if we come back to the same position, we experience the same force). We can therefore apply Equation 7.2, which for one dimension reduces to Equation 7.2a,

$$\Delta U = U(x_2) - U(x_1) = -\int_{x_1}^{x_2} F(x)\, dx$$

For part (b), note that the total mechanical energy at the turning points is just the potential energy because the kinetic energy is zero at these points (object is reversing direction). Thus, find the points on the graph where the potential energy is equivalent to the total mechanical energy, which is given to be –1 J.

EVALUATE   (a) Inserting the expression for force and performing the integration gives

$$\Delta U = -\int_{x_1}^{x_2}\left(ax - bx^3\right)dx = \left(\frac{ax^2}{2} - \frac{bx^4}{4}\right)\Bigg|_{x_1}^{x_2} = -\frac{a}{2}\left(x_2^2 - x_1^2\right) + \frac{b}{4}\left(x_2^4 - x_1^4\right)$$

Without loss of generality, we can define the potential energy at x = 0 to be zero, so

$$U(x) = -\frac{a}{2}x^2 + \frac{b}{4}x^4$$

(**b**) A graph of $U(x)$ for $x \geq 0$, when $a = 5$ N/m, $b = 2$ N/m^3, and x is in meters, is shown. (Note that the potential energy is symmetric, $U(-x) = U(x)$, but that only positive displacements are considered in this problem.) The conservation of energy can be written in terms of the total energy, $E = \frac{1}{2}m(dx/dt)^2 + U(x)$, so that $dx/dt = \pm\sqrt{2\left[E - U(x)\right]/m}$. The maximum speed occurs when $U(x)$ is a minimum; $dU/dx = 0$ and $d^2U/dx^2 > 0$. Taking the derivative, one finds $0 = -ax + bx^3$, which has solutions $x = 0$ and $x = \pm\sqrt{a/b} = \pm\sqrt{5/2}$ m $= \pm 1.58$ m. The second derivative $d^2U/dx^2 = -a + 3bx^2$ is negative for $x = 0$, which is a local maximum, but is positive for $x = \pm\sqrt{a/b}$, which are minima with $U_{min} = U(\pm\sqrt{a/b}) = -a^2/4b =$ $-(25/8)$ J $= -3.13$ J. There is real physical motion $(K \geq 0)$ for total energy $E \geq U_{min}$. The turning points (where $dx/dt = 0$) can be found from the equation $U(x) = E$; there are four solutions (two positive) for energies with $U_{min} < E < 0$, and two solutions (one positive) for $E > 0$. The equation $U(x) - E = 0$ is equivalent to $x^4 - 2(a/b)x^2 - 4(E/b) = 0$. The quadratic formula gives $x = \pm\{(a/b) \pm [(a/b)^2 + 4(E/b)]^{1/2}\}^{1/2}$ for $U_{min} < E < 0$, and $x = \pm\{(a/b) + [(a/b)^2 + 4(E/b)]^{1/2}\}^{1/2}$ for $E > 0$. For the particular values given $(E = -1$ J$)$, the positive turning points are $x = \left[\left(5 \pm \sqrt{17}\right)\big/(2 \text{ m})\right]^{1/2} = 0.662$ m and 2.14 m, as can be seen in the graph below. Retaining one a single significant figure gives x = 0.7 and 2 m.

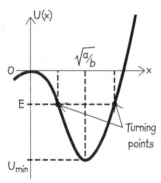

ASSESS   The graph is identical for negative displacements.

51.   **INTERPRET**   In this problem we are asked to find the speed of the skier at two different locations, given that the downward slope has a coefficient of friction $\mu_k = 0.11$. Because friction is a nonconservative force, we cannot apply conservation of total mechanical energy. Instead, we must use the concept of work done by a force combined with total mechanical energy.

DEVELOP   We find the work done by the friction force and subtract this work from the total energy to find the energy remaining after each slope The work done by friction skiing down a straight slope of length L is

$$W_f = -f_k L = -\mu_k n L = -\mu_k \left(mg\cos\theta\right)\left(\frac{h}{\sin\theta}\right) = -\mu_k mgh\cot\theta$$

where $h = L\sin\theta$ is the vertical drop of the slope. Conservation of energy applied between the start and the first level section now gives $\Delta K_{AB} + \Delta U_{AB} = W_{f,AB}$ or

$$\frac{1}{2}mv_B^2 = mg\left(y_A - y_B\right) - \mu_k mg\left(y_A - y_B\right)\cot\theta_{AB}$$

Similarly, for the motion between the top and the second level, we must include all the work done by friction, so

$$\Delta K_{AC} + \Delta U_{AC} = W_{f,AB} + W_{f,BC}$$

or

$$\frac{1}{2}mv_C^2 = mg\left(y_A - y_C\right) - \mu_k mg\left(y_A - y_B\right)\cot\theta_{AB} - \mu_k mg\left(y_B - y_C\right)\cot\theta_{BC}$$

EVALUATE   Solving the equation for v_B, we obtain

$$v_B = \sqrt{2g\left(y_A - y_B\right)\left(1 - \mu_k\cot\theta_{AB}\right)} = \sqrt{2\left(9.8\text{ m/s}^2\right)\left(25\text{ m}\right)\left[1 - 0.11\cot\left(32°\right)\right]} = 20\text{ m/s}$$

Similarly, for v_C, we have

$$v_C = \sqrt{2g\left[\left(y_A - y_C\right) - \mu_k\left(y_A - y_B\right)\cot\theta_{AB} - \mu_k\left(y_B - y_C\right)\cot\theta_{BC}\right]}$$
$$= \sqrt{2\left(9.8\text{ m/s}^2\right)\left[63\text{ m} - \left(0.11\right)\left(25\text{ m}\right)\cot\left(32°\right) - \left(0.11\right)\left(38\text{ m}\right)\cot\left(20°\right)\right]}$$
$$= 30\text{ m/s}$$

ASSESS   Let's consider the case where $\mu_k = 0$. In this limit, the results become

$$v_B = \sqrt{2g\left(y_A - y_B\right)} = \sqrt{2\left(9.8\text{ m/s}^2\right)\left(25\text{ m}\right)} = 22\text{ m/s}$$
$$v_C = \sqrt{2g\left(y_A - y_C\right)} = \sqrt{2\left(9.8\text{ m/s}^2\right)\left(63\text{ m}\right)} = 35\text{ m/s}$$

which are the same as the result of Problem 5.19 for the frictionless case.

53.   **INTERPRET**   In this problem we want to find the distance a block slides on surface with friction after being launched by a compressed spring. The force of friction is not conservative, so we will apply the principle that the work done by friction accounts force the change in the mechanical energy (Equation 7.5).

DEVELOP   Suppose the block comes to rest at point C (see figure below), which is a distance L from its initial position at rest against the compressed spring at point A. Use Equation 7.5, $\Delta K + \Delta U = W_{nc}$, where W_{nc} is the work done by nonconservative force (i.e., friction in this problem). This leads to

$$W_{nc} = -\mu_k mgL = \overbrace{\Delta K}^{=0} + \Delta U = -\frac{1}{2}kx^2$$

because the kinetic energies at A and C and the change in gravitational potential energy are zero.

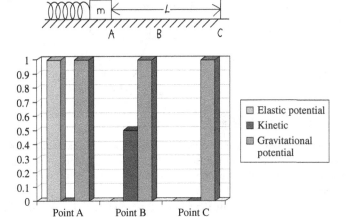

EVALUATE   Solving the above equation, we obtain

$$L = \frac{kx^2}{2\mu_k mg} = \frac{\left(340\text{ N/m}\right)\left(0.18\text{ m}\right)^2}{2\left(0.27\right)\left(1.5\text{ kg}\right)\left(9.8\text{ m/s}^2\right)} = 1.4\text{ m}$$

ASSESS   The distance moved by the block is proportional to the spring's potential energy, $kx^2/2$. In addition, it is inversely proportional to the coefficient of friction, μ_k. In the limit $\mu_k = 0$, the surface is frictionless and we expect the mass to travel indefinitely. The exchange between elastic potential, kinetic, and gravitational potential energy can be seen from the bar chart in the figure above. Because the block slides on a horizontal surface, the gravitational potential energy does not change (and we define it arbitrarily to be 1 J). At position A, the block-spring system's energy is entirely elastic potential energy. At point B, the system's energy is kinetic, but friction has already consumed some of the energy, so the kinetic energy is not equal to the initial elastic potential energy. At point C, the block stops, and all its initial elastic potential energy has been consumed.

55. **INTERPRET** In this problem we want to find the final position of a block after being launched from a compressed spring. Its path involves a frictional surface followed by a frictionless curve. There forces acting on the block are conservative (gravity and the elastic force) and nonconservative (friction). We will define the block's initial position as the zero of gravitational potential energy.

DEVELOP The energy of the block when it first encounters friction is completely kinetic and, by conservation of total mechanical energy (Equation 7.7) it is equal to the initial elastic potential energy of the block/spring system:

$$K_0 = \frac{1}{2}kx^2$$

Upon crossing the friction zone, the work done by the friction is

$$W_{nc} = -\mu_k mgL$$

Depending on the ratio of $K_0/|W_{nc}|$, the block will move back and forth several times before losing all its energy and coming to rest.

EVALUATE Initially the block has an energy

$$K_0 = \frac{1}{2}kx^2 = \frac{1}{2}(200 \text{ N/m})(0.15 \text{ m})^2 = 2.25 \text{ J}$$

The work done by the friction is

$$\Delta E = W_{nc} = -\mu_k mgL = -(0.27)(0.19 \text{ kg})(9.8 \text{ m/s}^2)(0.85 \text{ m}) = -0.427 \text{ J}$$

Because $K_0/|W_{nc}| = 5.27$, five complete crossings are made, leaving the block with energy

$K = K_0 - 5|W_{nc}| = 0.113\text{J}$ on the curved side. This remaining energy is sufficient to move the block a distance

$$s = \frac{K}{\mu_k mg} = \frac{0.113 \text{ J}}{(0.27)(0.19 \text{ kg})(9.8 \text{ m/s}^2)} = 0.225 \text{ m}$$

so the block comes to rest 85 cm – 22.5 cm = 62.5 cm to the right of the beginning of the friction patch.

ASSESS Because $K_0 > |W_{nc}|$, the block does not lose all its energy the first time when it moves across the frictional zone. No energy is lost while it moves along the frictionless curve. The number of times the block moves back and forth across the frictional zone depends on the ratio $K_0/|W_{nc}|$.

57. **INTERPRET** The object of interest is the roller coaster that, after being launched from a compressed spring, moves along a frictionless circular loop. The physical quantity we are asked about is the minimum compression of the spring that allows the car to stay on the track. To find this, we will need to apply Newton's second law as well as conservation of mechanical energy.

DEVELOP If the car stays on the track, the normal force applied by the track must be greater than zero and the radial component of the cars acceleration is $a = v^2/R$. Applying Newton's second law to the roller coaster gives

$$n = \frac{mv^2}{R} + mg\cos\theta \geq 0 \quad \rightarrow \quad v^2 \geq -gR\cos\theta$$

The function $-\cos\theta$ is maximal at the top of the loop ($\theta = 180°$, see figure below), so $v_B^2 \geq gR$ is the condition for the car to stay on the track all the way around. This is the result obtained in Example 5.7.

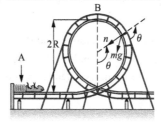

With the minimum speed at point B determined, apply conservation of total mechanical energy (Equation 7.7), to find the minimum compression length of the spring.

EVALUATE   In the absence of friction, conservation of total mechanical energy requires

$$K_A + U_A = K_B + U_B \quad \rightarrow \quad 0 + \frac{1}{2}kx^2 + mgy_A = \frac{1}{2}mv_B^2 + mgy_B$$

Solving for x, we obtain

$$x^2 = \frac{m}{k}\left[v_B^2 + 2g\left(y_B - y_A\right)\right] \geq \frac{5mgR}{k}$$

or

$$x \geq \pm\sqrt{\frac{5mgR}{k}} = \sqrt{\frac{5(840 \text{ kg})(9.8 \text{ m/s}^2)(6.2 \text{ m})}{31{,}000 \text{ N/m}}} = 2.9 \text{ m}$$

ASSESS   Our result indicates that if the radius of the loop increases, then the amount of spring compression must increase in proportion to the square root of the radius for the car to stay on the track. On the other hand, when a stiffer spring (with larger k) is used, then less compression would be required. Notice that we retained the positive sign in our solution because we defined x to be a compression. The negative sign therefore corresponds to an extension of the spring, which gives the spring the same elastic potential energy, but would accelerate the roller coaster in the opposite direction.

59.   **INTERPRET**   In this problem we want to find the distance a child can move across a frictional surface after sliding down a frictionless incline. The problem involves the conservative force of gravity for the first part (the incline) and the nonconservative force of friction for the second part (the level). Thus, we can apply conservation of total mechanical energy to the incline, and the concept of work done by friction to the level section. We will take the zero of gravitational potential energy to be the bottom of the incline.

DEVELOP   At the top of the hill, the child's mechanical energy is entirely gravitational potential energy, so $U_0 + K_0 = mgh$, where $h = 7.2$ m. At the bottom of the hill, just before starting across the rough surface, all this energy is converted to kinetic energy, so $U + K = K$. By conservation of total mechanical energy, we can equate these two expressions, which gives

$$K = mgh$$

where K is the kinetic (and total) energy at the beginning of the rough section. As the child progresses across the rough surface, this energy is consumed by the work done by friction, and the sled stops when the energy supply is exhausted. This is expressed by Equation 7.5, $\Delta U + \Delta K = W_{nc}$. Because the rough section is level, $\Delta U = 0$, and the work done by friction is $W_{nc} = \vec{f}_k \cdot \vec{x} = -\mu_k mgx$, so we have

$$\Delta K = K_{final} - K = 0 - mgh = -\mu_k mgx$$

EVALUATE   Solving the above equation for x, we obtain

$$x = \frac{h}{\mu_k} = \frac{7.2 \text{ m}}{0.51} = 14 \text{ m}$$

ASSESS   As expected, the distance the child travels is proportional to h, because the greater is h; the more gravitational potential energy there is to convert to kinetic energy. On the other hand, we expect x to be inversely proportional to the coefficient of friction, μ_k. A smaller μ_k will allow the child to travel a much further distance before losing all its kinetic energy. In the limit that $\mu_k \to 0$, the child will slide forever, as expected from Newton's second law.

61.   **INTERPRET**   This problem deals with a conservative, one-dimensional force F(x). We know that this force is conservative because it depends only on position. We can therefore apply the fact that the work done by this force on the particle equates to the loss in the particles potential energy (see Equation 7.2).

DEVELOP   Applying Equation 7.2 gives

$$\Delta U = -\int_{x_1}^{x_2} \vec{F}(x) \cdot d\vec{r} = -\int_{x_1}^{x_2} F(x)\, dx$$

Because the force F(x) is conservative, we can apply conservation of total mechanical energy (Equation 7.7), $U_0 + K_0 = U + K$. The subscript 0 refers to the initial state, so $K_0 = 0$ because the particle is initially at rest. Once $\Delta U = U - U_0$ is known, the speed of the particle can be calculated from

$$K = \frac{1}{2}mv^2 = -\Delta U$$

EVALUATE   Integrating $F(x) = a\sqrt{x}$, we obtain

$$\Delta U = U - U_0 = -\int_0^x a\sqrt{x'}\, dx' = \frac{2a}{3}(x')^{3/2}\bigg|_0^x = \frac{2a}{3}x^{3/2}$$

Therefore, the speed of the particle as a function of x is

$$v = \sqrt{\frac{2K}{m}} = \sqrt{\frac{-2\Delta U}{m}} = \left(\frac{4a}{3m}x^{3/2}\right)^{1/2} = 2x^{3/4}\sqrt{\frac{a}{3m}}$$

ASSESS   We can check our answer by substituting the result back to the expression for K. This leads to

$$K = \frac{1}{2}mv^2 = \frac{1}{2}m\left(\frac{4a}{3m}x^{3/2}\right) = \frac{2a}{3}x^{3/2}$$

Indeed, we see that $K = -\Delta U$, as required by the principle of conservation of energy.

63.   **INTERPRET**   We find whether a spring-launched block makes it to the top of an incline with friction, and how much kinetic energy it has when it gets there (if it gets there.) We can use energy methods to solve this problem, but friction is a factor so mechanical energy is not conserved.

DEVELOP   The initial energy of the system is spring potential energy: $U_i = \frac{1}{2}kx^2$. This energy is converted to kinetic energy and gravitational potential energy, but some energy is lost to (non-conservative) friction. From Equation 7.5, we can write this as:

$$U_i = K + U_g + W_{nc}$$

The gravitational potential energy is related to the height, h, the block reaches: $U_g = F_g h$. However, the work done by friction is related to the distance, Δs, the block travels along the incline: $W_{nc} = f\Delta s$, where the friction force is given by $f = \mu F_g \cos\theta$. The two distances are related by: $h = \Delta s \sin\theta$.

EVALUATE   We can determine whether the block makes it to the top of the incline by setting $\Delta s = L$, the length of the incline, and solving $K = U_i - U_g - W_{nc}$. If $K \geq 0$, then the block does reach the top. Otherwise it doesn't, and we can solve for the distance where its kinetic energy drops to zero. So solving for the kinetic energy at the top:

$$K = \frac{1}{2}kx^2 - F_g L(\sin\theta + \mu\cos\theta)$$
$$= \frac{1}{2}(2.0 \text{ kN/m})(10 \text{ cm})^2 - (4.5 \text{ N})(2.0 \text{ m})(\sin 30° + 0.50\cos 30°) = 1.6 \text{ J}$$

So yes, the block reaches the top with 1.6 J of kinetic energy.

ASSESS   Another way to solve this problem is to set $K = 0$ and then solve for the distance that the block would have to travel, Δs, before coming to rest. If $\Delta s \geq L$, then the block reaches the top of the incline with kinetic energy to spare.

65.   **INTERPRET**   This problem involves the conservative forces of a spring and gravity (of the Moon, in this case). We can therefore apply conservation of total mechanical energy to find the requisite spring constant. We will take the surface of the Moon to the zero of gravitational potential energy.

DEVELOP   To apply conservation of total mechanical energy, we need to express the initial energy, when the spring is compressed, and the final energy, just when a bin is separating from the spring. The initial energy is $U_0 + K_0 = ky^2/2$, and the final energy is $U + K = mg_M y + mv^2/2$. Equating the two allows us to solve for the spring constant k.

EVALUATE

$$\frac{1}{2}ky^2 = mg_M y + \frac{1}{2}mv^2$$

$$k = \frac{2mg_M}{y} + \frac{mv^2}{y^2} = \frac{2(1000\ \text{kg})(1.6\ \text{m/s}^2)}{15\ \text{m}} + \frac{(1000\ \text{kg})(2400\ \text{m/s})^2}{(15\ \text{m})^2} = 2.1\times10^2\ \text{N/m} + 2.6\times10^7\ \text{N/m} = 2.6\times10^7\ \text{N/m}$$

where we have used the acceleration of gravity on the Moon's surface from Appendix E.

ASSESS This is an extraordinarily strong spring—some two orders of magnitude larger than the effective spring constant in a carbon monoxide molecule (cf. Problem 32). Note also that the contribution of the Moon's gravity in our calculation is negligible, being five orders of magnitude less than the contribution of the spring.

67. **INTERPRET** The energy stored in the artificial tendon is given by Equation 7.2a for a one-dimensional force. What's unique in this problem is that the force suddenly changes when the second spring is engaged.

DEVELOP When only one spring is engaged, the force exerted by the artificial tendon is $F = -kx$ for $x = 0$ to $x = x_1$. But then for $x > x_1$, the second spring engages, thus increasing the force to $F = -(k + ak)x$. Therefore, in doing the integral of Equation 7.2a, we should divide it into two parts:

$$\Delta U = -\int_0^{x_1} F(x)\,dx - \int_{x_1}^{x_2} F(x)\,dx$$

EVALUATE Performing the integration, we find

$$\Delta U = \frac{1}{2}kx^2\Big|_0^{x_1} + \frac{1}{2}k(1+a)x^2\Big|_{x_1}^{x_2} = \frac{1}{2}kx_2^2 + \frac{1}{2}ka\left(x_2^2 - x_1^2\right)$$

ASSESS The first term in our result is the energy stored in the first spring, and the second term is the energy stored in the second spring.

69. **INTERPRET** We are asked to analyze a graph characterizing the potential energy between two deuterons.

DEVELOP When the deuterons are far apart, their potential energy is zero, $U_0 = 0$, but they will have some initial kinetic energy, K_0. We assume the deuterons are moving towards each other. As the distance between the deuterons shrinks, the graph shows that the potential energy increases. According to the conservation of mechanical energy (Equation 7.7), the kinetic energy will correspondingly decrease: $K(x) = K_0 - U(x)$. The question, then, is will the kinetic energy go to zero before the deuterons reach the well in the potential at $x \approx 1$ fm, where they will be fused (bound) together?

EVALUATE The deuterons won't be able to fuse if they run out of kinetic energy before reaching the peak at 5 fm in the potential energy curve. In other words, the initial kinetic energy has to be greater or equal to this "energy barrier" (i.e. $K_0 \geq U_{\text{peak}}$). The potential energy at the peak is equal to about 0.3 MeV. The answer is (d).

ASSESS We recall once again the analogy to a ball and a hill. If the ball starts at the bottom of the hill, it will only be able to reach the peak if its initial kinetic energy is greater or equal to the potential energy separating the top and bottom of the hill.

71. **INTERPRET** We are asked to analyze a graph characterizing the potential energy between two deuterons.

DEVELOP The force is given by Equation 7.8: $F_x = -dU/dx$.

EVALUATE The slope of the curve at 4 fm is positive, so the force is negative. That means the force points toward smaller x, which means it is pulling the deuterons closer together. This is an attractive force. The answer is (b).

ASSESS We shouldn't confuse the magnitude of the force with the fact that the potential, $U(x)$ is zero at $x = 4$ fm. An attractive force is consistent with the notion that the deuterons are bound to each other inside the potential well. The force at $x = 4$ fm is pulling the deuterons back to the equilibrium position at $x \approx 1$ fm, like a stretched spring.

8

GRAVITY

EXERCISES

Section 8.2 Universal Gravitation

11. **INTERPRET** This problem involves Newton's law of universal gravitation. We can use this law to find the radius of the planet given that we weigh twice the amount we do on Earth.

    DEVELOP Newton's law of universal gravitation (Equation 8.1) is

    $$F = \frac{GM_1M_2}{r^2}$$

    On a spherical Earth, this gives $F_E = GM_E m / R_E^2$, where m is the mass of the explorer. On the spherical planet, this gives $F_P = GM_P m / R_P^2$. We are told that the planet has the same mass as the Earth, so $M_E = M_P$, and that the space explorers weigh twice as much on the new planet, so $2F_E = F_P$.

    EVALUATE Taking the ratio of these expressions for force on each planet and solving for the radius R_P of the new planet gives

    $$\frac{F_E}{F_P} = \frac{GM_E m}{R_E^2} \frac{R_P^2}{GM_P m}$$

    $$\frac{1}{2} = \frac{R_P^2}{R_E^2}$$

    $$R_P = \frac{R_E}{\sqrt{2}}$$

    ASSESS Notice that the force due to gravity is not linear in the radius of the planet.

13. **INTERPRET** This problem involves Newton's law of universal gravitation. We are to find the radius of the Earth that would result in gravity tripling at the surface of the Earth.

    DEVELOP We shrink the Earth to a radius R such that the force due to gravity at its surface is three times the actual value. For this situation, Newton's law of universal gravitation gives

    $$F = \frac{GM_E m}{R^2} = \frac{3GM_E m}{R_E^2}$$

    where R_E is the normal radius of the Earth.

    EVALUATE Solving for the ratio of R to R_E, we find $R/R_E = 1/\sqrt{3} = 57.7\%$.

    ASSESS Thus, the reduced Earth would have about half the diameter of the actual Earth.

15. **INTERPRET** This involves using the gravitational force between two identical spheres to calculate their mass.

    DEVELOP According to Newton's law of universal gravitation (Equation 8.1), the identical spheres $(m_1 = m_2 = m)$ generate a force between them of $F = Gm^2 / r^2$.

    EVALUATE Rearranging the gravitational force equation, the mass of each sphere is

$$m = r\sqrt{\frac{F}{G}} = (0.14 \text{ m})\sqrt{\frac{0.25 \times 10^{-6} \text{ N}}{6.67 \times 10^{-11} \text{ N} \cdot \text{m}^2/\text{kg}^2}} = 8.6 \text{ kg}$$

ASSESS   Does this mass make sense given the small separation between the spheres? The density of lead is 11.34 g/cm^3, so the radius of each sphere is:

$$r = \sqrt[3]{\frac{3m}{4\pi\rho}} = \sqrt[3]{\frac{3(8600 \text{ g})}{4\pi(11.34 \text{ g/cm}^3)}} = 5.7 \text{ cm}$$

So yes, this is consistent with the fact that the centers of the two spheres are 14 cm apart, since there's about 3 cm of space between the closest edges of the spheres.

17.   **INTERPRET**   We're asked to find the height of the building by using the difference in the gravitational acceleration at the top and bottom of the building. The change in the acceleration is due to the change in the distance to the center of the Earth.

DEVELOP   In general, the acceleration due to gravity is given in Equation 8.2: $a = GM/r^2$. The acceleration is measured at street level, where $r = R_E$, and compared to reading at the top of the Willis Tower, where $r = R_E + h$. Here, R_E is the radius of the Earth, and h is the height of the building. The difference in the acceleration measurements should equal:

$$\Delta a = \frac{GM_E}{R_E^2} - \frac{GM_E}{(R_E + h)^2} = \frac{GM_E}{R_E^2}\left[1 - \frac{1}{(1 + h/R_E)^2}\right]$$

Since $h \ll R_E$, we can use the binomial approximation from Appendix A: $(1 + h/R_E)^{-2} \approx 1 - 2h/R_E$. The above expression reduces to: $\Delta a \approx 2gh/R_E$, where we have used $g = GM_E/R_E^2$ for the average value of the gravitational acceleration on the Earth's surface.

EVALUATE   Using the above expression for the acceleration difference, we solve for the height of the tower:

$$h \approx R_E \frac{\Delta a}{2g} = (6.37 \times 10^6 \text{ m})\frac{(1.36 \text{ mm/s}^2)}{2(9.8 \text{ m/s}^2)} = 442 \text{ m}$$

ASSESS   The 108-story Willis Tower is indeed 442 m tall. Note that present gravimeters can measure differences in the gravitational acceleration as small as a few tenths of a milligal, where 1 milligal = 10^{-5} m/s^2 is the unit used to measure gravity anomalies by geologists.

Section 8.3 Orbital Motion

19.   **INTERPRET**   This problem involves using Newton's second law and Newton's universal law of gravitation to find the speed of a satellite in geosynchronous orbit (which means that the satellite completes one orbit in 24 hours, so it stays above the same place on the Earth).

DEVELOP   By Newton's second law we have $F = ma_c$, where $a_c = v^2/r$ is the centripetal acceleration of the satellite with r being the radius of its orbit and v being the orbital speed. The gravitational force between the Earth and the satellite provides the centripetal force to keep the orbit circular. Thus,

$$\frac{GM_E m_s}{r^2} = \frac{m_s v^2}{r}$$

The orbital speed can be expressed as the circumference divided by the period T, or $v = (2\pi r)/T$, which we can use to eliminate the radius of the orbit so we can solve for the velocity.

EVALUATE   Solving the above equation for v with $T = 24$ h = 86,400 s, we obtain

$$v = \sqrt[3]{\frac{2\pi GM_E}{T}} = \sqrt[3]{\frac{2\pi(6.67 \times 10^{-11} \text{ N} \cdot \text{m}^2/\text{kg}^2)(5.97 \times 10^{24} \text{ kg})}{86,400 \text{ s}}} = 3070 \text{ m/s}$$

ASSESS   Dividing the circumference by this velocity and solving for the orbital radius gives

$$r = \frac{vT}{2\pi} = \frac{(3070 \text{ m/s})(86,400 \text{ s})}{2\pi} = 4.22 \times 10^7 \text{ m}$$

which agrees with the result of Example 8.3.

21. INTERPRET   The problem involves finding the orbital period of one of Jupiter's moons.
    DEVELOP   We'll assume the orbit is circular, in which case Equation 8.4 gives the period:
    $T^2 = 4\pi^2 r^3 / GM$. We're told the radius of the orbit, and we can find the mass of Jupiter from Appendix E:
    $M = 1.90 \times 10^{27}$ kg.
    EVALUATE   Io's orbital period is

$$T = \sqrt{\frac{4\pi^2 \left(4.22 \times 10^8 \text{ m}\right)^3}{\left(6.67 \times 10^{-11} \text{ N} \cdot \text{m}^2/\text{kg}^2\right)\left(1.90 \times 10^{27} \text{ kg}\right)}} = 1.53 \times 10^5 \text{ s} = 1.77 \text{ d}$$

    ASSESS   The answer agrees with the rotation period given in Appendix E for the moon Io.

23. INTERPRET   We're asked to find the altitude of a spacecraft orbiting Mars given its orbital period.
    DEVELOP   Equation 8.4 relates the period and radius of an orbiting body to the mass of the object it is orbiting
    around: $T^2 = 4\pi^2 r^3 / GM$. The mass of Mars from Appendix E is: $M_M = 6.42 \times 10^{23}$ kg. Once we solve for the
    orbital radius, we will have to subtract the radius of Mars $\left(R_M = 3.38 \times 10^6 \text{ m}\right)$ to find the altitude: $h = r - R_M$.

    EVALUATE   The distance between the Mars Renaissance Orbiter and the center of the planet Mars is

$$r = \sqrt[3]{\frac{1}{4\pi^2} GMT^2} = \sqrt[3]{\frac{1}{4\pi^2}\left(6.67 \times 10^{-11} \tfrac{\text{N}\cdot\text{m}^2}{\text{kg}^2}\right)\left(6.42 \times 10^{23} \text{ kg}\right)(112 \cdot 60 \text{ s})^2} = 3.659 \times 10^6 \text{ m}$$

    This implies that the altitude of the spacecraft is
$$h = r - R_M = 3.659 \times 10^6 \text{ m} - 3.38 \times 10^6 \text{ m} = 0.28 \times 10^6 \text{ m}$$

    ASSESS   This is 280 km. Compare this to Example 8.2, where it was shown that a low Earth orbit with an altitude
    of 380 km has a period of about 90 min. Since Mars has less mass, spacecrafts must orbit at a smaller radius in
    order to have roughly the same orbital period.

Section 8.4 Gravitational Energy

25. INTERPRET   This problem deals with the gravitational potential energy of an object. We are asked to find the
    energy required to raise an object to a given height in the Earth's gravitational field.
    DEVELOP   If we neglect any kinetic energy differences associated with the orbital or rotational motion of the
    Earth or package, the required energy is just the difference in gravitational potential energy given by Equation 8.5,
    $\Delta U = GM_E m \left[R_E^{-1} - (R_E + h)^{-1} \right]$, where $h = 1800$ km $= 1.8 \times 10^6$ m.
    EVALUATE   Evaluating the expression above with the data from Appendix E gives

$$\Delta U = \left(6.67 \times 10^{-11} \text{ N} \cdot \text{m}^2/\text{kg}^2\right)\left(5.97 \times 10^{24} \text{ kg}\right)(230 \text{ kg})\left[\frac{1}{6.37 \times 10^6 \text{ m}} - \frac{1}{6.37 \times 10^6 \text{ m} + 1.8 \times 10^6 \text{ m}}\right] = 3.17 \text{ GJ}$$

    ASSESS   In terms of the more convenient combination of constants $GM_E = gR_E^2$,
    $\Delta U = mgR_E h / (R_E + h) = 3.17$ GJ

27. INTERPRET   This problem involves conservation of total mechanical energy, as in Example 8.5, except that for
    this problem, we are given the final altitude and need to find the launch speed.
    DEVELOP   Apply conservation of total mechanical energy (Equation 7.7), $U_0 + K_0 = U + K$, where

$$U_0 + K_0 = -\frac{GM_E m}{R_E} + \frac{1}{2}mv^2$$

and

$$U + K = -\frac{GM_E m}{R_E + h} + 0$$

EVALUATE   Solving the expression derived from conservation of total mechanical energy for the initial velocity v gives

$$v = \pm\sqrt{2GM_E\left(\frac{1}{R_E} - \frac{1}{R_E + h}\right)}$$

$$= \sqrt{2\left(6.67\times10^{-11}\ \text{N}\cdot\text{m}^2/\text{kg}^2\right)\left(5.97\times10^{24}\ \text{kg}\right)\left(\frac{1}{\left(6.37\times10^6\ \text{m}\right)} - \frac{1}{\left(6.37+1.1\right)\times10^6\ \text{m}}\right)}$$

$$= 4.29\ \text{km/s}$$

ASSESS   The positive square root was chosen because we are interested in the magnitude of the speed. Notice that our result is larger than the initial speed of 3.1 km/s for Example 8.5, which makes sense because the altitude attained (1100 km) is higher than that attained (530 km) in Example 8.5.

29.   **INTERPRET**   This problem involves finding the total mechanical energy associated with the Earth's orbit about the Sun. We will assume the orbit is circular.

DEVELOP   Apply Equation 8.8b, using data from Appendix E, to find the total mechanical energy. The two masses involved are the mass M_E of the Earth and the mass M_S of the Sun.

EVALUATE   The total mechanical energy associated with the Earth's orbital motion is

$$E_{\text{Tot}} = \frac{1}{2}U = -\frac{GM_S M_E}{2r}$$

$$= -\frac{\left(6.67\times10^{-11}\ \text{N}\cdot\text{m}^2/\text{kg}^2\right)\left(1.99\times10^{30}\ \text{kg}\right)\left(5.97\times10^{24}\ \text{kg}\right)}{2\left(1.50\times10^{11}\ \text{m}\right)} = -2.64\times10^{33}\ \text{J}$$

ASSESS   Alternatively, from Equation 8.8a, $E_{\text{tot}} = -K = -\frac{1}{2}M_E\left(2\pi r/T\right)^2 = -2.66\times10^{33}\ \text{J}$, consistent with the accuracy of the data used in Appendix E.

31.   **INTERPRET**   This problem involves calculating the escape speed from two celestial bodies with different characteristics.

DEVELOP   Solve Equation 8.7, $v_{\text{esc}} = \sqrt{2GM/r}$, for the escape speed v_{esc} for each body. Use data from Appendix E as needed.

EVALUATE   (a) For Jupiter's moon Callisto, the escape speed is

$$v_{\text{esc}} = \sqrt{2\left(6.67\times10^{-11}\ \text{N}\cdot\text{m}^2/\text{kg}^2\right)\left(1.07\times10^{23}\ \text{kg}\right)/\left(2.40\times10^6\ \text{m}\right)} = 2.44\ \text{km/s}.$$

(b) For a neutron star, $v_{\text{esc}} = \sqrt{2\left(6.67\times10^{-11}\ \text{N}\cdot\text{m}^2/\text{kg}^2\right)\left(1.99\times10^{30}\ \text{kg}\right)/\left(6\times10^3\ \text{m}\right)} = 2.10\times10^8\ \text{m/s}.$

ASSESS   The escape speed from a neutron star is about 70% of the speed of light.

PROBLEMS

33.   **INTERPRET**   This problem involves Newton's second law (F = ma) and Newton's law of universal gravitation. We are to find the acceleration at an altitude equal to half the planet's radius, given the gravitational acceleration on the planet's surface.

DEVELOP   By Newton's law of universal gravitation, Equation 8.1, we see that the force due to gravity on an object is $F = GMm/r^2$. Newtons' second law $F = ma$ relates force to acceleration, so we can express the acceleration on the surface of the planet and at an altitude above the surface equal to one-half of the planet's radius.

EVALUATE   On the planet's surface, the acceleration due to gravity is

$$F_{net} = \frac{GMm}{r^2} = mg$$

$$g = \frac{GM}{r^2}$$

At the given altitude h, we substitute $3r/2$ for r, then solve the system of equations for g_h. This gives

$$g_h = \frac{GM}{(3r/2)^2} = \left(\frac{2}{3}\right)^2 \frac{GM}{r^2} = \frac{4}{9} g = \frac{4}{9}(22.5 \text{ m/s}^2) = 10.0 \text{ m/s}^2$$

ASSESS   The acceleration scales with the inverse of the radial distance squared, to increasing the radius by a factor $3/2$ results in a decrease in the acceleration by a factor $(2/3)^2$.

35.   INTERPRET   This problem explores the gravitational acceleration of a gravitating body as a function of altitude h.

DEVELOP   Using Equation 8.1, the gravitational force between a mass m and a planet of mass M_p is $F = GM^p m/r^2$ where r is their separation, measured from the center of the planet. From Newton's second law (for constant mass), $F = ma$, the acceleration of gravity at any altitude $h = r - R_p$ above the surface of a spherical planet of radius R_p, is

$$g(h) = \frac{GM_p}{\left(R_p + h\right)^2} = \frac{GM_p}{R_p^2}\left(\frac{R_p}{R_p + h}\right)^2 = g(0)\left(\frac{R_p}{R_p + h}\right)^2$$

where $g(0)$ is the value at the surface. Once the ratio $g(h)/g(0)$ is known, we can find the altitude h in terms of R_p.

EVALUATE   Solving for h, we find

$$\frac{h}{R_p} = \sqrt{\frac{g(0)}{g(h)}} - 1$$

Therefore, for $g(h)/g(0) = 1/2$, we have $h/R_p = \sqrt{2} - 1 = 0.414$.

ASSESS   To see if the result makes sense, we take the limit $h = 0$, where the object rests on the surface of the planet. In this limit, we recover $g(0)$ as the gravitational acceleration. The equation also shows that $g(h)$ decreases as the altitude h is increased, and $g(h)$ approaches zero as $h \to \infty$.

37.   INTERPRET   In this problem we want to find the Moon's acceleration in its circular orbit about the Earth. In addition, we want to use the result to confirm the inverse-square law for the gravitational force.

DEVELOP   Using Newton's second law ($F_{net} = ma$), and the Equation 5.1 for centripetal acceleration, we find that the centripetal force that keeps the Moon's orbit circular is:

$$F_{net} = M_M a_c = \frac{M_M v^2}{r_{ME}} = \frac{4\pi^2 M_M r_{ME}}{T^2}$$

The equation allows us to compute the acceleration of the Moon in its circular orbit.

EVALUATE   Substituting the values given in the problem statement, we find

$$a_c = \frac{v^2}{r_{mE}} = \frac{4\pi^2 r_{mE}}{T^2} = \frac{4\pi^2 \left(3.84 \times 10^8 \text{ m}\right)}{\left(27.3 \text{ d} \times 86,400 \text{ s/d}\right)^2} = 2.73 \times 10^{-3} \text{ m/s}^2$$

As a fraction of the acceleration due to gravity on the surface of the Earth, this is

$$\frac{a_c}{g} = \frac{2.73 \times 10^{-3} \text{ m/s}^2}{9.81 \text{ m/s}^2} = 2.78 \times 10^{-4}$$

Comparing this result with the ratio of the radius of the Earth to the radius of the Moon's orbit gives

$$\left(\frac{R_E}{r_{ME}}\right)^2 = \left(\frac{6.37 \times 10^6 \text{ m}}{3.85 \times 10^8 \text{ m}}\right)^2 = 2.74 \times 10^{-4}$$

which suggests that Newton's inverse-square law for gravity is valid.

ASSESS   Why do the answers differ by 0.04/2.78 = 1.4%? The problem is not the data used, nor its precision (all data is accurate to 3 significant figures). However, we assumed that the Moon's orbit was circular, which is not exactly true. At its nearest, the Moon is some 364×10^3 km from the Earth, whereas at its farthest, it is some 407×10^3 km from the Earth. This assumption is responsible for the difference in the results obtained above.

39.   INTERPRET   This problem involves Newton's law of universal gravitation, which is used to find the period of a circular orbit (Equation 8.4). We are asked to find the half-period of a circular orbit 130 m above the surface of the Moon.

DEVELOP   Equation 8.4 gives the period of a circular orbit to be $T^2 = 4\pi^2 r^3/(GM)$, where M is the mass of the Moon and r is the radius of the orbit. For an orbit at a height $h = 130$ m above the surface of the Moon, $r = R_M + h$. Use the data available in Appendix E to evaluate the half period $T/2$.

EVALUATE   The half period of the astronaut's orbit was

$$\frac{T}{2} = \pi\sqrt{\frac{(R_M + h)^3}{GM}} = \pi\sqrt{\frac{(1.74\times10^6 \text{ m} + 0.13\times10^6 \text{ m})^3}{(6.67\times10^{-11} \text{ N}\cdot\text{m}^2/\text{kg}^2)(7.35\times10^{22} \text{ kg})}} = 3.63\times10^3 \text{ s} = 60.5 \text{ min}$$

or about an hour.

ASSESS   During this hour, the astronaut could not communicate with the Earth.

41.   INTERPRET   We will be estimating the mass of the galaxy by using the Sun's orbit around the galaxy. This is similar to measuring the mass of the Earth by the orbit of the moon.

DEVELOP   Equation 8.4 relates mass of a central object to the period and radius of an orbiting object: $T^2 = 4\pi^2 r^3 / GM$. However, the central object in this case is a point or a sphere, so we will have to assume that the galaxy is spherical and that most of its mass is located interior to the orbit of the Sun.

EVALUATE   Using the radius and period given for the Sun's orbit, the mass of the galaxy is approximately

$$M = \frac{4\pi^2 r^3}{GT^2} = \frac{4\pi^2 (2.6\times10^{20} \text{ m})^3}{(6.67\times10^{-11} \frac{\text{N}\cdot\text{m}^2}{\text{kg}^2})(2\times10^8 \times \pi \times 10^7 \text{s})^2} = 2.6\times10^{41} \text{ kg}$$

ASSESS   If we divide our result by the mass of the Sun $(1.99\times10^{30} \text{ kg})$, we find that it is equivalent to about 100 billion Suns, which is a reasonable estimate for the number of stars in the galaxy. Astronomers plot the orbital velocity of objects (such as stars, clusters of stars, or clouds of hydrogen atoms) versus their distance from the galactic center to obtain "the rotation curve" for our galaxy and others. What's surprising about these curves is that they are flat (i.e., nearly constant) out to distances far beyond the central bright region of most galaxies. One would have expected the velocity of orbiting objects to drop off at large radii, as indicated in Equation 8.3. The fact that it doesn't seems to imply some sort of "dark matter," which doesn't emit or scatter light and yet accounts for over 80% of the mass in a galaxy. Dark matter is currently a topic of great interest in astronomy.

43.   INTERPRET   We're asked to solve a three-body problem, in which three identical stars are situated on the vertices of an equilateral triangle.

DEVELOP   We're told that the system rotates. In order for the configuration to remain stable, each star must rotate with the same speed. Let's assume the rotational direction is clockwise, as shown in the figure below.

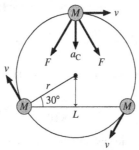

This is uniform circular motion about a radius $r = L/2\cos 30°$. The centripetal acceleration $\left(a_c = v^2/r\right)$ is provided by gravity. Specifically, each star is pulled toward the two other stars. Taken separately, the magnitude of the force, F, between two stars is: $F = GM^2/L^2$ (Equation 8.1). Added together, the net force points toward the center of the triangle with a magnitude of

$$F_{net} = F\cos 30° + F\cos 30° = \frac{2GM^2\cos 30°}{L^2}$$

It's this force that supplies the centripetal acceleration: $F_{net} = Ma_c$.

EVALUATE Pulling together all the information above, we can find an expression for the speed of the stars' rotation:

$$v = \sqrt{a_c r} = \sqrt{\left(\frac{F_{net}}{M}\right)\left(\frac{L}{2\cos 30°}\right)} = \sqrt{\frac{GM}{L}}$$

Notice how this has a similar form to Equation 8.3: $v = \sqrt{GM/r}$, for the orbital speed of a two-body system. The period in the three-body system is:

$$T = \frac{2\pi r}{v} = \frac{\pi L}{\cos 30°}\sqrt{\frac{L}{GM}}$$

To draw some comparison with Equation 8.4, we square the above equation and use $\cos^2 30° = \frac{3}{4}$,

$$T^2 = \frac{4\pi^2 L^3}{3GM}$$

ASSESS This says the period becomes longer, the farther the stars are separated, which makes sense. The system rotates faster (shorter period) when the mass of the stars is larger, which also makes sense.

45. **INTERPRET** This problem requires us to find the semimajor axis of an asteroid's orbit by using Kepler's third law (Equation 8.4). We are to express the results in terms of the Earth's orbital period and radius about the Sun.

DEVELOP Kepler's third law is

$$T^2 = \frac{4\pi^2 r^3}{GM}$$

where r is the semimajor axis of the elliptical orbit. To find the semimajor axis of the asteroid's orbit, we take the ratio of Kepler's third law applied to both objects:

$$\left(\frac{T_{ast}}{T_E}\right)^2 = \left(\frac{r_{ast}}{r_E}\right)^3 \quad \Rightarrow \quad r_{ast} = r_E\left(\frac{T_{ast}}{T_E}\right)^{2/3}$$

EVALUATE Taking one unit of distance to be the Earth's orbit ($\equiv 1$ AU) and one unit of time to be the Earth's period ($= 1$ y), we have

$$r_{ast} = (1\,\text{AU})\left[\left(\frac{1417\,\text{d}}{1\,\text{y}}\right)\left(\frac{1\,\text{y}}{363\,\text{d}}\right)\right]^{2/3} = 2.47\,\text{AU}$$

ASSESS Converting this distance to meters gives $r_{ast} = (2.47\,\text{AU})(1.5 \times 10^{11}\,\text{m/AU}) = 3.71 \times 10^8\,\text{m}$.

47. **INTERPRET** This problem involves conservation of mechanical energy, which we can use to find the speed of Comet Halley when it reaches Neptune's orbit, given its speed at the perihelion.

DEVELOP At perihelion, the point of closest approach to the Sun, the comet's distance from the Sun is $r_i = 8.79 \times 10^{10}\,\text{m}$ and its speed is $v = 54.6 \times 10^3\,\text{m/s}$. The distance of Neptune's orbit is at $r_f = 4.50 \times 10^{12}\,\text{m}$. We use conservation of energy: $U_i + K_i = U_f + K_f$.

EVALUATE Inserting the given quantities into the formula above gives

$$U_i + K_i = U_f + K_f$$

$$-G\frac{Mm}{r_i} + \frac{1}{2}mv_i^2 = -G\frac{Mm}{r_f} + \frac{1}{2}mv_f^2$$

$$v_f = \pm\sqrt{2GM\left(\frac{1}{r_f} - \frac{1}{r_i}\right) + v_i^2}$$

$$= \pm\sqrt{2\left(6.67\times10^{-11}\ \text{N}\cdot\text{m}^2/\text{kg}^2\right)\left(1.99\times10^{30}\ \text{kg}\right)\left(\frac{1}{4.50\times10^{12}\ \text{m}} - \frac{1}{8.79\times10^{10}\ \text{m}}\right) + \left(54.6\times10^3\ \text{m/s}\right)^2}$$

$$= 4.48\ \text{km/s}$$

where we take the positive square root because we are interested in the speed, not in the direction at which the asteroid is orbiting.

ASSESS   The speed changes by over 90%, but the energy does not change.

49.   **INTERPRET**   We want to show that an object lands on Earth with essentially the escape speed when it starts from rest far away from the Earth.

DEVELOP   Starting from rest means the initial kinetic energy is zero. The initial gravitational potential energy ($U_0 = -GMm/r_0$ from Equation 8.6) is nearly zero, given that $r_0 \gg R_E$. By conservation of energy, the object falls to the Earth's surface with kinetic and potential energy satisfying:

$$K + U = \tfrac{1}{2}mv^2 - \frac{GMm}{R_E} = K_0 + U_0 = -\frac{GMm}{r_0}$$

EVALUATE   Solving for the velocity in the above equation gives:

$$v = \sqrt{2GM\left(\frac{1}{R_E} - \frac{1}{r_0}\right)} = \sqrt{\frac{2GM}{R_E}\left(1 - \frac{R_E}{r_0}\right)} \approx v_{esc}\left(1 - \frac{R_E}{2r_0}\right) \approx v_{esc}$$

where we have used the definition of the escape velocity from Equation 8.7, as well as the binomial approximation from Appendix A, since $R_E/r_0 \ll 1$.

ASSESS   Since gravity is a conservative force, the scenario where an object falls to Earth from a great distance is just the time-reversal of the scenario where the object leaves Earth with the essentially escape velocity. It's like one movie played either forwards or backwards. So the landing velocity in the falling scenario should be the same as the take-off velocity in the escaping scenario.

51.   **INTERPRET**   The question boils down to: is the comet's orbit open or closed? Will it orbit around the Sun multiple times (and therefore pass by the Earth again), or is it destined to escape our solar system?

DEVELOP   The comet's orbit is open if the given velocity is greater than the escape velocity: $v = \sqrt{2GM/r}$. In this case, the mass is the Sun and the radius is the Earth's distance from the Sun.

EVALUATE   The escape velocity from the Sun at Earth's orbital radius is

$$v_{esc} = \sqrt{\frac{2GM_S}{r_E}} = \sqrt{\frac{2\left(6.67\times10^{-11}\ \frac{\text{N}\cdot\text{m}^2}{\text{kg}^2}\right)\left(1.99\times10^{30}\ \text{kg}\right)}{\left(150\times10^9\ \text{m}\right)}} = 42.1\ \text{km/s}$$

The comet is going faster than this escape velocity, so it is on a open (hyperbola) orbit, see Figure 8.9. It will not return to Earth's vicinity.

ASSESS   The escape velocity calculated here is much smaller than the 618 km/s escape velocity given in Appendix E for the Sun. But the larger value is the escape velocity from the Sun's surface. Farther away at the Earth's orbital radius the escape velocity doesn't need to be so high. Also note that the comet velocity is much greater than the Earth's escape velocity of 11.2 km/s, which just means that there's no danger of the comet getting captured in a closed orbit around Earth.

53.   **INTERPRET**   This problem involves the energy contained in orbital motion and position, which we will use to calculate the energy needed to put a satellite into circular orbit a height h above the surface of the Earth.

DEVELOP The total energy of a circular orbit is given by Equation 8.8b, $E_{\text{orbit}} = -GM_E m/2r$. In orbit, the radius $r = R_E + h$. On the ground, the total mechanical energy of the satellite is $E_{\text{surface}} = U_0 + K_0 = -GM_E m/R_E + mv^2/2$, where v is the velocity of the surface of the Earth as it rotates on its axis. However, we are instructed to neglect this ($mv^2/2 \sim 0.34\%$ of $|U_0|$), so we have $E_{\text{surface}} = -GM_E m/R_E$. Equate these two expressions for total mechanical energy to find the energy required to place a satellite in orbit.

EVALUATE The energy required to put a satellite into an orbit at a height h is therefore approximated by

$$E_{\text{orbit}} - E_{\text{surface}} = -\frac{GM_E m}{2(R_E + h)} + \frac{GM_E m}{R_E} = \left(\frac{GM_E m}{R_E}\right)\left(\frac{-1}{2(1 + h/R_E)} + \frac{2(1 + h/R_E)}{2(1 + h/R_E)}\right) = \left(\frac{GM_E m}{R_E}\right)\left[\frac{R_E + 2h}{2(R_E + h)}\right]$$

which agrees with the formula in the problem statement.

ASSESS Notice that the second factor in the result is dimensionless, and the first factor has units of energy, so the units work out to units of energy, as required. If we let h → 0, we find

$$E_{\text{orbit}} - E_{\text{surface}} = \left(\frac{GM_E m}{2R_E}\right)$$

which is the energy of a satellite orbiting around the Earth at zero altitude (i.e., an object sitting on the surface of the Earth). This is as expected, and just represents the neglected kinetic energy of the satellite due to the Earth's rotation. Had we included this term, the difference would be zero.

55. **INTERPRET** This problem involves using conservation of total mechanical energy to find the speed of a satellite for several different orbits. It also requires applying Kepler's third law to relate orbital radii to orbital periods.

DEVELOP The speed of a satellite in a circular orbit is given by Equation 8.3, $v^2 = GM/r$, where r is the distance to the center of the Earth. If the speed is to change to v' where $v' = 1.1v$, then the orbital radius will satisfy $v'^2 = (1.1)^2 v^2 = GM/r'$, which gives $r' = (1.1)^{-2} r$. From this, we can solve for the difference in orbital height $\Delta h = r - r'$. For part (b), take the ratio of Kepler's third law (Equation 8.4) applied to each orbit. This gives

$$\left(\frac{T}{T'}\right)^2 = \left(\frac{r}{r'}\right)^3$$

where the primed quantities are for the new orbit. Given that $T' = 0.9T$, we can again solve for $\Delta h = r - r'$

EVALUATE (a) For a 10% increase in orbital speed, the orbital height decreases by

$$\Delta h = r - r' = r\left[1 - (1.1)^{-2}\right] = (R_E + h)\left[1 - (1.1)^{-2}\right] = (6.37 \times 10^6 \text{ m} + 5.50 \times 10^6 \text{ m})\left[1 - (1.1)^{-2}\right] = 2.06 \times 10^6 \text{ m}$$

(b) For a 10% decrease in orbital period,

$$\left(\frac{T}{T'}\right)^2 = \left(\frac{1}{0.9}\right)^2 = \left(\frac{r}{r'}\right)^3$$

$$r' = (0.9)^{2/3} r$$

$$\Delta h = r - r' = r\left[1 - (0.9)^{2/3}\right] = (R_E + h)\left[1 - (0.9)^{2/3}\right] = (6.37 \times 10^6 \text{ m} + 5.50 \times 10^6 \text{ m})\left[1 - (0.9)^{2/3}\right] = 0.805 \times 10^6 \text{ m}$$

ASSESS We find that the orbital height is more sensitive to orbital speed than it is to orbital period.

57. **INTERPRET** This problem again involves conservation of total mechanical energy, this time applied to two rockets so that we can find their speeds when they cross the Moon's orbit.

DEVELOP Conservation of energy applied to the rockets gives:

$$K_0 + U_0 = K + U \quad \Rightarrow \quad \frac{1}{2}mv_0^2 - \frac{GM_E m}{r_0} = \frac{1}{2}mv^2 - \frac{GM_E m}{r}$$

Solving for v, we obtain

$$v = \pm\sqrt{v_0^2 + 2GM_E\left(\frac{1}{r} - \frac{1}{r_0}\right)}$$

The initial position of the rockets is $r_0 = R_E = $, and the final position is $r = R_M = 3.85 \times 10^8$ m (Appendix E).

EVALUATE   Evaluating the above expression for the final speed v gives

$$v_1 = \sqrt{(12,000 \text{ m/s})^2 + 2\left(6.67 \times 10^{-11} \text{ N} \cdot \text{m}^2/\text{kg}^2\right)\left(5.97 \times 10^{24} \text{ kg}\right)\left(\frac{1}{3.85 \times 10^8 \text{ m}} - \frac{1}{\left(6.37 \times 10^6 \text{ m}\right)}\right)} = 4.59 \text{ km/s}$$

and

$$v_2 = \sqrt{(18,000 \text{ m/s})^2 + 2\left(6.67 \times 10^{-11} \text{ N} \cdot \text{m}^2/\text{kg}^2\right)\left(5.97 \times 10^{24} \text{ kg}\right)\left(\frac{1}{3.85 \times 10^8 \text{ m}} - \frac{1}{\left(6.37 \times 10^6 \text{ m}\right)}\right)} = 14.2 \text{ km/s}$$

where we have taken the positive square root because we are interested in speed, not in direction.

ASSESS   As $r \to \infty$, $v \to 0$ as expected because the second term under the radical becomes the escape speed, which would be equal to the first term for launching a rocket infinitely far from Earth.

59.   **INTERPRET**   This problem involves conservation of total mechanical energy, which we can use to find the speed of the missile at the apex of its trajectory. We ignore nonconservative forces such as air resistance, so that only gravity is considered to act on the missile.

DEVELOP   Applying conservation of total mechanical energy (Equation 7.7) gives

$$K_0 + U_0 = K + U \quad \Rightarrow \quad \frac{1}{2}mv_0^2 - \frac{GM_E m}{R_E} = \frac{1}{2}mv^2 - \frac{GM_E m}{r} \quad \text{or} \quad v(r) = \pm\sqrt{v_0^2 + 2GM_E\left(\frac{1}{r} - \frac{1}{R_E}\right)}$$

We will take the positive square root because we are interested in the missile's speed, not its direction.

EVALUATE   Inserting the $r = R_E + 1200$ km and $v_0 = 6.1$ km/s into the above expression, we find the speed at the apex of the trajectory is

$$v = \sqrt{(6100 \text{ m/s})^2 + 2\left(6.67 \times 10^{-11} \text{ N} \cdot \text{m}^2/\text{kg}^2\right)\left(5.97 \times 10^{24} \text{ kg}\right)\left(\frac{1}{\left(6.37 \times 10^6 \text{ m} + 1.20 \times 10^6 \text{ m}\right)} - \frac{1}{6.37 \times 10^6 \text{ m}}\right)}$$

$$= 4.17 \text{ km/s}$$

ASSESS   If the missile was launched directly upward, it would reach a height of

$$-\frac{GM_E m}{R_E} + \frac{1}{2}mv_0^2 = -\frac{GM_E m}{R_E + h}$$

$$h = \frac{GM_E}{GM_E/R_E - v_0^2/2} - R_E = 2.70 \times 10^6 \text{ m}$$

which is just over twice the height given in the problem statement.

61.   **INTERPRET**   Only conservative forces (i.e., gravity) act on Mercury. Therefore, we can apply conservation of total mechanical energy to find Mercury's perihelion distance.

DEVELOP   Conservation of total mechanical energy (Equation 7.7) gives

$$K_0 + U_0 = K + U$$

$$\frac{1}{2}mv_a^2 - \frac{GM_S m}{r_a} = \frac{1}{2}mv_p^2 - \frac{GM_S m}{r_p}$$

which we can solve for the perihelion distance r_p.

EVALUATE   Solving the expression above for r_p and inserting the given quantities gives

$$r_p = \frac{2GM_S}{v_p^2 - v_a^2 + 2GM_S/r_a}$$

$$= \frac{2\left(6.67 \times 10^{-11} \text{ N} \cdot \text{m}^2/\text{kg}^2\right)\left(1.99 \times 10^{30} \text{ kg}\right)}{\left(59.0 \times 10^3 \text{ m/s}\right) - \left(38.8 \times 10^3 \text{ m/s}\right) + 2\left(6.67 \times 10^{-11} \text{ N} \cdot \text{m}^2/\text{kg}^2\right)\left(1.99 \times 10^{30} \text{ kg}\right)/\left(6.99 \times 10^{10} \text{ m}\right)}$$

$$= 4.60 \times 10^{10} \text{ m}$$

ASSESS   Kepler's second law provides a more direct solution:

$r_p = r_a \left(v_a/v_p\right) = \left(6.99 \times 10^{10} \text{ m}\right)\left(38.8/59.0\right) = 4.60 \times 10^{10} \text{ m}$ (see the solution to Problem 58).

63.   **INTERPRET**   This problem involves Kepler's third law, which we can apply to find the orbital periods of the satellites in their various orbits.

DEVELOP   In a lower circular orbit (smaller r) the orbital speed is faster (see Equation 8.3). The time for 10 complete orbits of the faster satellite must equal the time for 9.5 geosynchronous orbits. Thus, $10T' = 9.5T$, where the prime indicates the faster, lower orbiting satellite. Thus, the ratio of the orbital periods is $T'/T = 0.95$. Knowing the ratio of the periods, we can find the radius of the lower orbit by applying Kepler's third law to both orbits and taking the ratio.

EVALUATE   The ratio of Kepler's third law applied to both orbits gives

$$\left(\frac{T'}{T}\right)^2 = (0.95)^2 = \left(\frac{r'}{r}\right)^3$$

$$r' = r(0.95)^{2/3}$$

$$r - r' = r\left[1 - (0.95)^{2/3}\right]\left(42.2 \times 10^6 \text{ m}\right) = 1.42 \times 10^3 \text{ km}$$

where we have used the orbital radius $r = 42{,}200$ km from Example 8.3.

ASSESS   To catch up with the other satellite in a single orbit, we would have to descend a distance

$$r - r' = r\left[1 - (0.5)^{2/3}\right]\left(42.2 \times 10^6 \text{ m}\right) = 15.6 \times 10^3 \text{ km}$$

which would leave the satellite at a height of 42.2 Mm –15.6 Mm = 26.6×10^3 km.

65.   **INTERPRET**   This problem involves Kepler's third law, which we will apply to convert the rate of change in the orbital period of the Moon to the rate of change in its orbital distance (i.e., its radial speed). We assume that the Moon's orbit is approximately circular for this calculation.

DEVELOP   Kepler's third law relates the orbital period to the semimajor axis of an elliptical orbit (of which a circular orbit is a special case): $T^2 = 4\pi^2 r^3/(GM)$. We are told the rate of change is

$$\frac{dT}{dt} = \left(35 \times 10^{-3} \frac{\text{s}}{100 \text{ y}}\right)\left(\frac{1 \text{ y}}{365 \text{ d}}\right)\left(\frac{1 \text{ d}}{86{,}400 \text{ s}}\right) = 1.11 \times 10^{-11}$$

so we can differentiate Kepler's law to find the rate of change in the orbital radius r.

EVALUATE   Differentiating Kepler's law gives

$$\frac{dT}{dt} = \pm \frac{d}{dT}\sqrt{\frac{4\pi^2 r^3}{GM}} = \frac{3}{2}\sqrt{\frac{4\pi^2 r}{GM}}\frac{dr}{dt}$$

$$\frac{dr}{dt} = \frac{dT}{dt}\left(\frac{2}{3}\sqrt{\frac{GM}{4\pi^2 r}}\right) = \left(1.11 \times 10^{-11}\right)\left(\frac{2}{3}\sqrt{\frac{\left(6.67 \times 10^{-11} \text{ N} \cdot \text{m}^2/\text{kg}^2\right)\left(7.35 \times 10^{22} \text{ kg}\right)}{4\pi^2\left(3.85 \times 10^8 \text{ m}\right)}}\right)$$

$$= \left(1.33 \times 10^{-10} \text{ m/s}\right)\left(\frac{86{,}400 \text{ s}}{1 \text{ d}}\right)\left(\frac{36500 \text{ d}}{1 \text{ c}}\right)\left(\frac{100 \text{ cm}}{1 \text{ m}}\right) = 41.9 \text{ cm/c}$$

where we have taken the positive square root because the orbital radius is increasing, not decreasing.

ASSESS   At this speed, we don't have to worry about the Moon leaving any time soon!

67.   **INTERPRET**   You need to determine if a hockey puck hit at the maximum known speed could somehow enter orbit around the Moon.

DEVELOP   Let's assume the puck is hit in a direction more or less parallel to the Moon's surface. The easiest orbit to reach would be a circular orbit with radius just slightly greater than the Moon's surface radius: $r \simeq R_M$. From Equations 8.8a and b, this lowest energy orbit has kinetic energy, $K = GM_M m / 2R_M$.

EVALUATE   What speed would a hockey puck have to be hit at to reach the lowest energy orbit from above?

$$v = \sqrt{\frac{2K}{m}} = \sqrt{\frac{GM_M}{R_M}} = \frac{1}{\sqrt{2}} v_{esc,M} = \frac{1}{\sqrt{2}} (2.38 \text{ km/s}) = 6100 \text{ km/h}$$

Notice that we have used the escape velocity from the Moon, given in Appendix E. In any case, there's no danger that a 168 km/h puck will go into lunar orbit.

ASSESS   Notice that the minimum orbital velocity is only 30% smaller than the escape velocity.

69.   **INTERPRET**   We're asked to calculate the position of the L1 Lagrange point, where the gravity of the Earth and the Sun combine to give a period of 1 year around the Sun. It's worth noting that when only the Sun's gravity is considered, the only place with a 1-year period would be at the Earth's orbital radius:

$$T_E = \sqrt{\frac{4\pi^2 r_E^3}{GM_S}} = 1 \text{ y}$$

where we have used Equation 8.4 with r_E the Earth's distance from the Sun, and M_S the mass of the Sun.

DEVELOP   Let's assume that the point L1 is at a distance r_{LE} from the Earth and a distance r_{LS} from the Sun. From the remarks in the text, we know that L1 is between the Earth and the Sun, so $r_{LE} + r_{LS} = r_E$. The sum of the gravitational attraction from both bodies is

$$F_{net} = \frac{GM_S m}{r_{LS}^2} - \frac{GM_E m}{r_{LE}^2}$$

This sum supplies a centripetal force that keeps any object there in uniform circular motion around the Sun: $F_{net} = mv^2 / r_{LS}$. The orbital speed results in a period of $T = 2\pi r_{LS} / v$, which by definition is equal to one year. Combining all this information, we have:

$$\frac{GM_S}{r_{LS}^2} - \frac{GM_E}{r_{LE}^2} = \frac{4\pi^2 r_{LS}}{T^2}$$

We will now substitute $r_{LS} = r_E - r_{LE}$, as well as introduce the variables $x = r_{LE} / r_E$ and $y = M_E / M_S$, in order to obtain:

$$\frac{1}{(1-x)^3} - \frac{y}{x^2(1-x)} = \frac{4\pi^2 r_E^3}{GM_S T^2} = \frac{T_E^2}{T^2} = 1$$

Notice how we were able to substitute the Earth's orbital period, T_E, into the right-hand side of the equation. Both the Earth's period and the period at L1 are equal to 1 year, so they cancel.

EVALUATE   We are now faced with a rather difficult equation to solve:

$$\frac{1}{(1-x)^2} - \frac{y}{x^2} = 1 - x$$

But we can assume that the Lagrange point is much closer to Earth than to the Sun, so $x \ll 1$. In which case, the first term on the left can be reduced using the binomial approximation: $(1-x)^{-2} \approx 1 + 2x$ (see Appendix A). We then have

$$x \approx \sqrt[3]{\frac{y}{3}} = \sqrt[3]{\frac{M_E}{3M_S}} = \sqrt[3]{\frac{\left(5.97 \times 10^{24}\,\text{kg}\right)}{3\left(1.99 \times 10^{30}\,\text{kg}\right)}} = 0.01$$

This implies that L1 is 1% of the distance between the Earth and the Sun. Relative to the Earth, L1 is at a distance of

$$r_{LE} \approx 0.01 r_E = 0.01\left(150 \times 10^6\,\text{km}\right) = 1.5 \times 10^6\,\text{km}$$

ASSESS   One can check in an outside reference that indeed the L1 Lagrange point is around 1.5 million km from Earth. There are 4 other Lagrange points, called L2, L3, L4 and L5. Like L1, they are all stationary points, meaning an object situated there will not move relative to the Earth and Sun.

71.   **INTERPRET**   We consider the characteristics of the Global Positioning System.
DEVELOP   The satellite speed can be found from Equation 8.3: $v = \sqrt{GM_E / \left(R_E + h\right)}$, where as in the previous problem we write the orbital radius as the sum of the radius of the Earth, R_E, and the altitude, h.
EVALUATE   Plugging in the known values, the speed of one of the satellites is

$$v = \sqrt{\frac{GM_E}{\left(R_E + h\right)}} = \sqrt{\frac{\left(6.67 \times 10^{-11}\,\frac{\text{N·m}^2}{\text{kg}^2}\right)\left(5.97 \times 10^{24}\,\text{kg}\right)}{\left(6.37 \times 10^6\,\text{m} + 20.2 \times 10^6\,\text{m}\right)}} = 3.9\ \text{km/s}$$

The answer is (d).
ASSESS   This is slower than the International Space Station, which orbits at 7.7 km/s (see Example 8.2). However, the station is much closer to Earth at an altitude of 380 km. It completes an orbit in 90 min, as compared to 12 hours.

73.   **INTERPRET**   We consider the characteristics of the Global Positioning System.
DEVELOP   The total energy of an object in a circular orbit is given in Equation 8.8b: $E = -GM_E m / 2\left(R_E + h\right)$.
EVALUATE   Using the mass for the next generation of GPS satellites, the total energy is

$$E = -\frac{GM_E m}{2\left(R_E + h\right)} = -\frac{\left(6.67 \times 10^{-11}\,\frac{\text{N·m}^2}{\text{kg}^2}\right)\left(5.97 \times 10^{24}\,\text{kg}\right)\left(844\ \text{kg}\right)}{2\left(6.37 \times 10^6\,\text{m} + 20.2 \times 10^6\,\text{m}\right)} = -6.3\ \text{GJ}$$

The answer is (d).
ASSESS   The total energy is negative because the satellites are in bound orbits around the Earth.

SYSTEMS OF PARTICLES

EXERCISES

Section 9.1 Center of Mass

13. **INTERPRET**   This is a two-dimensional problem about the center of mass. Our system consists of three masses located at the vertices of an equilateral triangle. Two masses are known and the location of the center of mass is given, so we can find the location of the third mass.

DEVELOP   The center of mass of a system of particles is given by Equation 9.2:

$$\vec{r}_{cm} = \frac{\sum_i m_i \vec{r}_i}{\sum_i m_i} = \frac{\sum_i m_i \vec{r}_i}{M}$$

We shall choose x-y coordinates with origin $(0,0)$ at the midpoint of the base. With this arrangement, the center of the mass is located at $x_{cm} = 0$ and $y_{cm} = y_3/2$, where y_3 is the position of the third mass (and, of course, $y_1 = y_2 = 0$ for the equal masses $m_1 = m_2 = m$ on the base).

EVALUATE   Using Equation 9.2, the y coordinate of the center of mass is

$$y_{cm} = \frac{m_1 y_1 + m_2 y_2 + m_3 y_3}{m_1 + m_2 + m_3} = \frac{m(0) + m(0) + m_3 y_3}{m + m + m_3} = \frac{m_3 (2y_{cm})}{2m + m_3}$$

Solving for m_3, we have $2m + m_3 = 2m_3$, or $m_3 = 2m$.

ASSESS   From symmetry consideration, it is apparent that $x_{cm} = 0$. However, we have $m + m = 2m$ at the bottom two vertices of the triangle. Because $y_{cm} = y_3/2$ (i.e., y_{cm} is halfway to the top vertex), we expect the mass there to be $2m$ (See Example 9.2).

15. **INTERPRET**   This two-dimensional problem is about locating the center of mass. Our system consists of three equal masses located at the vertices of an equilateral triangle of side L.

DEVELOP   We take x-y coordinates with the origin at the center of one side as shown in the figure below. The center of mass of a system of particles is given by Equation 9.2:

$$\vec{r}_{cm} = \frac{\sum_i m_i \vec{r}_i}{\sum_i m_i} = \frac{\sum_i m_i \vec{r}_i}{M}$$

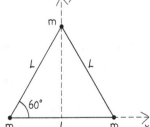

EVALUATE   From the symmetry (for every mass at x, there is an equal mass at $-x$) we have $x_{cm} = 0$. As for y_{cm}, because $y = 0$ for the two masses on the x-axis, and $y_3 = L\sin(60°) = L\sqrt{3}/2$ for the third mass, Equation 9.2 gives

$$y_{cm} = \frac{m_1 y_1 + m_2 y_2 + m_3 y_3}{m_1 + m_2 + m_3} = \frac{m(0) + m(0) + mL\sqrt{3}/2}{m + m + m} = \frac{\sqrt{3}}{6}L = 0.289L$$

ASSESS   From symmetry consideration, it is apparent that $x_{cm} = 0$. On the other hand, we have $m + m = 2m$ at the bottom two vertices of the triangle, and m at the top of the vertex. Therefore, we should expect y_{cm} to be one third of y_3. This indeed is the case, as y_{cm} can be rewritten as $y_{cm} = y_3/3$.

Section 9.2 Momentum

17.   **INTERPRET**   This problem involves conservation of linear momentum (Equation 9.7), which we can apply to find the speed of one out of two particles that separate after an explosion, given the speed of the other particle and mass of both particles.

DEVELOP   Before the explosion, the popcorn kernel has zero momentum, $(m_1 + m_2)\vec{v} = 0$. After the explosion, the total momentum of the two particles must still sum to zero, so we have $m_1\vec{v}_1 + m_2\vec{v}_2 = 0$. Thus,

$$\vec{v}_2 = \frac{m_1}{m_2}\vec{v}_1$$

so we can solve for v_2 given $m_1 = 91$ mg, $m_2 = 64$ mg, and $v_1 = (47 \text{ m/s})\hat{i}$.

EVALUATE   Inserting the given quantities gives $\vec{v}_2 = -(91 \text{ mg}/64 \text{ mg})(47 \text{ cm/s})\hat{i} = (-67 \text{ cm/s})\hat{i}$.

ASSESS   Notice that the smaller piece moves faster than the larger piece. Also notice that total mechanical energy is not conserved because $K_0 = (m_1 + m_2)v_2 = 0$ and $K = m_1 v_1^2 + m_2 v_2^2 \neq 0$.

19.   **INTERPRET**   This problem involves conservation of linear momentum (Equation 9.7), which we can apply to find the speed of one out of two particles that separate after an explosion, given the speed of the other particle and mass of both particles.

DEVELOP   Before the explosion, the uranium atom has zero momentum, so $(m_1 + m_2)\vec{v} = 0$. After fission, the total momentum of the two particles must still sum to zero, so we have $m_\alpha\vec{v}_\alpha + m_{U^{235}}\vec{v}_{U^{235}} = 0$. Thus,

$$\vec{v}_{U^{235}} = \frac{m_\alpha}{m_{U^{235}}}\vec{v}_\alpha$$

The initial speed can be obtained from the kinetic energy, $\vec{v}_\alpha = \pm\sqrt{2K_\alpha/m_\alpha}\hat{i} \equiv \sqrt{2K_\alpha/m_\alpha}\hat{i}$, so we can solve for $v_{U^{235}}$ using data from Appendix D for the masses of the particles.

EVALUATE   Solving for $v_{U^{235}}$ gives

$$\vec{v}_{U^{235}} = -\frac{\sqrt{2m_\alpha K_\alpha}}{m_{U^{235}}}\hat{i} = -\left[\frac{2(4 \text{ u})(5.15 \text{ MeV})(1.60 \times 10^{-3} \text{ J/MeV})}{(235 \text{ u})^2 1(.66 \times 10^{-27} \text{ kg/u})}\right]^{1/2}\hat{i} = (-2.68 \times 10^5 \text{ m/s})\hat{i}$$

We're asked for the speed, a scalar quantity, so we report 2.68×10^5 m/s or 0.268 Mm/s.

ASSESS   Because $K_\alpha = 5.15 \text{ MeV} \ll m_\alpha c^2 = 3.73$ Gev, relativity can be ignored.

Section 9.3 Kinetic Energy of a System

21.   **INTERPRET**   In this problem we are asked about the energy gained by the baseball pieces after the baseball explodes. We can apply conservation of linear momentum to solve this problem.

DEVELOP   Applying conservation of linear momentum to the baseball gives

$$\vec{P}_i = \vec{P}_f \quad \Rightarrow \quad (m_1 + m_2)\vec{v}_0 = m_1\vec{v}_1 + m_2\vec{v}_2$$

The initial kinetic energy of the system is $K_i = \frac{1}{2}(m_1 + m_2)v_0^2$, and the total final kinetic energy is $K_f = \frac{1}{2}m_1 v_1^2 + \frac{1}{2}m_2 v_2^2$. Therefore, the change in kinetic energy is

$$\Delta K = K_f - K_i = \frac{1}{2}m_1 v_1^2 + \frac{1}{2}m_2 v_2^2 - \frac{1}{2}(m_1 + m_2)v_0^2 = \frac{1}{2}m_1\left(v_1^2 - v_0^2\right) + \frac{1}{2}m_2\left(v_2^2 - v_0^2\right)$$

EVALUATE Let the forward direction be positive. By conservation of momentum, the velocity of the second piece, with mass $m_2 = m - m_1 = 150 \text{ g} - 38 \text{ g} = 112 \text{ g}$, is

$$v_2 = \frac{(m_1 + m_2)v_0 - m_1 v_1}{m_2} = \frac{(150 \text{ g})(60 \text{ km/h}) - (38 \text{ g})(85 \text{ km/h})}{112 \text{ g}} = 51.5 \text{ km/h} = 14.3 \text{ m/s}$$

In SI units $v_0 = 16.67 \text{ m/s}$ and $v_1 = 23.6 \text{ m/s}$, so the difference in kinetic energy is

$$\Delta K = \Delta K_1 + \Delta K_2 = \frac{1}{2}m_1\left(v_1^2 - v_0^2\right) + \frac{1}{2}m_2\left(v_2^2 - v_0^2\right)$$

$$= \frac{1}{2}\left(38 \times 10^{-3} \text{ kg}\right)\left[(23.6 \text{ m/s})^2 - (16.7 \text{ m/s})^2\right] + \frac{1}{2}\left(112 \times 10^{-3} \text{ kg}\right)\left[(14.3 \text{ m/s})^2 - (16.7 \text{ m/s})0^2\right]$$

$$= 1.21 \text{ J}$$

ASSESS The change in kinetic energy for the first piece (ΔK_1) is positive because $v_1 > v_0$, but negative for the second ($\Delta K_2 < 0$ because $v_2 < v_0$).

Section 9.4 Collisions

23. **INTERPRET** Your asked to compare the impulse during a collision to the impulse of gravity over the same time period.

    DEVELOP An impulse is a change in momentum produced by a force acting on an object over a specific time period. The gravitational force will be constant over the time of the collision, so we can find the gravity's impulse on each spacecraft using Equation 9.10a: $\vec{J} = \vec{F}_g \Delta t = \Delta \vec{p}$.

    EVALUATE We're not concerned with the direction of the impulse, but just the magnitude:

    $$J = mg\Delta t = (140 \text{ kg})(8.7 \text{ m/s}^2)(120 \times 10^{-3} \text{ s}) = 146 \text{ N} \cdot \text{s}$$

    The impulse imparted by gravity is 0.08% of the collision impulse.

    ASSESS During the collision, the influence of gravity can be neglected. That's because the average force from the collision is very large: $\overline{\vec{F}} = \vec{J}/\Delta t = 1.5 \text{ MN}$.

25. **INTERPRET** You need to determine how to fire a rocket to obtain the needed impulse.

    DEVELOP We're given the average thrust, so the needed time comes from Equation 9.9a: $\Delta t = J/\overline{F}$.

    EVALUATE For the required impulse, the space probes rocket must fire for

    $$\Delta t = \frac{J}{\overline{F}} = \frac{5.64 \text{ N} \cdot \text{s}}{135 \times 10^{-3} \text{ N}} = 41.8 \text{ s}$$

    ASSESS This might seem like a long time for such a small impulse. But the rocket exerts a tiny force on the space probe. Often, spacecraft need precision thrusters like the one here to make small adjustments in their trajectory or orientation.

Section 9.5 Totally Inelastic Collisions

27. **INTERPRET** In this problem, we are asked to show that half of the initial kinetic energy of a system is lost in a totally inelastic collision between two equal masses.

    DEVELOP Suppose we have two masses m_1 and m_2 moving with velocities $\vec{v}_1$ and $\vec{v}_2$, respectively. After undergoing a totally inelastic collision, the two masses stick together and move with final velocity $\vec{v}_f$. Although the collision is totally inelastic, momentum conservation still applies, and we have (Equation 9.11):

    $$m_1 \vec{v}_1 + m_2 \vec{v}_2 = (m_1 + m_2)\vec{v}_f \implies \vec{v}_f = \frac{m_1 \vec{v}_1 + m_2 \vec{v}_2}{m_1 + m_2}$$

    The initial total kinetic energy of the two-particle system is $K_i = \frac{1}{2}m_1 v_1^2 + \frac{1}{2}m_2 v_2^2$, whereas the final kinetic energy of the system after collision is $K_f = \frac{1}{2}(m_1 + m_2)v_f^2$. Therefore, the change in kinetic energy is given by

    $$\Delta K = K_f - K_i = \frac{1}{2}(m_1 + m_2)v_f^2 - \frac{1}{2}m_1 v_1^2 - \frac{1}{2}m_2 v_2^2 = \frac{1}{2}m_1\left(v_f^2 - v_1^2\right) + \frac{1}{2}m_2\left(v_f^2 - v_2^2\right)$$

EVALUATE   In our case, we have $m_1 = m_2 = m$, $v_1 = v$, and $v_2 = 0$. The initial kinetic energy of the system is therefore $K_i = \frac{1}{2}mv^2$. The final speed is

$$v_f = \frac{m_1 v_1 + m_2 v_2}{m_1 + m_2} = \frac{mv}{m+m} = \frac{1}{2}v$$

Therefore, the change in total kinetic energy is

$$\Delta K = \frac{1}{2}m_1\left(v_f^2 - v_1^2\right) + \frac{1}{2}m_2\left(v_f^2 - v_2^2\right) = \frac{1}{2}m\left(\frac{v^2}{4} - v^2\right) + \frac{1}{2}m\left(\frac{v^2}{4} - 0\right) = -\frac{1}{4}mv^2$$

Thus, we see that half of the total initial kinetic energy is lost in the collision process.

ASSESS   For a totally inelastic collision, one may show that the general expression for ΔK is

$$\Delta K = \frac{1}{2}\frac{\left(m_1 v_1 + m_2 v_2\right)^2}{m_1 + m_2} - \frac{1}{2}m_1 v_1^2 - \frac{1}{2}m_2 v_2^2 = -\frac{m_1 m_2}{2\left(m_1 + m_2\right)}\left(v_1 - v_2\right)^2$$

Clearly, ΔK is always negative, and it depends on the relative speed between m_1 and m_2.

29.   **INTERPRET**   This is a totally inelastic collision, since the trucks move together as one after the collision. You can find the mass of the second truck using conservation of momentum.

DEVELOP   According to Equation 9.11, conservation of momentum in the truck collision implies

$$m_1\vec{v}_1 + m_2\vec{v}_2 = (m_1 + m_2)\vec{v}_f$$

EVALUATE   The first truck is at rest ($\vec{v}_1 = 0$), which means the final velocity must be in the same direction as $\vec{v}_2$, so we don't need to work with the full vector notation. Given that the first truck has a mass of $m_1 = 5500 \text{ kg} + 3800 \text{ kg} = 9300 \text{ kg}$, we can solve for the mass of the second:

$$m_2 = m_1\frac{v_2 - v_f}{v_f} = (9300 \text{ kg})\frac{65 \text{ km/h} - 27 \text{ km/h}}{27 \text{ km/h}} = 13,089 \text{ kg}$$

Subtracting the 5500-kg mass of the truck itself leaves a load of 7589 kg, so the second truck was about 410 kg under the legal limit.

ASSESS   Note that because the velocities only appear in a ratio, it isn't necessary to convert to m/s.

Section 9.6 Elastic Collisions

31.   **INTERPRET**   This problem is about head-on (i.e. one-dimensional) elastic collisions. We want to find the speed of the ball after it rebounds elastically from a moving car.

DEVELOP   Both mechanical energy and linear momentum are conserved in an elastic collision. In this one-dimensional case, conservation of linear momentum gives

$$m_1 v_{1i} + m_2 v_{2i} = m_1 v_{1f} + m_2 v_{2f}$$

Conservation of energy gives

$$\frac{1}{2}m_1 v_{1i}^2 + \frac{1}{2}m_2 v_{2i}^2 = \frac{1}{2}m_1 v_{1f}^2 + \frac{1}{2}m_2 v_{2f}^2$$

Using the two conservation equations, the final speeds of m_1 and m_2 are (see Equations 9.15a and 9.15b):

$$v_{1f} = \frac{m_1 - m_2}{m_1 + m_2}v_{1i} + \frac{2m_2}{m_1 + m_2}v_{2i}$$

$$v_{2f} = \frac{2m_1}{m_1 + m_2}v_{1i} + \frac{m_2 - m_1}{m_1 + m_2}v_{2i}$$

EVALUATE   Let the subscripts 1 and 2 be for the car and the ball, respectively. We choose positive velocities in the direction of the car. The speed of the ball after it rebounds is

$$v_{2f} = \frac{2m_1}{m_1 + m_2}v_{1i} + \frac{m_2 - m_1}{m_1 + m_2}v_{2i} \approx 2v_{1i} - v_{2i} = 2(14 \text{ m/s}) - (-18 \text{ m/s}) = 46 \text{ m/s}$$

where we have used $m_1 \gg m_2$.

ASSESS   Similarly, the final speed of the car is

$$v_{1f} = \frac{m_1 - m_2}{m_1 + m_2} v_{1i} + \frac{2m_2}{m_1 + m_2} v_{2i} \approx v_{1i} = 14 \text{ m/s}$$

We do not expect the speed of the car to change much after colliding with a ball. However, the ball rebounds with a much greater speed than before. If the car were stationary with $v_{1i} = 0$, then we would find $v_{2f} = -v_{2i} = 18$ m/s.

33.   **INTERPRET**   In this problem we are asked to find the speeds of the protons after they collide elastically head-on. The problem is thus one-dimensional and involved conservation of mechanical energy and linear momentum.

DEVELOP   Consider the general situation where two masses m_1 and m_2 moving with velocities $\vec{v}_1$ and $\vec{v}_2$, undergo elastic collision. Both momentum and energy are conserved in this process. Using the conservation equations, the final speeds of m_1 and m_2 are (see Equations 9.15a and 9.15b):

$$v_{1f} = \frac{m_1 - m_2}{m_1 + m_2} v_{1i} + \frac{2m_2}{m_1 + m_2} v_{2i}$$

$$v_{2f} = \frac{2m_1}{m_1 + m_2} v_{1i} + \frac{m_2 - m_1}{m_1 + m_2} v_{2i}$$

EVALUATE   We choose positive velocities to be in the direction of $\vec{v}_1$. With $m_1 = m_2 = m$, and $v_1 = v = 11$ Mm/s, and $v_2 = -v_1 = -v = -11$ Mm/s, the final speeds are

$$v_{1f} = \frac{m_1 - m_2}{m_1 + m_2} v_{1i} + \frac{2m_2}{m_1 + m_2} v_{2i} = v_{2i} = -v = -11 \text{ Mm/s}$$

$$v_{2f} = \frac{2m_1}{m_1 + m_2} v_{1i} + \frac{m_2 - m_1}{m_1 + m_2} v_{2i} = v_{1i} = v = 11 \text{ Mm/s}$$

ASSESS   In this case, the protons simply exchange places—the final speed of the first proton is equal to the initial speed of the second proton, while the final speed of the second proton is equal to the initial speed of the first proton.

PROBLEMS

35.   **INTERPRET**   In this problem we want to find the center of mass of a pentagon of side a with one trianglular section missing.

DEVELOP   We choose coordinates as shown in the figure below. If the fifth isosceles triangle (with the same uniform density as the others) were present, the center of mass of the whole pentagon would be at the origin, so

$$0 = \frac{my_5 + 4my_{cm}}{5m} = \frac{y_5 + 4y_{cm}}{5}$$

where y_{cm} gives the position of the center of mass of the figure we want to find, and y_5 is the position of the center of mass of the fifth triangle. Of course, the mass of the figure is four times the mass of the triangle.

EVALUATE   From symmetry, the x coordinate of the center of mass is $x_{cm} = 0$. Now, to calculate y_{cm}, we make use of the result obtained in Example 9.3 where the center of mass of an isosceles triangle is calculated. This gives $y_5 = -2L/3$. In addition, from the geometry of a pentagon, we have $\tan(36°) = a/(2L)$. Therefore, the y coordinate of the center of mass is

$$y_{cm} = -\frac{1}{4} y_5 = \frac{L}{6} = \frac{a}{12} \cot(36°) = 0.115\, a$$

ASSESS   From symmetry argument, the center of mass must lie along the line that bisects the figure. With the missing triangle, we expect it to be located above $y = 0$, which would have been the center of mass for a complete pentagon.

37.   **INTERPRET**   We are asked to calculate the center-of-mass motion of a three body system.

DEVELOP   The position of the center of mass is given by Equation 9.2: $\vec{r}_{cm} = \sum m_i \vec{r}_i / M$. In this case, the masses are all equal, $m_i = m$, so the total mass is $M = 3m$. Once we find the center-of-mass position, the velocity and acceleration can be found through differentiating.

EVALUATE   The mass term divides out, so the center-of-mass position is the sum of the three given vectors:

$$\vec{r}_{cm} = \frac{1}{3}\sum \vec{r}_i = \frac{1}{3}\left[\sum a_i \hat{i} + \sum b_i \hat{j}\right] = \left(t^2 + \frac{10}{3}t + \frac{7}{3}\right)\hat{i} + \left(\frac{2}{3}t + \frac{8}{3}\right)\hat{j}$$

The center-of-mass velocity is the first derivative:

$$\vec{v}_{cm} = \frac{d\vec{r}_{cm}}{dt} = \left(2t + \frac{10}{3}\right)\hat{i} + \left(\frac{2}{3}\right)\hat{j}$$

The center-of-mass acceleration is the second derivative:

$$\vec{a}_{cm} = \frac{d\vec{v}_{cm}}{dt} = (2)\hat{i}$$

ASSESS   The acceleration is constant and in the x-direction. This is due to the t^2-term in the position of the particle 1. There must be a force that accelerates this particle, and correspondingly accelerates the center of mass.

39.   **INTERPRET**   This problem involves the center of mass of a two-body system, which remains stationary in the absence of external horizontal forces.

DEVELOP   When the mouse starts at the rim, the center of mass of the mouse-bowl system has x component:

$$x_{cm} = (m_b x_b + m_m x_m)/(m_b + m_m) = m_m R/(m_b + m_m)$$

since initially $x_b = 0$ and $x_m = R$. Because there is no external horizontal force (no friction), x_{cm} remains constant as the mouse descends. When it reaches the center of the bowl, the center of mass of the system is

$$x_{cm} = (m_b x'_b + m_m x'_m)/(m_b + m_m) = (m_b d/10 + m_m d/10)/(m_b + m_m)$$

Because the center of mass does not move, we can equate these two expressions for the center of mass to find the ratio of m_b to m_m.

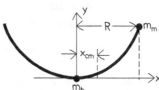

EVALUATE   Using the fact that $2R = d$, we find

$$m_m R/(m_b + m_m) = (m_b R/5 + m_m R/5)/(m_b + m_m)$$
$$m_b = 4m_m$$

ASSESS   The bowl is 4 times more massive than the mouse, which makes sense because the bowl has been horizontally displaced.

41.   **INTERPRET**   In this problem we are asked to find the center of mass of a uniform solid cone. We will need to integrate thin slices of the cone to find the answer.

DEVELOP   Choose the z axis along the axis of the cone, with the origin at the center of the base (see figure below). Because the cone is symmetric about the z axis, the center of mass is on the z axis [for each mass element at position (x, y, z) there is an equal mass element at position $(-x, -y, z)$, so the integral over x and y gives zero]. Thus, we only need to find the z coordinate of the center of mass, so Equation 9.4 reduces to

$$z_{cm} = \frac{\int z\,dm}{M}$$

For the mass element dm, take a disk at height z and of radius $r = R(1 - z/h)$ that is parallel to the base. Then $dm = \rho \pi r^2 \, dz = \rho \pi R^2 (1 - z/h)^2 \, dz$ where ρ is the density of the cone, and $M = \frac{1}{3}\rho \pi R^2 h$ is the total mass of the cone.

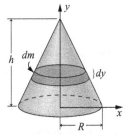

EVALUATE For the z coordinate of the center of mass, the integral above gives

$$z_{cm} = \frac{1}{M}\int_0^h z \, dm = \frac{3}{\rho \pi R^2 h}\int_0^h z \rho \pi R^2 (1 - z/h)^2 \, dz$$

$$= \frac{3}{h}\int_0^h \left(z - \frac{2z^2}{h} + \frac{z^3}{h^2} \right) dz = \frac{3}{h}\left(\frac{h^2}{2} - \frac{2h^2}{3} + \frac{h^2}{4} \right) = \frac{1}{4}h$$

so the complete center of mass coordinate is $(0, 0, h/4)$.

ASSESS The result makes sense because we expect z_{cm} to be closer to the bottom of the cone because more mass is distributed in this region.

43. INTERPRET We are asked about the compression of the spring due to a totally inelastic collision.

DEVELOP Since the total momentum of the system is conserved in the process, we have
$$P_i = P_f \quad \Rightarrow \quad m_1 v_1 = (m_1 + m_2)v_f$$

The potential energy of the spring at maximum compression equals the kinetic energy of the two-car system prior to contact with the spring: $\frac{1}{2}kx_{max}^2 = \frac{1}{2}(m_1 + m_2)v_f^2$.

For (b), we note that when the cars rebound, they are coupled together and both have the same velocity. Since the spring is ideal (by assumption), its maximum potential energy, $\frac{1}{2}kx_{max}^2$, is transformed back into kinetic energy of the cars.

EVALUATE (a) The second car is initially at rest so $v_2 = 0$. By momentum conservation, the speed of the cars after collision is

$$v_f = \frac{m_1 v_1}{m_1 + m_2} = \frac{(9,400 \text{ kg})(8.5 \text{ m/s})}{11,000 \text{ kg} + 9,400 \text{ kg}} = 3.92 \text{ m/s}$$

which leads to

$$x_{max} = v_f \sqrt{\frac{m_1 + m_2}{k}} = (3.92 \text{ m/s})\sqrt{\frac{11,000 \text{ kg} + 9,400 \text{ kg}}{0.32 \times 10^6 \text{ N/m}}} = 0.99 \text{ m}$$

(b) The spring's potential energy is converted back into the kinetic energy of the cars, so the rebound speed should be the same (only in the opposite direction) as the speed prior to the spring being compressed:
$$v_{reb} = v_f = 3.9 \text{ m/s}$$

where we only keep the significant figures.

ASSESS During the collision in the first part of the motion, the momentum is conserved but energy is not. However, during the spring compression and release in the second part, energy is conserved. Therefore, the cars rebound with the same speed as that before coming into contact with the spring.

45. INTERPRET This problem involves the Newton's second law in the form of Equation 9.6. We can use this to find an expression for the initial acceleration of the car due to the water jet that bounces off its rear window, and to find the final speed of the car.

DEVELOP Draw a diagram of the situation (see figure below). Consider the initial situation, when the car is at rest. From Equation 9.6, we know that the force exerted on the car by the water is the negative of the rate of change of momentum of the water:

$$\vec{F}_c = -\left(\frac{d\vec{p}_w}{dt}\right)$$

Let the water momentum be $p_w = mv_0$, where v_0 is the speed of the water with respect to the road. When the car is at rest, this speed is the same before and after reflecting off the car's rear window. In this case,

$$\left(\frac{d\vec{p}_w}{dt}\right) = \frac{d}{dt}\left(p_w\hat{j} - p_w\hat{i}\right) = \frac{dp_w}{dt}\left(\hat{j} - \hat{i}\right) = \frac{d}{dt}(mv_0)\left(\hat{j} - \hat{i}\right) = v_0\frac{dm}{dt}\left(\hat{j} - \hat{i}\right)$$

Thus, the initial force exerted on the car by the water jet is

$$\vec{F}_c = -\left(\frac{d\vec{p}_w}{dt}\right) = v_0\frac{dm}{dt}\left(\hat{i} - \hat{j}\right)$$

If we apply Newton's second law $\vec{F}_{net} = d\vec{p}_c/dt$ to the car, it reduces to $\vec{F}_{net} = M\vec{a}_c$, because the car's mass does not change. The net force acting on the car is simply the horizontal component of $\vec{F}_c$, because its vertical component is simply canceled by an increase in the normal force exerted on the car by the road. Thus,

$$\vec{F}_{net} = \vec{F}_c \cdot \hat{i} = v_0\frac{dm}{dt}\hat{i} = M\vec{a}_c$$

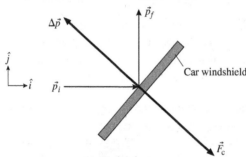

EVALUATE (a) Solving Newton's second law for the acceleration a_c of the car gives

$$\vec{a}_c = \frac{v_0}{M}\left(\frac{dm}{dt}\right)\hat{i}$$

(b) When the car starts moving, the change in the water's momentum is reduced because the speed v' of the water in the frame of reference of the car is reduced according to $v' = v_0 - v_c$, where v_c is the car's speed. The acceleration thus becomes

$$\vec{a}_c = \frac{v'}{M}\left(\frac{dm}{dt}\right)\hat{i} = \frac{v_0 - v_c}{M}\left(\frac{dm}{dt}\right)\hat{i}$$

Thus, when $v_c = v_0$, the car will no longer accelerate, so the final velocity of the car is v_0.

ASSESS As expected, the acceleration of the car increases with the water speed v_0 and the water mass rate dm/dt, but decreases with M, the mass of the car.

47. **INTERPRET** This is a head-on elastic collision where the initial speed, v, is the same for both objects.

DEVELOP For a one-dimensional collision like this, Equation 9.14 applies: $v_{1i} - v_{2i} = v_{2f} - v_{1f}$. In this case, $v_{1i} = v = -v_{2i}$, so we have $v_{2f} - v_{1f} = 2v$.

EVALUATE Plugging $v_{2f} = 2v + v_{1f}$ into the one-dimensional conservation of momentum equation (Equation 9.12a):

$$\left(m_1 - m_2\right)v = m_1v_{1f} + m_2\left(2v + v_{1f}\right) \;\rightarrow\; \left(m_1 + m_2\right)v = -\left(m_1 + m_2\right)v_{1f} \;\rightarrow\; v_{1f} = -v$$

And $v_{2f} = v$, using the previous relation. So the final speed of each object is equal (but opposite to the initial speed).

ASSESS Do the minus signs makes sense in the derivation? Let's assume the first object approaches from the left with positive velocity, and it bounces off to the left in the negative direction. The second object does the exact opposite, approaching from the right with negative velocity and bouncing back with positive velocity.

49.   **INTERPRET**   This problem involves finding the center of mass of an object composed of several parts (walls, base, and silage).

DEVELOP   Here it's convenient to find the centers of mass of sub-parts and then treat these parts as point particles to find the center of mass of the entire object. With no information about the geometry of the base, we will assume its an infinitely thin disk with the same diameter as the silo. Use a coordinate system with the origin at the center of the cylinder's base and the z axis running along the center of the silo cylinder. Because the system is symmetric about the z axis, the different centers of mass must lie along the z axis.

EVALUATE   (a) To find the center of mass when the silo is empty, find the center of mass of the cylindrical wall and base separately, and then treat the problem as if the mass of each object were concentrated at their respective center-of-mass points. By symmetry, the z coordinate of the center of mass of the wall must be at half the height, or $z_{cm}^{wall} = 15$ m . The z coordinate of the center of mass of the base is at $z_{cm}^{base} = 0$ m because it is infinitely thin. Inserting these results into Equation 9.2 to find the center of mass of the empty silo gives

$$z_{cm}^{silo} = \frac{m_{wall} z_{cm}^{wall} + m_{base} z_{cm}^{base}}{M} = \frac{\left(3.8 \times 10^4 \text{ kg}\right)\left(15 \text{ m}\right) + \left(6 \times 10^3 \text{ kg}\right)\left(0\right)}{3.8 \times 10^4 \text{ kg} + 6 \times 10^3 \text{ kg}} = 13 \text{ m}$$

so the complete coordinates of the center of mass are (0, 0, 13 m), to two significant figures.

(b) Treat the empty silo and silage as if their entire mass were concentrated at their respective center-of-mass points. We found the center of mass of the empty silo in part (a). The z coordinate of the center of mass of the silage is halfway up its 20-meter height, so $z_{cm}^{silage} = 10$ m and the silage's mass is

$$m_{silage} = \rho_{silage} V_{silage} = \rho_{silage} \left(\pi r^2 h\right) = \left(800 \text{ kg/m}^3\right)\left(\pi \times 4 \text{ m}^2 \times 20 \text{ m}\right) = 2.01 \times 10^5 \text{ kg}$$

Inserting these results into Equation 9.2 gives

$$z_{cm}^{silage} = \frac{m_{silo} z_{cm}^{silo} + m_{silage} z_{cm}^{silage}}{M} = \frac{\left(4.4 \times 10^4 \text{ kg}\right)\left(13 \text{ m}\right) + \left(\left(2.01 \times 10^5 \text{ kg}\right)\left(10 \text{ m}\right)\right)}{4.4 \times 10^4 \text{ kg} + 2.01 \times 10^5 \text{ kg}} = 11 \text{ m}$$

to two significant figures. Thus, the complete coordinates of the center of mass are (0, 0, 11 m).

ASSESS   When the silage is added, the center of mass is lowered, as expected because the silage fills from the bottom of the silo.

51.   **INTERPRET**   No external forces act on the three-body system, so total linear momentum is conserved. We can use this to find the velocity of the camera discarded by the astronaut.

DEVELOP   In the rest frame of the astronaut (i.e., in the inertial frame of reference in which the astronaut is at rest), the total momentum of the three-body system is zero. After the astronaut discards the two items, the total momentum must still be zero, so

$$m_1 \vec{v}_1 + m_2 \vec{v}_2 + m_3 \vec{v}_3 = 0$$

where the subscripts 1, 2, and 3 refer to the astronaut, the air canister, and the camera, respectively. Decomposing this vector equation into two scalar equations gives

$$m_1 v_1 \cos\left(200°\right) + m_2 v_2 \cos\left(0°\right) + m_3 v_{3,x} = 0$$
$$m_1 v_1 \sin\left(200°\right) + m_2 v_2 \sin\left(0°\right) + m_3 v_{3,y} = 0$$

which we can solve for $\vec{v}_3$.

EVALUATE   Solving first for the x component of the camera's velocity, we find

$$v_{3x} = -\frac{(60\ \text{kg})(0.85\ \text{m/s})\cos(200°) + (14\ \text{kg})(1.6\ \text{m/s})}{5.8\ \text{kg}} = 4.4\ \text{m/s}$$

Similarly, the y component is

$$v_{3y} = \frac{-(60\ \text{kg})(0.85\ \text{m/s})\sin(200°)}{5.8\ \text{kg}} = 3.0\ \text{m/s}$$

So the velocity of the camera is $\vec{v}_3 = (4.4\ \text{m/s})\hat{i} + (3.0\ \text{m/s})\hat{j}$

ASSESS   Alternatively, we can express the result in terms of the magnitude and direction of the velocity. This gives $v_3 = \sqrt{4.4^2 + 3.0^2}\ \text{m/s} = 5.3\ \text{m/s}$ and $\theta_3 = \text{atan}(3.0\ \text{ms}/4.4\ \text{ms}) = 34°$ counter-clockwise from the x axis.

53.   **INTERPRET**   This one-dimensional problem involves conservation of linear momentum and relative motion. We can use the former to find the speed of the sprinter with respect to the cart and the latter to find her speed relative to the ground.

DEVELOP   We choose a coordinate system in which the cart moves in the $-\hat{i}$ direction, and the sprinter runs in the $\hat{i}$ direction. The initial momentum of the system is

$$p = (m_s + m_c)v_{cm}$$

The final momentum of the system is

$$p = m_s v_s + m_c v_c = m_c v_c$$

because she has zero velocity with respect to the ground ($v_s = 0$). Equating these two expressions for total linear momentum (by conservation of total linear momentum), we have

$$(m_s + m_c)v_{cm} = m_s v_s + m_c v_c = m_c v_c$$

Using Equation 3.7 to express the sprinter's speed relative to the cart, we have

$$v_s = v_{rel} + v_c$$
$$v_{rel} = -v_c$$

because $v_s = 0$. We are given $v_{cm} = -7.6\ \text{m/s}$, $m_c = 240\ \text{kg}$, and $m_s = 55\ \text{kg}$.

EVALUATE   Solving the equations above for v_{rel}, we find

$$(m_s + m_c)v_{cm} = m_c v_c = -m_c v_{rel}$$
$$v_{rel} = -\frac{(m_s + m_c)v_{cm}}{m_c} = -\frac{(55\ \text{kg} + 240\ \text{kg})(-7.6\ \text{m/s})}{240\ \text{kg}} = 9.3\ \text{m/s}$$

ASSESS   The fact that the sprinter accelerates by pushing against the cart accelerates the cart from –7.6 m/s to –9.3 m/s, which is reasonable.

55.   **INTERPRET**   We're asked to find the speeds of two objects following their head-on elastic collision.

DEVELOP   The collision is one-dimensional, so Equations 9.15(a) and 9.15(b) are relevant. The information that we're given is that $m_1 = m$, $v_{1i} = 2v$, $m_2 = 4m$, and $v_{2i} = v$. Notice that both objects are initially moving in the same (positive) direction.

EVALUATE   Plugging the parameters into Equations 9.15,

$$v_{1f} = \frac{m - 4m}{m + 4m}(2v) + \frac{2(4m)}{m + 4m}(v) = \left(\frac{-6}{5} + \frac{8}{5}\right)v = \frac{2}{5}v$$

$$v_{2f} = \frac{2m}{m + 4m}(2v) + \frac{4m - m}{m + 4m}(v) = \left(\frac{4}{5} + \frac{3}{5}\right)v = \frac{7}{5}v$$

ASSESS   The first object loses some momentum from the collision $(v_{1i} < v_{1f})$, whereas the second object gets a "push" from the collision $(v_{2i} > v_{2f})$. Notice that $v_{1f} < v_{2f}$, otherwise it wouldn't make sense how the first object got ahead of the second object.

57.   **INTERPRET**   This problem involves conservation of momentum applied to a two-body system. The center of mass of this system does not move because of conservation of momentum. We will apply these principles to find the initial angle at which we threw the rock and the speed at which you must be moving.

DEVELOP   We choose a coordinate system in which your initial position is at the origin (see figure below). Apply Equation 3.15 to find the angle θ at which you throw the rock,

$$x_1 = \frac{v_0^2}{g}\sin(2\theta)$$

with $x_1 = 15.2 \text{ m} - x_2$ and $v_0 = 12.0$ m/s. We can find x_2 because we know the center of mass of the two-body system does not change since there are no horizontal forces acting on it. Thus,

$$x_{cm} = 0 = \frac{m_1 x_1 + m_2 x_2}{m_1 + m_2}$$

$$x_2 = -\frac{m_1 x_1}{m_2}$$

so

$$x_1 = 15.2 \text{ m} - \frac{m_1 x_1}{m_2}$$

$$x_1 = \frac{15.2 \text{ m}}{1 + m_1/m_2}$$

which allows us to solve Equation 3.15 for the angle θ. To find the speed at which you move after throwing the rock, apply conservation of linear momentum. Your initial horizontal momentum is zero, so your final momentum must also be zero, or

$$m_1 v_{1x} + m_2 v_{2x} = 0$$

where $v_{1x} = v_0 \cos\theta$, with θ being the angle with respect to the horizontal at which you throw the rock. Solve this equation for v_{2x}, which is the speed at which you move as a result of throwing the rock.

EVALUATE   (a) Inserting the known quantities into Equation 3.15 and solving for θ gives

$$\theta = \frac{1}{2}\text{asin}\left(\frac{x_1 g}{v_0^2}\right) = \frac{1}{2}\text{asin}\left(\frac{(15.2 \text{ m})g}{(1 + m_1/m_2)v_0^2}\right) = \frac{1}{2}\text{asin}\left(\frac{(15.2 \text{ m})(9.8 \text{ m/s}^2)}{\left[1 + (4.50 \text{ kg})/(65.0 \text{ kg})\right](12.0 \text{ m/s})^2}\right) = 37.7°$$

(b) The speed at which you recoil after throwing the rock is

$$v_{2x} = -\frac{m_1 v_{1x}}{m_2} = -\frac{m_1 v_0 \cos\theta}{m_2} = -\frac{(4.50 \text{ kg})(12.0 \text{ m/s})\cos(37.7°)}{65.0 \text{ kg}} = -65.8 \text{ cm/s}$$

ASSESS   The horizontal speed of the rock is $v_0\cos(\theta) = 9.50$ m/s. Thus, your recoil speed is much less than the rock's horizontal speed, as expected.

59.   **INTERPRET**   This two-body problem involves kinematics (Chapters 2 and 3) and motion of the center of mass. The rockets explodes into two equal-mass fragments at its peak height, which we can calculate from the kinematic equations of Chapter 2. We are given the time it takes from the explosion for one fragment to hit the ground and are asked to find the time at which the second fragment hits the ground.

DEVELOP   At the peak of the rocket's trajectory (just before the explosion), its center-of-mass y velocity is zero, or $v_{cm} = 0$. The motion of the center of mass is unaffected by the explosion, so just after the explosion, the velocity $v'_{cm} = 0$. Expressing the center of mass velocity in terms of the velocities v_1 and v_2 of fragments 1 and 2 gives

$$v'_{cm} = 0 = \frac{mv_1 + mv_2}{m + m}$$

$$v_1 = -v_2$$

where the equal-mass fragments each have mass m. We can now use the kinematic equations to find v_1. The height h at which the rocket explodes may be found using Equation 2.11, which gives (with a slight change in notation)

$$\overbrace{v_{cm}}^{=0} = v_0^2 - 2gh$$

$$h = \frac{v_0^2}{2g}$$

where v is the velocity at the peak of the trajectory and $v_0 = 40$ m/s. Knowing the height and the time t_1 for fragment 1 to hit the ground, we can find its initial velocity from Equation 3.13. This gives

$$\overbrace{y - y_0}^{=-h} = v_1 t_1 - \frac{1}{2}gt_1^2$$

$$v_1 = \frac{-h + gt_1^2/2}{t_1} = \frac{-v_0^2/2g + gt_1^2/2}{t_1} = \frac{-v_0^2}{2gt_1} + \frac{gt_1}{2}$$

Knowing v_1 (and thus v_2), we can find the time for fragment 2 to hit the ground by using the same Equation (i.e., 3.13), but solving for the time instead of the velocity. This gives

$$-h = v_2 t_2 - \frac{1}{2}gt_2^2$$

$$t_2 = \frac{v_2 \pm \sqrt{v_2^2 + 2gh}}{g} = \frac{-v_1 \pm \sqrt{v_1^2 + 2gh}}{g} = \frac{-v_1 \pm \sqrt{v_1^2 + v_0^2}}{g}$$

EVALUATE   Evaluating first v_1, we find

$$v_1 = \frac{-v_0^2}{2gt_1} + \frac{gt_1}{2} = \frac{-(40 \text{ m/s})^2}{2(9.81 \text{ m/s}^2)(2.87 \text{ s})} + \frac{(9.81 \text{ m/s}^2)(2.87 \text{ s})}{2} = -14.38 \text{ m/s}$$

where we have retained one extra significant figure because this is an intermediate result. Inserting this result into the expression for t_2 gives

$$t_2 = \frac{-v_1 \pm \sqrt{v_1^2 + v_0^2}}{g} = \frac{-(-14.38 \text{ m/s}) \pm \sqrt{(-14.38 \text{ m/s})^2 + (40 \text{ m/s})^2}}{(9.81 \text{ m/s}^2)} = 5.80 \text{ s}, -2.87 \text{ s}$$

The physically significant result is $t_2 = 5.80$ s.

ASSESS   The time for fragment 2 to reach the ground will increase with increasing $|v_1|$, which is reasonable because if fragment 1 has a larger downward velocity initially, then fragment 2 has a larger initial upward velocity. Also, note that if $v_1, v_2 \to 0$, then $t_2 = t_1 = 4.08$ s, which is intermediate between 2.87 s and 5.80 s, as expected.

61.   **INTERPRET**   In this two-body problem we are asked to find the relative speed between the satellite and the booster after the given impulse. We can apply conservation of momentum because there are no external forces acting on the system. Finally, this will be a one dimensional problem because we are dealing with only two bodies, so their relative motion must be linear to satisfy conservation of momentum.

DEVELOP   By Newton's third law, the explosion applies a force of equal magnitude to each body (satellite and booster), but in the opposite direction. We therefore have $\vec{J}_s = -\vec{J}_b$, where the subscripts s and b refer to the satellite and booster, respectively, and $J_s = J_b = 350$ N·s. As explained for Equation 9.10b, an impulse $\vec{J}$ is equal to the change of momentum: $\vec{J} = \Delta\vec{p} = m\Delta\vec{v}$, which allows us to find the speed of the satellite and the booster in the (stationary) center-of-mass frame. The relative speed of separation is $v_{rel} = |\vec{v}_s - \vec{v}_b|$.

EVALUATE   Initially both the satellite and the booster are at rest. After explosion, their velocities are

$$\vec{v}_s = \frac{\Delta\vec{p}_s}{m_s} = \frac{\vec{J}_s}{m_s} \qquad \vec{v}_b = \frac{\Delta\vec{p}_b}{m_b} = \frac{\vec{J}_b}{m_b} = \frac{-\vec{J}_s}{m_b}$$

Thus, the relative speed of separation is

$$|\vec{v}_s - \vec{v}_b| = \left|\frac{\vec{J}_s}{m_s} + \frac{\vec{J}_s}{m_b}\right| = J_s\left(\frac{1}{m_s} + \frac{1}{m_b}\right) = (350\ \text{N}\cdot\text{s})\left(\frac{1}{950\ \text{kg}} + \frac{1}{640\ \text{kg}}\right) = 0.92\ \text{m/s}$$

ASSESS   The relative speed is shown to depend on J_s, the magnitude of the impulse. The greater the impulse, the faster the satellite and the booster separate from each other.

63.   **INTERPRET**   This two-dimensional problem asks for the speed of one of two vehicles just before its totally inelastic collision with the second vehicle. Given the road condition (i.e., the coefficient of kinetic friction), we want to show that the speed of one of the cars exceeded 25 km/h. Energy is not conserved in this process, but momentum is. Furthermore, because work is done by friction, this problem involves the work-energy theorem.

DEVELOP   If the wreckage skidded on a horizontal road, the work-energy theorem requires that the work done by friction be equal to the change of the kinetic energy of both cars. $W_{nc} = \Delta K$ (see Equation 7.5). Because $W_{nc} = -f_k x = -\mu_k nx = -\mu_k(m_1 + m_2)gx$, and $\Delta K = K_f - K_i = 0 - \frac{1}{2}(m_1 + m_2)v^2$, where v is the speed of the wreckage immediately after collision, we are led to

$$\mu_k gx = \frac{1}{2}v^2$$

Therefore, the speed of the wreckage just after the collision is $v = \pm\sqrt{2\mu_k gx}$. Next, momentum conservation requires that the initial and final momentum are the same, so

$$m_1\vec{v}_1 + m_2\vec{v}_2 = (m_1 + m_2)\vec{v}$$
$$\vec{v} = \frac{m_1\vec{v}_1 + m_2\vec{v}_2}{m_1 + m_2}$$

where $\vec{v}$ is the initial velocity of the wreckage. To find the change in kinetic energy, we need to calculate the scalar product $\vec{v}\cdot\vec{v}$:

$$v^2 = \vec{v}\cdot\vec{v} = \left(\frac{m_1\vec{v}_1 + m_2\vec{v}_2}{m_1 + m_2}\right)\cdot\left(\frac{m_1\vec{v}_1 + m_2\vec{v}_2}{m_1 + m_2}\right)$$
$$= \frac{m_1^2 v_1^2 + m_2^2 v_2^2}{(m_1 + m_2)^2} + \frac{2m_1 m_2 \overbrace{\vec{v}_1\cdot\vec{v}_2}^{=0}}{(m_1 + m_2)^2}$$
$$= \frac{m_1^2 v_1^2 + m_2^2 v_2^2}{(m_1 + m_2)^2}$$

where we have used the fact that the scalar product $\vec{v}_1\cdot\vec{v}_2 = 0$ because the initial velocities are perpendicular to each other. In the next step, we insert the maximum speed for one car to find the minimum speed for the other car.

EVALUATE   Inserting $v = \pm\sqrt{2\mu_k gx}$ into the above expression for the initial velocity of the wreckage leads to

$$v^2 = \frac{m_1^2 v_1^2 + m_2^2 v_2^2}{(m_1 + m_2)^2} = 2\mu_k gx$$

Solving for v_1 gives

$$v_1 = \sqrt{\frac{2\mu_k gx(m_1 + m_2)^2 - m_2^2 v_2^2}{m_1^2}}$$

where we have taken the positive square root. Consider now the following situations:

Let subscript 1 correspond to the Toyota 2 to the Buick. If the speed of the Buick is $v_2 = 25$ km/h $= 6.94$ m/s, then the speed of the Toyota would be

$$v_1 = \sqrt{\frac{2\mu_k gx\left(m_1 + m_2\right)^2 - m_2^2 v_2^2}{m_1^2}}$$

$$= \sqrt{\frac{2(0.91)(9.8 \text{ m/s}^2)(22 \text{ m})(1200 \text{ kg} + 2200 \text{ kg})^2 - (2200 \text{ kg})^2 (6.94 \text{ m/s})^2}{(1200 \text{ kg})^2}}$$

$$= 55 \text{ m/s} = 200 \text{ km/h}$$

Thus, we conclude that the speed of the Toyota exceeded 25 km/h.

(2) Here, we reverse the assignment of the subscripts 1 and 2. If the speed of the Toyota is $v_2 = 25$ km/h $= 6.94$ m/s, then the speed of the Buick would be

$$v_1 = \sqrt{\frac{2\mu_k gx\left(m_1 + m_2\right)^2 - m_2^2 v_2^2}{m_1^2}}$$

$$= \sqrt{\frac{2(0.91)(9.8 \text{ m/s}^2)(22 \text{ m})(2200 \text{ kg} + 1200 \text{ kg})^2 - (1200 \text{ kg})^2 (6.94 \text{ m/s})^2}{(2200 \text{ kg})^2}}$$

$$= 30 \text{ m/s} = 110 \text{ km/h}$$

Thus, we conclude that the speed of the Buick exceeded 25 km/h.

From the analysis above, we conclude that if one car is going at 25 km/h, then the other one must have been speeding, so at least one car must have been speeding.

ASSESS   If we knew the direction of the wreckage velocity, we could easily find the car that was speeding.

65.   **INTERPRET**   This two-dimensional problem involves a totally inelastic collision, so momentum is conserved but kinetic energy is not conserved. We can use conservation of momentum to find the angle between the initial velocities before a the collision.

DEVELOP   The collision between the two masses is totally inelastic. Conservation of momentum tells us that

$$m_1\vec{v}_1 + m_2\vec{v}_2 = \left(m_1 + m_2\right)\vec{v}_f$$

$$\vec{v}_f = \frac{m_1\vec{v}_1 + m_2\vec{v}_2}{m_1 + m_2}$$

We known the magnitude of all the velocities involved, but not their relative orientation. We can find this by taking the scalar product of the final velocity with itself:

$$\vec{v}_f \cdot \vec{v}_f = v_f^2 = \left(\frac{m_1\vec{v}_1 + m_2\vec{v}_2}{m_1 + m_2}\right) \cdot \left(\frac{m_1\vec{v}_1 + m_2\vec{v}_2}{m_1 + m_2}\right) = \frac{m_1^2 v_1^2 + m_2^2 v_2^2 + 2m_1 m_2 \vec{v}_1 \cdot \vec{v}_2}{\left(m_1 + m_2\right)^2}$$

Using the definition of a scalar product, $\vec{v}_1 \cdot \vec{v}_2 = v_1 v_2 \cos\theta$, the angle between $\vec{v}_1$ and $\vec{v}_2$ can be found.

EVALUATE   With $m_1 = m_2 = m$ and $v_1 = v_2 = v = 2v_f$, the above equation can be simplified to

$$\frac{v^2}{4} = \frac{m^2 v^2 + m^2 v^2 + 2m^2 v^2 \cos\theta}{4m^2} = \frac{1}{2}v^2\left(1 + \cos\theta\right)$$

Therefore, the angle between the two initial velocities is

$$\theta = \text{ac os}\left(\frac{-1}{2}\right) = 120°$$

ASSESS   To see that the result makes sense, suppose $\vec{v}_1$ makes an angle $-60°$ with $+x$ and $\vec{v}_2$ makes an angle $+60°$ with $+x$. The y component of the total momentum cancels. But for the x component, we have

$$mv\cos\left(-60°\right) + mv\cos\left(60°\right) = \left(m + m\right)v_f$$

Solving for v_f, we get $v_f = v/2$, which confirms the result obtained above.

67. **INTERPRET**   The one-dimensional collision in this problem is elastic, so both momentum and energy are conserved. We are asked to find the ratio of the two masses if the one that is initially at rest acquires, after the collision, half of the kinetic energy that the other had before the collision.

DEVELOP   Momentum is conserved in this process. In this one-dimensional case, we may write

$$m_1 v_{1i} + m_2 v_{2i} = m_1 v_{1f} + m_2 v_{2f}$$

Since the collision is completely elastic, energy is conserved:

$$\frac{1}{2} m_1 v_{1i}^2 + \frac{1}{2} m_2 \overset{=0}{\overbrace{v_{2i}^2}} = \frac{1}{2} m_1 v_{1f}^2 + \frac{1}{2} m_2 v_{2f}^2$$

Using the two conservation equations, the final speeds of m_1 and m_2 are (see Equations 9.15a and 9.15b):

$$v_{1f} = \frac{m_1 - m_2}{m_1 + m_2} v_{1i} + \frac{2m_2}{m_1 + m_2} v_{2i} \text{ and } v_{2f} = \frac{2m_1}{m_1 + m_2} v_{1i} + \frac{m_2 - m_1}{m_1 + m_2} v_{2i}$$

Given that $v_{2i} = 0$, the above expressions may be simplified to

$$v_{1f} = \frac{m_1 - m_2}{m_1 + m_2} v_{1i} \qquad v_{2f} = \frac{2m_1}{m_1 + m_2} v_{1i}$$

Now, if half of the kinetic energy of the first object is transferred to the second, then

$$K_{2f} = \frac{1}{2} K_{1i} \implies \frac{1}{2} m_2 \left(\frac{2m_1 v_{1i}}{m_1 + m_2} \right)^2 = \frac{1}{4} m_1 v_{1i}^2$$

EVALUATE   The above equation can be further simplified to

$$8 m_1 m_2 = (m_1 + m_2)^2 \rightarrow 8 \left(\frac{m_1}{m_2} \right) = \left(\frac{m_1}{m_2} + 1 \right)^2$$

The resulting quadratic equation, $m_1^2 - 6m_1 m_2 + m_2^2 = 0$ has two solutions:

$$m_1 = \left(3 \pm \sqrt{8} \right) m_2 = \begin{cases} 5.83 m_2 \\ (5.83)^{-1} m_2 \end{cases}$$

Because the quadratic equation is symmetric in m_1 and m_2, one solution equals the other with m_1 and m_2 interchanged. Thus, one object is 5.83 times more massive than the other.

ASSESS   To check that our answer is correct, let's calculate the kinetic energy of the particles after the collision. Using $m_1 = 5.83 m_2$, we find

$$K_{2f} = \frac{1}{2} m_2 v_{2f}^2 = \frac{1}{2} m_2 \left(\frac{2m_1 v_{1i}}{m_1 + m_2} \right)^2 = \left(\frac{1}{2} \right) \frac{m_1}{5.83} \left(\frac{2m_1}{m_1 + m_1/5.83} \right)^2 v_{1i}^2$$

$$= \frac{1}{2} \frac{m_1}{5.83} \left(\frac{2}{1 + 1/5.83} \right)^2 v_{1i}^2 = \frac{1}{4} m_1 v_{1i}^2 = \frac{1}{2} K_{1i}$$

as expected.

69. **INTERPRET**   We are asked to derive Equation 9.15b which we can do using conservation of momentum. In addition, since Equation 9.15b describes an elastic collision, conservation of kinetic energy also applies.

DEVELOP   Use conservation of momentum,

$$m_1 \vec{v}_{1i} + m_2 \vec{v}_{2i} = m_1 \vec{v}_{1f} + m_2 \vec{v}_{2f}.$$

Because this is an elastic collision, kinetic energy is also conserved, so

$$\frac{1}{2} m_1 v_{1i}^2 + \frac{1}{2} m_2 v_{2i}^2 = \frac{1}{2} m_1 v_{1f}^2 + \frac{1}{2} m_2 v_{2f}^2.$$

Use these two equations to solve for v_{2f}. Much of this problem is done already in Equations 9.12a through 9.14.

EVALUATE   First solve Equation 9.14 for v_{1f} to get $v_{1f} = v_{2f} + v_{2i} - v_{1i}$. When we substitute this result into Equation 9.12, using the sign of v to denote the direction, we obtain $m_1 v_{1i} + m_2 v_{2i} = m_1 \left(v_{2f} + v_{2i} - v_{1i} \right) + m_2 v_{2f}$. Solving this for v_{2f} gives

$$m_1 v_{1i} + m_2 v_{2i} = m_1 v_{2i} - m_1 v_{1i} + \left(m_1 + m_2 \right) v_{2f}$$

$$v_{2f} = \frac{2 m_1}{m_1 + m_2} v_{1i} + \frac{m_2 - m_1}{m_1 + m_2} v_{2i}$$

ASSESS   Our result agrees with Equation 9.15b, as expected.

71.   **INTERPRET**   The two-dimensional problem involves an elastic collision between a proton and an initially stationary deuteron. Given the angle between their final velocities, we are to find the fraction of kinetic energy transferred from the proton to the deuteron in the process.

DEVELOP   Using the coordinate system shown in the sketch below (the deuteron's recoil angle θ_{2f} is negative), the components of the conservation of momentum equations for the elastic collision become

$$m_p v_{pi} = m_p v_{pf} \cos \theta_{1f} + m_d v_{df} \cos \theta_{df}$$

$$0 = m_p v_{pf} \sin \theta_{pf} + m_d v_{df} \sin \theta_{df}$$

In addition, conservation of energy gives us

$$\frac{1}{2} m_p v_{pi}^2 = \frac{1}{2} m_p v_{pf}^2 + \frac{1}{2} m_d v_{df}^2$$

so the fraction of initial kinetic energy transferred to the deuteron is

$$\frac{K_{df}}{K_{pi}} = 1 - \frac{K_{pf}}{K_{pi}} \quad \rightarrow \quad \frac{m_d}{m_p} \left(\frac{v_{df}}{v_{pi}} \right)^2 = 1 - \left(\frac{v_{pf}}{v_{pi}} \right)^2$$

EVALUATE   With $m_d = 2 m_p$, the conservation equations become

$$v_{pi} = v_{pf} \cos \theta_{pf} + 2 v_{pf} \cos \theta_{pf}$$

$$0 = v_{pf} \sin \theta_{pf} + 2 v_{df} \sin \theta_{df}$$

$$v_{pi}^2 = v_{pf}^2 + 2 v_{df}^2.$$

To find the final velocities, eliminate θ_{2f} from the first and second equations and v_{2f} from the third to get

$$v_{pi}^2 - 2 v_{pi} v_{pf} \theta_{pf} + v_{pf}^2 = 4 v_{df}^2 \left(\sin^2 \theta_{df} + \cos^2 \theta_{df} \right) = 4 v_{df}^2 = 2 v_{pi}^2 - 2 v_{pf}^2$$

This results in a quadratic equation for v_{pf}: $3 v_{pf}^2 - 2 v_{pi} v_{pf} \cos \theta_{pf} - v_{pi}^2 = 0$, with positive solution

$$v_{pf} = \frac{1}{3} v_{pi} \left(\cos \theta_{pf} + \sqrt{\cos^2 \theta_{pf} + 3} \right) = 0.902 v_{pi}$$

where we have used $\theta_{1f} = 37°$. From the kinetic energy equation, we have $v_{df} = \sqrt{\frac{1}{2} \left(v_{pi}^2 - v_{pf}^2 \right)} = 0.305 v_{pi}$, and from the transverse momentum equation, we have

$$\theta_{df} = \sin^{-1} \left(\frac{-v_{pf} \sin 37°}{2 v_{df}} \right) = \sin^{-1} \left(\frac{-\left(0.902 v_{pi} \right) \sin 37°}{2 \left(0.305 v_{pi} \right)} \right) = -62.7°$$

From either v_{pf} or v_{df}, the fraction of transferred kinetic energy is found to be

$$\frac{K_{df}}{K_{pi}} = 1 - \frac{K_{df}}{K_{pi}} = 1 - \left(\frac{v_{pf}}{v_{pi}} \right)^2 = 1 - \left(0.902 \right)^2 = 18.6\%$$

ASSESS The fraction of energy transfer can also be obtained as

$$\frac{K_{df}}{K_{pi}} = \frac{m_d}{m_p}\left(\frac{v_{df}}{v_{pi}}\right)^2 = 2(0.305)^2 = 0.186 = 18.6\%$$

Here, one does not need both final velocities to answer this question, but a more complete analysis of this collision, including the deuteron recoil angle, is instructive.

73. **INTERPRET** This problem asks us to find an expression for the impulse imparted by a time-varying force.

DEVELOP The impulse of a variable force requires integration (Equation 9.10b): $\vec{J} = \int \vec{F}(t)\, dt$. In this case, we'll need to use $\int \sin x\, dx = -\cos x$.

EVALUATE The force and the impulse are one-dimensional, so we will neglect the vector formalism. Evaluating the integral over the given time interval:

$$J = \int_0^{\pi/a} F_0 \sin at\, dt = \frac{F_0}{a}\cos at \Big|_0^{\pi/a} = \frac{F_0}{a}\left[\cos\pi - \cos 0\right] = \frac{2F_0}{a}$$

Notice that the argument of the sine and cosine functions is not degrees but radians, so $\cos\pi = -1$.

ASSESS The units are $N \cdot s$, as they should be for the impulse.

75. **INTERPRET** This one-dimensional two-body problem involves an inelastic collision so we can apply conservation of momentum but not conservation of energy. We will also need to apply some kinematics to find the maximum height and the speed with which the combination hit the ground.

DEVELOP By conservation of momentum, we can equate the momentum of the two-body system before and after the Frisbee-mud collision. This gives

$$m_m v_i = (m_F + m_m) v_f$$

Using Equation 2.11, we find the velocity $v_{m,i}$ with which the mud hits the Frisbee to be

$$v_{m,i} = \sqrt{v_{m,0}^2 - 2g(y - y_0)} = \sqrt{(17.7\ \text{m/s})^2 - 2(9.8\ \text{m/s}^2)(7.65\ \text{m} - 1.23\ \text{m})} = 13.69\ \text{m/s}$$

Therefore, the initial velocity of the mud-Frisbee combination is

$$v_f = \frac{m_m v_i}{m_F + m_m} = \frac{(0.240\ \text{kg})(13.7\ \text{m/s})}{0.240\ \text{kg} + 0.114\ \text{kg}} = 9.288\ \text{m/s}$$

upward. Use this result in the kinematic Equation 2.11 to find the maximum height and the speed upon hitting the ground for the mud-Frisbee combination.

EVALUATE (a) The maximum height reached is

$$y = y_0 + \frac{v^2 - v_f^2}{-2g} = 7.65\ \text{m} - \frac{(0.00\ \text{m/s})^2 - (9.28\ \text{m/s})^2}{2(9.8\ \text{m/s}^2)} = 8.10\ \text{m}$$

(b) An object falling from this height, unimpeded by air resistance or other obstacles, would attain a speed of

$$v = \pm\sqrt{2gy} = -\sqrt{2(9.8\ \text{m/s}^2)(8.10\ \text{m})} = -12.6\ \text{m/s}$$

when it reaches the ground (where we have retained the negative sign to indicate the velocity is downward).

ASSESS Notice that we retained extra significant figures for the intermediate results.

77. **INTERPRET** This one-dimensional, two-body problem involves a collision that is neither elastic nor inelastic, so we can apply conservation of momentum. Given that we are told the amount of kinetic energy that is lost in the collision, we can apply conservation of energy as well. We can use these principles to find the velocity of the wreckage.

DEVELOP Let the initial velocity car 1 be v, and that of cars after the wreckage be v_1 and v_2. Conservation of momentum requires that

$$mv = mv_1 + mv_2$$

$$v = v_1 + v_2$$

and conservation of energy gives

$$\frac{1}{2}mv^2 = \frac{1}{2}mv_1^2 + \frac{1}{2}mv_2^2 + \frac{5}{18}\left(\frac{1}{2}mv_1^2\right)$$

$$\frac{13}{18}v^2 = v_1^2 + v_2^2$$

Solve these two equations for the final velocities.

EVALUATE   Solving the expression from conservation of energy for $v_{1,f}$ gives

$$v_1 = \pm\sqrt{\frac{13}{18}v^2 - v_2^2}$$

Taking the positive square root (the negative solution simply corresponds to the cars moving in the opposite direction) and using the result from conservation of momentum eliminate $v_{2,f}$ gives

$$v_1 = \sqrt{\frac{13}{18}v^2 - (v - v_1)^2}$$

$$v_1^2 = \frac{13}{18}v^2 - (v^2 - 2vv_1 + v_1^2)$$

$$v_1^2 + 2vv_1 + \frac{5}{18}v^2 = 0$$

$$2\left(v_1 - \frac{1}{6}v\right)\left(v_1 - \frac{5}{6}v\right) = 0$$

which has the solutions $v_1 = 5v/6$ or $v/6$. Because $v_1 < v_2$ (because car 1 is behind car 2 after the wreck), we chose $v_1 = v/6$, which leads to $v_2 = 5v/6$.

ASSESS   We can easily verify that conservation of momentum and energy (including the converted kinetic energy) is respected with this solution.

79.   **INTERPRET**   This two-dimensional two-body problem involves an inelastic collision. In the vertical direction, the motion is governed by the force of gravity, and in the horizontal direction, we can use conservation of momentum to find the horizontal velocity of the combined bodies after the collision. We will need to use kinematic equations to find the velocities and positions at the various points along the trajectories.

DEVELOP   The peak of the projectile 1 is at half its range, which is given by Equation 3.15. Thus, the two projectiles collide at a horizontal position, measured from the launch point of projectile 1, of

$$x = \frac{1}{2g}v_0^2 \sin(2\theta) = \frac{(380 \text{ m/s})^2}{2(9.8 \text{ m/s}^2)}\sin(110°) = 6.92 \text{ km}$$

Just before the collision, the horizontal velocity of projectile 1 is $v_{1,x} = v_1 \cos\theta = (380 \text{ m/s})\cos(55°) = 218 \text{ m/s}$. From conservation of linear momentum, we know that the horizontal velocity of the combined projectile is

$$m_1 v_{1,x} + m_2 v_{2,x} = (m_1 + m_2)v_x$$

$$m_2 = m_1 \frac{v_x - v_{1,x}}{v_{2,x} - v_x}$$

We also know that the combined projectile travels to 9.6 km from the launch point before hitting the ground, so

$$x = x_0 + v_x t$$

$$v_x = \frac{x - x_0}{t} = \frac{9.6 \text{ km} - 6.92 \text{ km}}{t}$$

Finally, we can find t and thus solve for m2 using the kinematic Equation 2.7 ($v = v_0 + at$, with $a = -g$ and $v = 0$). This gives

$$0 = v_0 - gt$$

$$t = \frac{v_0}{g}$$

EVALUATE Evaluating the expression for v_x gives

$$v_x = \frac{9.6 \text{ km} - 6.92 \text{ km}}{v_0} g = \frac{(9.6 \text{ km} - 6.92 \text{ km})(9.8 \text{ m/s}^2)}{380 \text{ m/s}} = 84.3 \text{ m/s}$$

Inserting this result into the expression for m_2 gives

$$m_2 = m_1 \frac{v_x - v_{1,x}}{v_{2,x} - v_x} = (14 \text{ kg}) \frac{84.3 \text{ m/s} - 218 \text{ m/s}}{-140 \text{ m/s} - 84.3 \text{ m/s}} = 8.3 \text{ kg}$$

ASSESS The time required for the two to fall to the ground is $t = v_0/g = (380 \text{ m/s})/(9.8 \text{ m/s}) = 31.8$ s. We find that $m_2 < m_1$, which makes sense because the combined projectile continues to travel in the direction at which m1 was initially traveling.

81. **INTERPRET** We are asked to find the peak in the force for the collision in the previous problem.
    DEVELOP Let's simplify the force equation slightly by substituting in $x = t / \Delta t$:
    $$F(x) = (-14.2x^4 + 26.2x^3 - 14.5x^2 + 2.50x) \text{ MN}$$

    This function equals zero at $x = 0$ and $x = 1$, which corresponds to the beginning and end of the car's collision with the barrier. The force is positive between these two points, so there is a peak in the force somewhere between $x = 0$ and $x = 1$, just as we'd expect.
    EVALUATE At the peak, the derivative should be zero:
    $$\frac{dF}{dx} = -56.8x^3 + 78.6x^2 - 29.0x + 2.50 = 0$$

    where we have dropped the units for the time being. Dividing through by the x^3-coefficient gives a function we'll call $f(x)$:
    $$f(x) = x^3 - 1.38x^2 + 0.511x - 0.0440 = 0$$

    Solving a cubic like this is rather involved, but we know that there is a root between $x = 0$ and $x = 1$, so we can use trial and error to find where the cubic crosses the x-axis. We first note that $f(x)$ is negative at $x = 0$ and positive at $x = 1$. If we try $x = 0.5$, we get a negative result, so the crossing point must be to the right of $x = 0.5$. So we check and see that $f(0.75) < 0$, but $f(0.85) > 0$. Therefore the root is in between these points. This can be done several times to hone in on $x = 0.825$, which is perhaps sufficiently accurate for our purposes. In terms of the original variables, the peak occurs approximately at $t = 165$ ms with a value of $F = 327$ kN.
    ASSESS In the previous problem, the average force was found to be 128 kN. The peak value we found above is 2.5 times the average, which seems reasonable.

83. **INTERPRET** This problem is like the previous problem where we looked at the transfer of kinetic energy from a moving block to an initially stationary block. Here, we show that the fraction of energy transferred is independent of which block is the moving one and which is the stationary one.
    DEVELOP We can use the result from the previous problem to write the fraction of the initial energy in the moving block that is transferred to the initially stationary block:
    $$\frac{K_{\text{stat}}}{K_{\text{mov}}} = \frac{4(m_1/m_2)}{(1 + m_1/m_2)^2}$$

    where we assume in this case that the moving block has mass m_1 and the stationary block has mass m_2.

EVALUATE   If instead m_1 is the stationary block, and m_2 is the moving block, then the fraction becomes

$$\frac{K_{stat}}{K_{mov}} = \frac{4(m_2/m_1)}{(1 + m_2/m_1)^2} = \frac{4(m_2/m_1)}{(1 + m_2/m_1)^2}\frac{(m_1/m_2)^2}{(m_1/m_2)^2} = \frac{4(m_1/m_2)}{(1 + m_1/m_2)^2}$$

But this is the same as before, so the energy transfer is independent of which mass is initially stationary.

ASSESS   As an example of this independence, one can imagine a light block colliding with a heavy stationary block (like the ping pong ball and bowling ball collision in Section 9.6). After the collision, the heavy block barely moves, whereas the light block ricochets backward with essentially the same speed. In other words, very little energy is transferred $(K_{stat}/K_{mov} \approx 0)$. Then imagine the blocks switch places, with the light block initially at rest. In this case, the heavy block plows into the light block and knocks it forward. But the heavy block keeps moving with pretty much its initial speed, so again not much energy is transferred $(K_{stat}/K_{mov} \approx 0)$.

85.   INTERPRET   We find the center of mass of a slice of pizza with central angle θ and radius R. We would expect that it's along the center of the slice, and closer to the crust than to the tip.

DEVELOP   The equation for center of mass is $\vec{r}_{cm} = \frac{1}{M}\int \vec{r}\,dm$, which in two dimensions can be written out as:

$$\vec{r}_{cm} = \frac{1}{M}\int(x\hat{i} + y\hat{j})\,dm = \frac{1}{M}\int x\,dm\,\hat{i} + \frac{1}{M}\int y\,dm\,\hat{j} = x_{cm}\hat{i} + y_{cm}\hat{j}$$

We set up our coordinate system such that the slice is symmetric about the y axis, see the figure below. In this case, $y_{cm} = 0$. As for x_{cm}, we use $x = r\cos\phi$ and integrate over r from 0 to R, and over ϕ from $-\theta/2$ to $\theta/2$, where the angles are in radians.

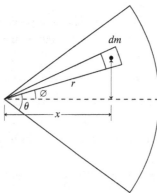

The infinitesimal mass element, dm, is equal to $\mu\,dA = \mu r\,dr\,d\phi$, where μ is the mass per unit area. Since the slice is uniform, the density is constant:

$$\mu = \frac{M}{A} = \frac{M}{(\theta/2\pi)(\pi R^2)} = \frac{2M}{\theta R^2}$$

One can check these values by verifying that $M = \int \mu\,dA$.

EVALUATE   Evaluating the integral for the x-component of the center of mass gives

$$x_{cm} = \frac{1}{M}\int_{-\theta/2}^{\theta/2}\int_0^R (r\cos\phi)\left(\frac{2M}{\theta R^2}\right)r\,dr\,d\phi$$

$$= \frac{2}{\theta R^2}\left[\tfrac{1}{3}r^3\right]_0^R \left[\sin\phi\right]_{-\theta/2}^{\theta/2} = \frac{4R}{3\theta}\sin(\theta/2)$$

ASSESS   We can check this answer by letting $\theta = \pi/4$, which corresponds to a 1/8th slice of pizza. In that case, $x_{cm} = 0.65R$, which matches our original prediction that the center of mass will be closer to the outer crust than to the tip. If instead $\theta = 2\pi$ (the full pizza), then the center of mass is at $x_{cm} = 0$, as we would expect.

87.   **INTERPRET**   We use conservation of momentum to find the speed of an astronaut after she throws her toolbox away, and use this speed and the given distance to determine whether she reaches safety before her oxygen runs out.

DEVELOP   The mass of the astronaut is $m_a = 80$ kg. The mass of the toolbox is $m_t = 10$ kg. The initial speed of both is zero, so the final speed of the toolbox is $v_{ta} = -8$ m/s *relative to the* astronaut. We can use conservation of momentum to find the speed of the astronaut: $0 = m_t v_t + m_a v_a$. Once we have this speed, we calculate how long it will take to travel a distance $x = 200$ m and hope that the answer is less than 4 minutes.

EVALUATE   First we find the speed of the toolbox relative to the rest frame: $v_t = v_a + v_{ta}$. Next we plug this into the conservation of momentum equation: $0 = m_t (v_a + v_{ta}) + m_a v_a$ and solve for the astronaut's speed:

$$0 = m_t (v_a + v_{ta}) + m_a v_a$$

$$v_a (m_t + m_a) = -m_t v_{ta}$$

$$v_a = v_{ta} \left(-\frac{m_t}{m_t + m_a} \right) = 0.89 \text{ m/s}$$

The time it takes is

$$t = \frac{x}{v_a} = \frac{200 \text{ m}}{0.89 \text{ m/s}} = 225 \text{ s} = 3.75 \text{ min}$$

ASSESS   She makes it with 15 seconds to spare.

89.   **INTERPRET**   We're asked to find the total mass and the center of mass for a thin rod with non-uniform density. The density increases from zero at one end to a maximum at the other end, so we'd expect the center of mass to be closer to the denser end.

DEVELOP   We will have to integrate to find the total mass: $M_{tot} = \int dm = \int \mu dx$. The limits of integration are between $x = 0$ and $x = L$. Since the mass is distributed along one-dimension, the center of mass integral takes a similar from: $r_{cm} = \frac{1}{M_{tot}} \int x \, dm = \frac{1}{M_{tot}} \int x \mu \, dx$.

EVALUATE   (a) The mass integral gives

$$M_{tot} = \int_0^L \mu \, dx = \int_0^L \frac{M x^a}{L^{1+a}} \, dx = \frac{M}{L^{1+a}} \left. \frac{x^{1+a}}{1+a} \right|_0^L = \frac{M}{1+a}$$

(b) Using the above result, the center of mass is

$$r_{cm} = \frac{1}{M_{tot}} \int_0^L x \mu \, dx = \frac{1+a}{M} \int_0^L \frac{M x^{1+a}}{L^{1+a}} \, dx = \frac{1+a}{L^{1+a}} \left. \frac{x^{2+a}}{2+a} \right|_0^L = \frac{1+a}{2+a} L$$

(c) If $a = 0$, the density is constant: $\mu = M / L$. The total mass is M, and the center of mass occurs at $\frac{1}{2} L$, just as we would expect for a rod with uniform density.

ASSESS   For $a = 1$, the center of mass occurs at $\frac{2}{3} L$, while for $a = 2$, it occurs at $\frac{3}{4} L$. This agrees with our premonition that having the density get larger toward $x = L$ will mean that the center of mass will be closer to that end.

91.   **INTERPRET**   We're asked to analyze the bouncing of a ball captured by a strobe camera.

DEVELOP   Right before the second collision, the ball has kinetic energy $K_i = \frac{1}{2} m v_{xi}^2 + \frac{1}{2} m v_{yi}^2$, while after the collision, it has $K_f = \frac{1}{2} m v_{xf}^2 + \frac{1}{2} m v_{yf}^2$. We argued in the previous problem that because the ball doesn't rebound to the same height, the vertical speed at ground level must be getting smaller after each collision $\left(v_y = \sqrt{2gh} \right)$. If we just consider the motion in the vertical direction, the fraction of energy lost is:

$$\left(\frac{\Delta K}{K_i} \right)_y = \frac{h_f - h_i}{h_i}$$

EVALUATE   With our fingers or with a small ruler, we can check that the peak height after the second collision is about 0.6 times the peak height before the collision. So by the equation above, the ball lost around 40% of its energy in the vertical direction. Assuming the loss in horizontal direction wasn't more than that, the fraction of the total energy lost is a little less than half.

The answer is (b).

ASSESS   We've treated the components of kinetic energy separately: $K_x = \frac{1}{2}mv_x^2$ and $K_y = \frac{1}{2}mv_y^2$. It should be noted that the two are not completely separate. If the ground were flat or if the ball were spinning, a collision could transfer energy in the vertical direction to energy in the horizontal direction, or vice versa.

93.   **INTERPRET**   We're asked to analyze the bouncing of a ball captured by a strobe camera.

DEVELOP   The way a strobe camera works is that it takes pictures at a set interval. So we can get a rough estimate for how long the ball was between collisions or in the midst of a collision by counting how many times the camera caught the ball in either setting.

EVALUATE   In the image, we count 7 times that the ball's picture was taken between the first and second collision. However, it appears that the ball's picture was taken only once during each collision. So the collision time is a tiny fraction of the time between collisions.

The answer is (a).

ASSESS   This matches our experience that collisions are very short-lived events.

10 ROTATIONAL MOTION

EXERCISES

Section 10.1 Angular Velocity and Acceleration

13. **INTERPRET** This problem involves calculating the angular speed of a variety of rotating objects.

    DEVELOP Apply Equation 10.1, $\bar{\omega} = \Delta\theta/\Delta t$, where $\Delta\theta$ is the rotation and Δt is the time interval for the rotation.

    EVALUATE (a) $\omega_E = (1 \text{ rev})/(1 \text{ d}) = 2\pi/(86,400 \text{ s}) = 7.27 \times 10^{-5} \text{ s}^{-1}$

    (b) $\omega_{min} = (1 \text{ rev})/(1 \text{ h}) = 2\pi/3600 \text{ s} = 1.75 \times 10^{-3} \text{ s}^{-1}$

    (c) $\omega_{hr} = (1 \text{ rev})/(12 \text{ h}) = \omega_{min}/12 = 1.45 \times 10^{-4} \text{ s}^{-1}$

    (d) $\omega = (300 \text{ rev})/\text{min} = 300 \times 2\pi/(60 \text{ s}) = 31.4 \text{ s}^{-1}$

    ASSESS Note that radians are a dimensionless angular measure, i.e., pure numbers; therefore angular speed can be expressed in units of inverse seconds.

15. **INTERPRET** This problem involves converting angular speed from various units to radians/s (which is the same as s^{-1}, or frequency).

    DEVELOP Use the appropriate conversion factors to convert each angular speed to units of rad/s.

    EVALUATE (a) $(720 \text{ rev/min})(2\pi \text{ rad/rev})(\text{min}/60 \text{ s}) = 24\pi \text{ rad/s} = 75 \text{ rad/s}$, to two significant figures.

    (b) $(50°/h)(\pi \text{ rad}/180°)(h/3600 \text{ s}) = 2.4 \times 10^{-4} \text{ rad/s}$, to two significant figures.

    (c) $(1000 \text{ rev/s})(2\pi \text{ rad/rev}) = 2000\pi \text{ s}^{-1} = 6 \times 10^3 \text{ rad/s}$ to a single significant figure.

    (d) $(1 \text{ rev/y}) = 2\pi \text{ rad}/(\pi \times 10^7 \text{ s}) = 2 \times 10^{-7} \text{ rad/s}$, to a single significant figure.

    ASSESS Note that radians are a dimensionless angular measure, i.e., pure numbers; therefore angular speed can be expressed in units of inverse seconds. The approximate value for 1 y used in part (d) is often handy for estimates, and is fairly accurate; see Chapter 1, Problem 20.)

17. **INTERPRET** For this problem, we are asked to find the average angular acceleration given the initial and final acceleration and the time interval.

    DEVELOP Apply Equation 10.4 (before the limit is taken), $\bar{\alpha} = \Delta\omega/\Delta t$. The change in the angular velocity is $\Delta\omega = \omega_f - \omega_i = 500 \text{ rpm} - 200 \text{ rpm} = 300 \text{ rpm}$ and the time interval is $\Delta t = (74 \text{ min})(60 \text{ s/min}) = 4400 \text{ s}$.

    EVALUATE Inserting the given quantities gives an average angular acceleration of

    $$\bar{\alpha} = \frac{300 \text{ rpm}}{4400 \text{ s}}\left(\frac{2\pi \text{ rad}}{\text{rev}}\right)\left(\frac{\text{min}}{60 \text{ s}}\right) = 7 \times 10^{-2} \text{ s}^{-1}$$

    ASSESS Note that the units cancel out to leave units of frequency, as expected.

19. **INTERPRET** This problem is an exercise in angular kinematics. We are given an angular acceleration and the acceleration period, and are asked to find the revolutions made in this time and the average angular speed.

    DEVELOP Apply the formulas in Table 10.1. To find the number of revolutions, we find the total angular displacement θ from Equation 10.8, then divide this by $2\pi (= 1 \text{ revolution})$ to find the number of revolutions. The linear analog to this can be thought of as finding a distance, then dividing it by a given distance (say, 10-km segments) to find the number of 10-km segments traveled. In both cases, we end up with a dimensionless number. To find the average angular speed, use Equation 10.1.

EVALUATE   (a) Inserting $\alpha = 0.010$ rad/s^2 and $t = 14$ s into Equation 10.8 gives a final rotational distance θ of

$$\Delta\theta = \theta - \theta_0 = \overbrace{\omega_0}^{=0} t + \frac{1}{2}\alpha t^2$$

$$\Delta\theta = \frac{1}{2}(0.010 \text{ rad/s}^2)(14 \text{ s})^2 = 0.98 \text{ rad}\left(\frac{1 \text{ rev}}{2\pi \text{ rad}}\right) = 0.16 \text{ rev}$$

(b) From Equation 10.7, with $\theta_0 = 0$, the final angular speed is

$$\bar{\omega} = \frac{\Delta\theta}{\Delta t} = \frac{\theta - \theta_0}{\Delta t} = \frac{0.98 \text{ rad}}{14 \text{ s}} = 0.07 \text{ rad/s}$$

ASSESS   The final angular speed of the merry-go-round is, from Equation 10.7,

$$\omega = \overbrace{\omega_0}^{=0} + \alpha t = (0.010 \text{ rad/s}^2)(14 \text{ s}) = 0.14 \text{ rad/s}$$

which is twice the average speed. This is expected because we start from zero speed and accelerate at a constant rate, so the average speed is attained at half the acceleration period, at which point the object in question is rotating at half the angular speed.

Section 10.2 Torque

21.   INTERPRET   This problem involves the concept of torque. We are given the force applied (by the child) and are asked to find the lever arm needed to produce a given torque.
DEVELOP   Solve Equation 10.10 for the lever arm r, assuming the child pushes perpendicular to the door (so $\theta = 90°$).
EVALUATE   Solving Equation 10.10 for r gives

$$r = \frac{\tau}{F\sin\theta} = \frac{110 \text{ N}\cdot\text{m}}{(90 \text{ N})\sin(90°)} = 1.2 \text{ m}$$

so the child must push 1.2 m from the center of the door.
ASSESS   This seems like a reasonable distance, given that the typical width of revolving doors is greater than 1.2 m.

23.   INTERPRET   This problem involves torque, which we are asked to calculate given a force (gravity on the mouse), the radial distance at which it is applied, and the angle at which it is applied (from the geometry of a clock).
DEVELOP   Draw a diagram of the situation (see figure below). From the geometry of a clock, we know that the angle the minute hand makes with the vertical is $\phi = 180°/3 = 60°$. The angle between the force and the radial position vector from the axis of rotation to the point where the force is applied is therefore $\theta = 180° - 60° = 120°$. The force applied by the mouse is simply its weight, so $F = mg$, and the lever arm is $r = 17$ cm.

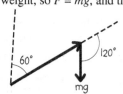

EVALUATE   Inserting the given quantities into Equation 10.10 gives

$$\tau = rF\sin\theta = (0.17 \text{ m})(0.055 \text{ kg})(9.8 \text{ m/s}^2)\sin(120°) = 7.9\times10^{-2} \text{ N}\cdot\text{m}$$

ASSESS   In 5 minutes, the torque applied by the mouse (assuming it doesn't move) will be $\tau = (0.17 \text{ m})(0.055$ kg$)(9.8 \text{ m/s}^2) = 9.2 \times 10^{-2}$ N·m, or 16% more torque.

Section 10.3 Rotational Inertia and the Analog of Newton's Law

25.   INTERPRET   This problem involves rotational inertia, which include both mass and the spatial distribution of that mass. Rotation inertia is the rotational analog of mass in linear motion. Because the spatial mass distribution enters into rotational inertia, the position of the axis of rotation is important. We are asked to find the rotational inertia of an arrangement of 4 masses about two different axes of rotation.

DEVELOP Draw a diagram of the situation (see figure below), and apply Equation 10.12.

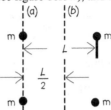

EVALUATE (a) For the axis labeled (a), two masses have $r = 0$, and the other two masses have $r = L$. Inserting these quantities into Equation 10.12 gives

$$I = \sum_i m_i r_i^2 = m(0)^2 + m(0)^2 + mL^2 + mL^2 = 2mL^2$$

(b) For the axis labeled (b), each mass has $r = L/2$, so Equation 10.12 gives

$$I = \sum_i m_i r_i^2 = m\left(\frac{L}{2}\right)^2 + m\left(\frac{L}{2}\right)^2 + m\left(\frac{L}{2}\right)^2 + m\left(\frac{L}{2}\right)^2 = mL^2$$

ASSESS Thus, there is more rotational inertia when the axis of rotation is at the edge of the object than when it is at the center of the object, as expected.

27. **INTERPRET** This problem involves combining the rotational inertia of several objects to find the overall rotation inertia of the combined object. In addition, we are asked to find the torque needed to give the object the given angular acceleration.

DEVELOP Because both the cylinder and the end caps rotate about the same axis, we can sum the rotational inertia of each object to find the total rotational inertia: $I_{tot} = I_{cyl} + 2I_{cap}$. The rotational inertia of the individual components are given in Table 10.2, and are $I_{cyl} = M_{cyl}R^2$ and $I_{cap} = M_{cap}R^2/2$. To find the torque, apply the rotational analog of Newton's second law (for constant mass), Equation 10.11: $\tau = I\alpha$.

EVALUATE (a) The total rotational inertia of the capped cylinder is

$$I_{tot} = I_{cyl} + 2I_{cap} = M_{cyl}R^2 + 2\left(\frac{1}{2}M_{cap}R^2\right) = (0.071\ \text{m})^2(0.065\ \text{kg} + 0.022\ \text{kg}) = 4.4 \times 10^{-4}\ \text{kg} \cdot \text{m}^2$$

(b) The torque needed to accelerate the capped cylinder is

$$\tau = I_{tot}\alpha = (4.39 \times 10^{-4}\ \text{kg} \cdot \text{m}^2)(3.4\ \text{rad/s}^2) = 1.5 \times 10^{-3}\ \text{N} \cdot \text{m}$$

ASSESS Notice that we used more significant figures for the total rotational inertia in part (b) because it was an intermediate result in this case.

29. **INTERPRET** We are asked to find the rotational inertia I of three point masses located at the corners of an equilateral triangle. We find I for the axis through the center perpendicular to the plane, and for the axis along the line through one corner and the midpoint of the opposite side, as shown in the figure below.

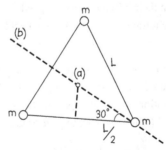

DEVELOP We use the equation for rotational inertia of a collection of point masses: $I = \sum m_i r_i^2$. Therefore, we will need the distances of the three particles from the appropriate axis. In part (a), the distance of each particle from the center of the triangle is $r_a = (L/2)/\cos 30°$. In part (b), one mass is on the axis where the distance is zero. The other two masses are at a distance of $r_b = L/2$.

EVALUATE   (a) The rotational inertia around the center of the triangle is

$$I = \sum m_i r_i^2 = 3\left[mr_a^2 \right] = \frac{3mL^2}{4\cos^2 30°} = mL^2$$

(b) The rotational inertia around a mid-line of the triangle is

$$I = \sum m_i r_i^2 = 2\left[mr_b^2 \right] = \frac{2mL^2}{4} = \tfrac{1}{2} mL^2$$

ASSESS   Both of these axes pass through the center of mass of the triangle. But the axis in part (a) maximizes the distance to the three masses, so it has a larger moment of inertia than the axis in part (b).

31.   INTERPRET   This problem involves finding the rotational inertia of a sphere of constant density that rotates about an axis through its geometric center. In addition, we are to find the torque needed to impart the given acceleration to this sphere.

DEVELOP   Use the formula $I = 2MR^2/5$ from Table 10.2 to calculate the rotational inertia of the neutron star. The radius R_{NS} of the neutron star is

$$M_{NS} = \rho V = \rho\left(\frac{4}{3}\pi R_{NS}^3 \right) = 1.8 M_S$$

$$R_{NS} = \sqrt[3]{\frac{3(1.8 M_S)}{4\pi\rho}}$$

where (from Appendix E) $M_S = 1.99 \times 10^{30}$ kg. The torque may be found by inserting the resulting inertia and the given angular acceleration into Equation 10.11.

EVALUATE   (a) The rotational inertia of the neutron star is

$$I_{NS} = \frac{2}{5}M_{NS}R_{NS}^2 = \frac{2}{5}(1.8 M_S)\left[\frac{3(1.8 M_S)}{4\pi\rho} \right]^{2/3}$$

$$= \frac{2}{5}(1.8 \times 1.99 \times 10^{30} \text{ kg})^{5/3}\left(\frac{3}{4\pi(1\times 10^{18} \text{ kg}\cdot\text{m}^{-3})} \right)^{2/3} = 1\times 10^{38} \text{ kg}\cdot\text{m}^2$$

(b) To achieve a spin-down rate of $-5 \times 10-5$ s-2, the torque needed is

$$\tau = I_{NS}\alpha = (1.29\times 10^{38} \text{ kg}\cdot\text{m}^2)(-5\times 10^{-5} \text{ rad/s}^2) = -6\times 10^{33} \text{ N}\cdot\text{m}$$

ASSESS   The results are reported to a single significant figure to reflect the precision of the data. Checking the units for part (a), we find $(\text{kg})^{5/3}(\text{m}^3)^{2/3}(\text{kg})^{-2/3} = (\text{kg})^{5/3-2/3}(\text{m})^{3\times 2/3} = \text{kg m}^2$, as expected.

33.   INTERPRET   This problem involves calculating the torque that results from a frictional force applied about a 41-cm shaft, and the angular acceleration this engenders. We are then asked to find the time it takes the shaft (and the accompanying flywheel) to stop, given their initial rotational speed.

DEVELOP   From Equation 10.10, the torque applied to the flywheel is

$$\tau = rF \sin\theta = R_{shaft} f_k$$

where $\theta = 90°$, $f_k = 34$ kN, and $R_{shaft} = (41 \text{ cm})/2 = 0.205$ m. Inserting this torque into the rotational analog of Newton's second law (for constant mass), we can find the angular acceleration. We find $\alpha = -\tau/I_{fw}$, where the negative sign indicates that the acceleration is directed opposite to the motion. Use Table 10.2 to find the formulas for the rotational inertia of the flywheel (which we take to be a solid disk). This is

$$I_{fw} = \frac{1}{2}M_{fw}R_{fw}^2$$

where $M_{fw} = 7.7 \times 10^4$ kg and $R_{fw} = 2.4$ m. The time it will take the flywheel to stop is, from Equation 10.7 with $\omega = 0$,

$$0 = \omega_0 + \alpha t$$

$$t = -\frac{\omega_0}{\alpha} = \frac{\omega_0 I_{fw}}{\tau} = \frac{\omega_0 M_{fw} R_{fw}^2}{2 R_{shaft} f_k}$$

EVALUATE Inserting the given quantities into the expression for the time gives

$$t = \frac{\omega_0 M_{fw} R_{fw}^2}{2 f_k R_{shaft}} = \frac{(360 \text{ rpm})(7.7 \times 10^4 \text{ kg})(2.4 \text{ m})^2}{2(34 \times 10^3 \text{ N})(0.205 \text{ m})}\left(\frac{2\pi \text{ rad}}{\text{rev}}\right)\left(\frac{\text{min}}{60 \text{ s}}\right) = 1200 \text{ s} = 20 \text{ min}$$

ASSESS The exact rotational inertia for the flywheel is $M\left(R_{shaft}^2 + R_{fw}^2\right)/2$, which is just 0.7% different from $MR_{fw}^2/2$ for the given radii.

Section 10.4 Rotational Energy

35. **INTERPRET** We're asked to imagine extracting energy from the Earth's rotational kinetic energy. We want to know how long it would take to slow the rotation rate enough for the day to increase by 1 minute.

    DEVELOP We imagine that rotational kinetic energy is extracted from the Earth at a rate of $P = 15 \times 10^{12}$ W. The rotational kinetic energy will correspondingly decrease, manifesting itself as a slowdown in the rotational speed. Over sufficient time, t, the rotational speed will decrease from its current value of $\omega_0 = 2\pi/1\text{d}$ to a value for which the day is one minute longer: $\omega_f = 2\pi/(1\text{d} + 1 \text{ min})$. Equating the change in rotational kinetic energy to the energy extracted gives:

    $$\tfrac{1}{2} I_E \left(\omega_0^2 - \omega_f^2\right) = Pt$$

    EVALUATE From Problem 30, the rotational inertia of the Earth can be estimated as $I_E = 9.69 \times 10^{37} \text{ kg} \cdot \text{m}^2$. Because $1 \text{ min} \ll 1\text{d}$, we can approximate the change in the rotational velocity squared as:

    $$\left(\omega_0^2 - \omega_f^2\right) = \left(\frac{2\pi}{1\text{d}}\right)^2\left[1 - \left(1 + \frac{1\text{min}}{1\text{d}}\right)^{-2}\right] \approx \left(\frac{2\pi}{1\text{d}}\right)^2\left[2\left(\frac{1\text{min}}{1\text{d}}\right)\right]$$

    Plugging this into the above energy equation and solving for the time gives:

    $$t = \frac{\tfrac{1}{2} I_E \left(\omega_0^2 - \omega_f^2\right)}{P} = \frac{\left(9.69 \times 10^{37} \text{ kg} \cdot \text{m}^2\right)(2\pi)^2\left(\frac{1}{24 \cdot 60}\right)}{\left(1.5 \times 10^{13} \text{ W}\right)(24 \cdot 60 \cdot 60 \text{ s})^2} = 2.4 \times 10^{13} \text{ s} = 750{,}000 \text{ y}$$

    ASSESS This is nearly one million years, which simply shows how much kinetic energy there is in the Earth's rotation. Of course, the computed time may be an underestimate, since humankind is continuously increasing its power consumption.

37. **INTERPRET** We are asked to find the energy stored in a the flywheel of Problem 10.33, so we will use the concepts of rotational inertia and kinetic energy of rotation. We also need to find the power output of a generator if the speed of the flywheel changes a given amount in a given time.

    DEVELOP Apply Equation 10.18, $K = I\omega^2/2$, to calculate the kinetic energy stored in the flywheel. We will need to convert the angular speed in rpm to rad/s, and calculate the rotational inertia of the flywheel disk using $I = mR^2/2$ (from Table 10.2). From the work-energy theorem (see Equation 10.19) and using $\bar{P} = W/\Delta t$, we have

    $$\bar{P} = \frac{W}{\Delta t} = \frac{\Delta K}{\Delta t}$$

    where $\omega_i = 360$ rpm, $\omega_f = 300$ rpm, and $\Delta t = 3$s.
    The mass of the flywheel is m = 7.7×10^4 kg, the radius is $R = 2.4$ m, and the initial rotation rate is 360 rpm.

EVALUATE

(a) The energy stored in the flywheel is

$$K = \frac{1}{2}I\omega^2 = \frac{1}{4}mR^2\omega^2$$

$$= \frac{1}{4}(7.7\times10^4 \text{ kg})(2.4 \text{ m})^2 \left(360 \frac{\text{rev}}{\text{min}}\right)^2 \left(\frac{2\pi \text{ rad}}{\text{rev}}\right)^2 \left(\frac{1 \text{ min}}{60 \text{ s}}\right)^2 = 1.6\times10^8 \text{ J}$$

(b) The average power output during the deceleration of the flywheel is

$$\bar{P} = \frac{\Delta K}{\Delta t} = \frac{mR^2}{4\Delta t}\left(\omega_f^2 - \omega_i^2\right)$$

$$= \frac{(7.7\times10^4 \text{ kg})(2.4 \text{ m})^2}{4(3 \text{ s})}\left[\left(300 \frac{\text{rev}}{\text{min}}\right)^2 - \left(360 \frac{\text{rev}}{\text{min}}\right)^2\right]\left(\frac{2\pi \text{ rad}}{\text{rev}}\right)^2 \left(\frac{1 \text{ min}}{60 \text{ s}}\right)^2 = 16 \text{ MW}$$

ASSESS   This is a good way of generating enormous power pulses.

Section 10.5 Rolling Motion

39.   **INTERPRET**   This problem involves comparing the rotational and translational kinetic energy, so we will use the relationship between ω and v ($v = r\omega$).

DEVELOP   The total kinetic energy is

$$K = \frac{1}{2}Mv^2 + \frac{1}{2}I\omega^2$$

where the first term is the translational kinetic energy and the second term is the rotational kinetic energy. From Table 10.2, we find that the rotational inertia of a solid disk is $I = mr^2/2$. Recalling that $v = r\omega$, we can calculate the ratio $f = K_{rot}/K$.

EVALUATE   The ratio of rotational kinetic energy to total kinetic energy is

$$f = \frac{\cancel{\frac{1}{2}}I\omega^2}{\cancel{\frac{1}{2}}Mv^2 + \cancel{\frac{1}{2}}I\omega^2} = \frac{\left(\frac{1}{2}MR^2\right)\omega^2}{M(\omega R)^2 + \left(\frac{1}{2}MR^2\right)\omega^2}$$

$$= \frac{\frac{1}{2}}{1+\frac{1}{2}} = \frac{1}{3}$$

ASSESS   This is consistent with what we noted in the previous problem: the rotational inertia is $mR^2/2$ so the rotational kinetic energy is ½ the translational kinetic energy when it rolls without slipping.

PROBLEMS

41.   **INTERPRET**   The problem is about the rotational motion of the wheel. By identifying the analogous situation for linear motion (see Table 10.1), we can apply the correct formula.

DEVELOP   We are given the angular displacement and the angular acceleration, and the initial angular speed (= 0). To find the final angular speed, we can apply Equation 10.9, which relates all these quantities:

$$\omega^2 = \omega_0^2 + 2\alpha(\theta - \theta_0)$$

To find the time it takes for the wheel to make 2 turns, apply Equation 10.8:

$$\theta = \theta_0 + \omega_0 t + \frac{1}{2}at^2$$

For the calculation, we will convert the angular acceleration to s^{-2}

$$\alpha = 18\left(\frac{\text{rev}}{\text{min}\cdot\text{s}}\right)\left(\frac{2\pi \text{ rad}}{\text{rev}}\right)\left(\frac{1 \text{ min}}{60 \text{ s}}\right) = \frac{6\pi}{10} \text{ rad}\cdot\text{s}^{-2}$$

EVALUATE (a) Inserting the angular acceleration and the angular displacement $\theta - \theta_0 = 4\pi$ into Equation 10.9, we find the final angular velocity is

$$\omega = \pm\sqrt{\omega_0^2 + 2\alpha(\theta - \theta_0)}$$
$$= \sqrt{0 + 2\left(\frac{6\pi}{10}\right)(4\pi \text{ rad})} = \pi\sqrt{4.8} = 6.9 \text{ rad/s}$$

where the two signs indicate that the wheel could turn either clockwise or counter clockwise (we arbitrarily chose the positive sign).

(b) Inserting the acceleration and the angular displacement $\theta - \theta_0 = 4\pi$ rad into Equation 10.8 gives

$$\theta - \theta_0 - \overset{=0}{\overbrace{\omega_0 t}} = \frac{1}{2}\alpha t^2$$

$$t = \pm\sqrt{\frac{2(\theta - \theta_0)}{\alpha}} = \sqrt{\frac{2(4\pi \text{ rad})}{6\pi/10 \text{ rad/s}^2}} = \sqrt{\frac{40}{3}} \text{ s} = 3.7 \text{ s}$$

ASSESS Another way to answer (b) is to use Equation 10.7:

$$\omega = \omega_0 + \alpha t \quad \Rightarrow \quad t = \frac{\omega - \overset{=0}{\overbrace{\omega_0}}}{\alpha} = \frac{\pi\sqrt{4.8} \text{ rad/s}}{6\pi/10 \text{ rad/s}^2} = 3.7 \text{ s}$$

where we have used $\omega = \pi\sqrt{4.8}$ rad/s because it is more precise than 6.9 rad/s, which is only precise to two significant figures.

43. **INTERPRET** We're asked to characterize one of the eagle's downstrokes, in which case the top of the stroke is θ_0 and the bottom of the stroke is θ, as shown in the figure below.

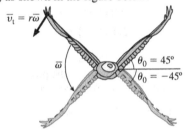

DEVELOP If the eagle flaps 20 times per minute, then it makes a full flap every 3 seconds. A full flap consists of an upstroke and a downstroke, so a single downstroke takes $\Delta t = 1.5$ s . We can plug this time into Equation 10.1 to determine the average angular velocity $\left(\bar{\omega} = \Delta\theta / \Delta t\right)$. The tangential velocity at the tip can be found using Equation 10.3, $\bar{v} = \bar{\omega} r$. For the radius, r, we assume it's roughly half the wingspan, which is by definition the distance between the two wing tips.

EVALUATE (a) Let's first convert the angles from degrees to radians:

$$\theta_0 = 45° \left(\tfrac{2\pi \text{ rad}}{360°}\right) = 0.785 \text{ rad} ; \quad \theta = -45° \left(\tfrac{2\pi \text{ rad}}{360°}\right) = -0.785 \text{ rad}$$

So the average angular velocity is:

$$\bar{\omega} = \frac{\Delta\theta}{\Delta t} = \frac{(0.785 \text{ rad}) - (-0.785 \text{ rad})}{1.5 \text{ s}} = 1.05 \text{ rad/s} \approx 1.1 \text{ rad/s}$$

(b) As for the tangential velocity at the tip of the wing:

$$\bar{v} = \bar{\omega} r = (1.05 \text{ rad/s})\left(\tfrac{1}{2} \cdot 2.1 \text{ m}\right) = 1.1 \text{ m/s}$$

ASSESS The eagle makes a single downstroke in 1.5s, which seems reasonable. And since its wings are about a meter long each, it makes sense that the tangential velocity is approximately one meter per second.

45. **INTERPRET**   This problem involves angular acceleration, which we shall assume is constant. We are provided the initial and final angular speed of the motor and the time interval over which the motor accelerates and are asked to find several characteristics of the rotational kinematics of the engine.

DEVELOP   Because we have no information about the variation in time of the acceleration, we can only calculate the average acceleration over the given time interval. This is given by Equation 10.4 in the form

$$\alpha = \frac{\Delta \omega}{\Delta t}$$

To find the tangential (i.e., linear) acceleration, differentiate Equation 10.3 with respect to time to find

$$a = \frac{dv}{dt} = \frac{d}{dt}(\omega r) = \frac{d\omega}{dt}r = \alpha r$$

(note that this result holds only for constant radius). Finally, knowing the angular acceleration and the initial and final angular velocities, we can apply Equation 10.9 to find the number of revolutions made during the given time interval.

EVALUATE   (a) The average angular acceleration is

$$\bar{\alpha} = \frac{\Delta \omega}{\Delta t} = \frac{(5500 - 1200)\ \text{rpm}}{(2.7\ \text{s})}\left(\frac{1\ \text{min}}{60\ \text{s}}\right) = 170\ \text{s}^{-2}$$

to two significant figures.

(b) With $d = 3.75$ cm, we find an average linear acceleration of

$$\bar{a} = \bar{\alpha}r = \left(167\ \text{s}^{-2}\right)\left(\frac{3.5\ \text{cm}}{2}\right) = 2.9\ \text{m/s}^2$$

(c) The engine makes

$$\Delta\theta = \frac{\omega^2 - \omega_0^2}{2\alpha} = \frac{(5500\ \text{rpm})^2 - (1200\ \text{rpm})^2}{2\left(167\ \text{s}^{-2}\right)\left(60\ \text{s/min}\right)^2} = 150\ \text{revolutions}$$

during this 2.7-s time interval (to two significant figures).

ASSESS   Note that dimensional analysis would lead us to the proper formula for part (b), where we needed to multiply the angular acceleration by a length to recover linear acceleration.

47. **INTERPRET**   The problem concerns the *E. coli* bacteria, whose linear motion is related directly to the rotational motion of its flagellum.

DEVELOP   The time that it takes for the bacteria to cross the microscope's field of view is simply the linear distance divided by the linear velocity: $t = \Delta x / v$. Over the same time, the flagellum completes a number of revolutions given by: $\Delta\theta = \omega t$.

EVALUATE   Combining the two equations from above gives

$$\Delta\theta = \frac{\omega \Delta x}{v} = \frac{(600\ \text{rad/s})(150\ \mu\text{m})}{(25\ \mu\text{m/s})}\left(\frac{1\ \text{rev}}{2\pi\ \text{rad}}\right) = 570\ \text{rev}$$

ASSESS   The units all work out, and the answer seems reasonable. Note that the v we use here is not necessarily the same as the v in Equation 10.3 $(v = \omega r)$, which is the tangential speed of the rotating object.

49. **INTERPRET**   This problem involves finding the rotational inertia of an object about several different axes. For some axes, the object can be decomposed into objects for which the rotational inertia is given in Table 10.2, whereas for the other axes we must apply Equation 10.13 to find the rotational inertia for the various axes.

DEVELOP   For part (a), the square frame can be decomposed into two rods parallel and two rods perpendicular to the axis. For the parallel rods, we can treat them as if all the mass were concentrated at the center of mass, so $I_{par} = Mr^2 = M(L/2)^2 = ML^2/4$. The rotational inertia of the perpendicular rods can be found from Table 10.2, and is $I_{per} = ML^2/12$. For part (b), we apply Equation 10.13 first to a single rod. Using the coordinate system drawn in the figure below, the integral of Equation 10.13 becomes

$$I = \int_0^{L\cos\theta} y^2\,dm = \frac{M}{L}\frac{\tan^2\theta}{\cos\theta}\int_0^{L\cos\theta} x^2\,dx = \frac{M}{L}\frac{\tan^2\theta}{\cos\theta}\left(\frac{x^3}{3}\right)_0^{L\cos\theta} = \frac{ML^2}{3}\sin^2\theta$$

Because all four rods are symmetric, the total rotational inertial will be four times this result.

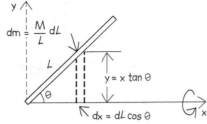

For part (c), apply the parallel axis theorem. From Table 10.2 we find the rotational inertia of a rod rotating about an axis through its center of mass is $I_{cm} = ML^2/12$. The parallel-axis theorem tells us the rotational inertia about a parallel axis a distance L/2 from the center-of-mass axis is

$$I = I_{cm} + M\left(\frac{L}{2}\right)^2 = ML^2\left(\frac{1}{12} + \frac{1}{4}\right) = \frac{1}{3}ML^2$$

EVALUATE   (a) Because we have two rods parallel to the axis and two rods perpendicular to the axis, the total rotational inertia is

$$I_a = 2I_{par} + 2I_{per} = 2\frac{ML^2}{4} + 2\frac{ML^2}{12} = \frac{2}{3}ML^2$$

(b) Given that we have four rods, each with the rotational inertia given by the expression above, we can sum them to find the total rotational inertia. The result is

$$I_b = 4I_{rod} = \frac{4}{3}ML^2\sin^2\left(\frac{\pi}{4}\right) = \frac{4}{3}ML^2\left(\frac{1}{\sqrt{2}}\right)^2 = \frac{2}{3}ML^2$$

(c) Again, we have four rods, each with the rotational inertial derived above. Therefore, the total rotational inertia is

$$I_c = \frac{4}{3}ML^2$$

ASSESS   Notice how we used symmetry to simplify the calculations.

51.   **INTERPRET**   This problem involves applying the parallel axis theorem to find the rotational inertia of an object. We can use Table 10.2 to find the expression for the rotational inertia for an axis through the center of mass of the object.
DEVELOP   The object is a flat plate that is rotating about one of its long edges (of length *b*). Therefore, if we displace the axis of rotation to go through the center of the plate, we have the situation depicted in the last entry of Table 10.2, so $I_{cm} = Ma^2/12$. The displacement of the axis of rotation is $d = a/2$.
EVALUATE   Applying the parallel-axis theorem (Equation 10.17), gives

$$I = I_{cm} + Md^2 = Ma^2\left(\frac{1}{12} + \frac{1}{4}\right) = \frac{1}{3}Ma^2$$

ASSESS   Notice the length b of the long side does not enter into the result. This makes sense because a longer plate will simply have more mass than a shorter one, but the distribution of the mass will not have changed.

53.   **INTERPRET**   The problem concerns the cellular motor that drives the flagellum of the *E. coli* bacteria. We are asked to find the force exerted by this motor, given the torque and the radius at which the force is applied.
DEVELOP   We're told that the force is applied tangentially, so $\theta = 90°$, and Equation 10.10 reduces to: $\tau = rF$.
EVALUATE   Solving for the motor's applied force:

$$F = \frac{\tau}{r} = \frac{400\ \text{pN}\cdot\text{nm}}{12\ \text{nm}} = 33\ \text{pN}$$

ASSESS   This is a very small force, but it's rather impressive that an *E. coli*, with a typical mass of about 10^{-15} kg, can exert a force that is over 1000 times its own weight.

55.   **INTERPRET**   You are asked to find the time it takes for the space station to start from rest and reach a certain angular speed, with a given thrust.
DEVELOP   The space station is essentially a ring with radius $R = 11$ m and rotational inertia $I = MR^2$ (from Table 10.1). The two rockets provide a net torque of $\tau = 2FR$, as can be seen from the figure below.

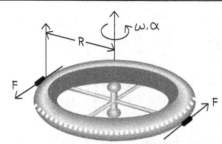

This torque causes an angular acceleration, $\alpha = \tau / I = 2F / MR$, that spins up the station from rest to an angular velocity, ω. This final rotation speed is chosen such that the centripetal acceleration at the rim is equal to the gravitational acceleration on the surface of the Earth:

$$a_c = \frac{v^2}{R} = \frac{(\omega R)^2}{R} = \omega^2 R = g \quad \rightarrow \quad \omega = \sqrt{\frac{g}{R}}$$

Your job is to determine how long the rockets must fire to reach this angular velocity and how many rotations does the station make during this time period.

EVALUATE   (a) The time can be found with Equation 10.7:

$$t = \frac{\omega}{\alpha} = \frac{\sqrt{g/R}}{2F/MR} = \frac{M\sqrt{gR}}{2F} = \frac{(5.0 \times 10^5 \text{ kg}) \sqrt{(9.8 \text{ m/s}^2)(11 \text{ m})}}{2(100 \text{ N})} = 2.60 \times 10^4 \text{ s} = 7.2 \text{ h}$$

(b) We could use Equation 10.8 to find the number of revolutions completed in this time, but Equation 10.9 provides a simple formula with the weight of the space station:

$$\Delta\theta = \frac{\omega^2}{2\alpha} = \frac{Mg}{4F} = \frac{(5.0 \times 10^5 \text{ kg})(9.8 \text{ m/s}^2)}{4(100 \text{ N})} = \frac{12{,}250 \text{ rad}}{2\pi \text{ rad/rev}} = 1900 \text{ rev}$$

ASSESS   These are relatively small rockets, so it takes a fair amount of time to reach the desired rotational velocity. Since $t \sim 1/F$, a larger thrust will shorten this spin up time.

57. **INTERPRET**   This problem involves Newton's second law in both linear and rotational form, which we can apply to find the coefficient of friction between block and slope, given the acceleration of the block. We will also need to consider the rotational inertia of the wheel in this problem.

DEVELOP   Draw a diagram of the situation (see figure below). Applying Newton's second law (Equation 4.3) to the mass gives

$$\left.\begin{array}{l} mg\sin\theta - f_k - T = ma \\ n - mg\cos\theta = 0 \end{array}\right\} mg\sin\theta - \mu_k mg\cos\theta - T = ma$$

where we have used the Equation 5.3 to express the force due to kinetic friction, $f_k = \mu_k n$.

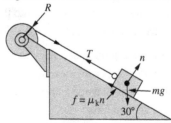

Likewise, applying the rotational analog of Newton's second law (Equation 10.11) to the wheel gives

$$\tau_{net} = I\alpha$$
$$TR = I\alpha$$

where $t_{net} = TR$ because the tension is the only torsional force acting on the wheel, $I = MR^2/2$ (from Table 10.2) and $a = \alpha R$ (Equation 10.4). These equations allow us to determine μ_k.

EVALUATE   Solving first for the tension from the rotational application of Newton's second law gives

$$T = \frac{I\alpha}{R} = \frac{\left(MR^2/2\right)\left(a/R\right)}{R} = \frac{1}{2}Ma$$

Insert this into the equation derived from Newton's second law applied to the block, and solve for μ_k:

$$\mu_k = \frac{mg\sin\theta - ma - Ma/2}{mg\cos\theta}$$

$$= \frac{\left(2.4\ \text{kg}\right)\left(9.8\ \text{m/s}^2\right)\sin\left(30°\right) - \left(2.4\ \text{kg} + 0.425\ \text{kg}\right)\left(1.6\ \text{m/s}^2\right)}{\left(2.4\ \text{kg}\right)\left(9.8\ \text{m/s}^2\right)\cos\left(30°\right)} = 0.36$$

ASSESS   To see that our expression for μ_k makes sense, let's check some limits: **(i)** If $a = 0$, then $\mu_k = \tan\theta$. This is precisely the equation we obtained in Chapter 5 (see Example 5.10). **(ii)** $M = 0$ and $\mu_k = 0$. The situation corresponds to a block of mass m sliding down a frictionless slope with acceleration $a = g\sin\theta$.

59. **INTERPRET**   In this problem we want to find the angular speed of the potter's wheel after he exerts a tangential force to the edge of the wheel. We can address this problem in several ways, either through the work-energy theorem, or through Newton's second law (Equation 10.11). The force produces a torque that causes the wheel to rotate.

DEVELOP   We will apply the work-energy theorem for constant torque (Equation 10.19). This gives

$$W = \tau\Delta\theta = \Delta K = K_f - K_0 = \frac{1}{2}I\omega^2$$

because the wheel starts from rest. The equation allows us to determine the angular velocity ω.

EVALUATE   Because the force acting on the wheel is tangential to the wheel circumference, $\theta = 90°$ in Equation 10.10; so $\tau = FR$. In addition, From Table 10.2, we know that the rotational inertia of a disk is I = MR2/2. Inserting $\Delta\theta = \frac{1}{8}$ rev $= \frac{\pi}{4}$ rad, we have

$$\omega^2 = \frac{2\tau\Delta\theta}{I} = \frac{2FR\Delta\theta}{MR^2/2} = \frac{4F\Delta\theta}{MR}$$

or

$$\omega = \pm\sqrt{\frac{4F\Delta\theta}{MR}} = \pm\sqrt{\frac{4\left(75\ \text{N}\right)\left(\pi\ \text{rad}/4\right)}{\left(120\ \text{kg}\right)\left(0.45\ \text{m}\right)}} = \pm 2.1\ \text{rad/s}$$

ASSESS   The two signs indicate that the potter may spin the wheel either clockwise or counter clockwise. The greater the force exerted on the wheel, the larger the angular speed. On the other hand, larger M and R result in a larger rotational inertia, and smaller angular speed (if the same force is applied). If we apply Newton's second law to this problem, we find

$$\tau_{net} = FR = Ia = MR^2\alpha/2$$

$$\alpha = \frac{2F}{MR}$$

Inserting this result into Equation 10.9 and solving for the final angular velocity gives

$$\omega^2 = \overbrace{\omega_0^2}^{=0} + 2\alpha\overbrace{\left(\theta - \theta_0\right)}^{\Delta\theta}$$

$$\omega = \pm\sqrt{\frac{4F\Delta\theta}{MR}}$$

which is the same expression found using the work-energy theorem.

61.   **INTERPRET**   This problem involves conservation of energy: gravitational potential energy is converted to center-of-mass kinetic energy and rotational kinetic energy.

DEVELOP   By conservation of energy, the sum of the gravitational potential energy and the total kinetic energy (Equation 10.20) is a constant. If we assume the gravitational potential is zero where the ball is at rest, then this constant is zero, or in other words:

$$K_{cm} + K_{rot} = \tfrac{1}{2}Mv^2 + \tfrac{1}{2}I\omega^2 = -U$$

As it rolls down the incline, the potential decreases: $U = -Mgh,$ where the height is related to the distance rolled down the incline by: $h = d\sin\theta.$ The ball is hollow, so its rotational inertia is $I = \tfrac{2}{3}MR^2,$ and we assume that it rolls without slipping, so $v = \omega R$ (Equation 10.21).

EVALUATE   Plugging in the various expressions into the energy conservation equation gives:

$$\tfrac{1}{2}Mv^2 + \tfrac{1}{3}Mv^2 = Mgd\sin\theta$$

Solving for the speed,

$$v = \sqrt{\tfrac{6}{5}gd\sin\theta}$$

ASSESS   If the ball were sliding down the incline without friction, the speed would have been $v = \sqrt{2gd\sin\theta}.$ The fact that the ball is rolling means it will go slower down the incline.

63.   **INTERPRET**   The kinetic energy of the wheel consists of two parts: the kinetic energy of the center of mass, $K_{cm},$ and the rotational kinetic energy, K_{rot} . We want to find how changing the moment of inertia and mass of the wheel affects the total kinetic energy.

DEVELOP   The total kinetic energy of the wheel consists of center-of-mass energy and internal rotational energy associated with the spin about the center of mass (see Equation 10.20):

$$K_{tot} = K_{cm} + K_{rot} = \frac{1}{2}Mv_{cm}^2 + \frac{1}{2}I_{cm}\omega^2$$

With the condition for rolling without slipping, $v = \omega R,$ the total kinetic energy can be rewritten as

$$K_{tot} = K_{cm} + K_{rot} = \frac{1}{2}Mv_{cm}^2 + \frac{1}{2}I_{cm}\left(\frac{v_{cm}}{R}\right)^2 = \frac{1}{2}Mv_{cm}^2\left(1 + \frac{I_{cm}}{MR^2}\right)$$

The initial condition is $I_{cm}/(MR^2) = 0.40 = 40\%.$ After the redesign,

$$\frac{I'_{cm}}{M'R^2} = \frac{0.9I_{cm}}{(0.8M)R^2} = 1.125\frac{I_{cm}}{MR^2} = (1.125)(0.40) = 0.45$$

EVALUATE   The fractional decrease in kinetic energy is

$$\frac{K - K'}{K} = 1 - \frac{K'}{K} = 1 - \frac{M'\left[1 + I'_{cm}/(M'R^2)\right]}{M\left[1 + I_{cm}/(MR^2)\right]} = 1 - \frac{0.8M}{M}\frac{(1+0.45)}{(1+0.40)} = 0.171 = 17\%$$

to two significant figures.

ASSESS   Initially, K_{cm} accounts for $1/14 = 71\%$ of the total kinetic energy, while K_{rot} accounts for the remaining $0.4/14 = 29\%.$ After the redesign $M \to M' = 0.8M$, so the translational kinetic energy decreases by 20%, while the rotational kinetic energy goes down by 10% $(I_{cm} \to I'_{cm} = 0.9I_{cm})$. Therefore, the total kinetic energy is now

$$(0.8)\frac{1}{1.4} + (0.9)\frac{0.4}{1.4} = 0.829 = 83\%$$

of the original. This is a 17% decrease.

65.   **INTERPRET**   This problem involves finding the rotational inertia of a circular disk after an off-center hole has been drilled through it. The parallel-axis theorem is likely to be useful here.

DEVELOP   Equation 10.12 shows that the rotational inertia of an object is the sum of the rotational inertias of its pieces, so

$$I_{disk} = I_{hole} + I_{remainder}$$

The hint expresses this fact as $I_{remainder} = I_{disk} - I_{hole}$. Here $I_{disk} = MR^2/2$ is the rotational inertia of the whole disk about an axis perpendicular to the disk and through the disk center (see Example 10.7). Use the parallel-axis theorem to find the rotational inertia of the hole, I_{hole}. This gives

$$I_{hole} = M_{hole}\left(\frac{R}{4}\right)^2 + I_{cm} = M_{hole}\frac{R^2}{16} + M_{hole}\frac{R^2}{32} = \frac{3}{32}M_{hole}R^2$$

where $R/4$ is the distance of the hole's center of mass from the axis of the disk, and we have $I_{cm} = M_{hole}(R/4)^2/2$ as the rotational inertia of the hole material about a parallel axis through its center of mass (see Example 10.7). With these equations, we can determine $I_{remainder}$.

EVALUATE   Because the planar mass density of the disk (assumed to be uniform) is $\sigma = M/\pi R^2$, the mass of the hole material is

$$M_{hole} = \sigma A_{hole} = \frac{M}{\pi R^2}\pi\left(\frac{R}{4}\right)^2 = \frac{M}{16}$$

Therefore, the rotational inertia of the hole is

$$I_{hole} = \left(\frac{3}{32}\right)\left(\frac{M}{16}\right)R^2 = \frac{3}{512}MR^2$$

and

$$I_{remainder} = I_{disk} - I_{hole} = \frac{1}{2}MR^2 - \frac{3}{512}MR^2 = \frac{253}{512}MR^2 = 0.494MR^2$$

ASSESS   If the hole drilled were concentric with the disk, we would have

$$I'_{hole} = I_{cm} = \frac{1}{2}\frac{M}{16}\left(\frac{R}{4}\right)^2 = \frac{1}{512}MR^2$$

and

$$I'_{remainder} = I_{disk} - I'_{hole} = \frac{1}{2}MR^2 - \frac{1}{512}MR^2 = \frac{255}{512}MR^2 = 0.498MR^2$$

The same result is obtained if we use the formula $M'(R_1^2 + R_2^2)/2$ derived in Problem 51, with $M' = \pi R^2 - \pi(R/4)^2 = (15/16)\pi R^2 = (15/16)M$, $R_1 = R$ and $R_2 = R/4$.

67.   **INTERPRET**   This problem involves conservation of total mechanical energy, which we can use to find how high up the hill the motorcyclist can go.

DEVELOP   If all possible losses are neglected, the total mechanical energy of the motorcycle and rider is conserved as it coasts uphill, so the total kinetic energy at the bottom equals the total potential energy at the highest point,

$$K_{trans} + K_{rot} = M_{tot}gh$$

The translation kinetic energy of the cycle and rider (including the wheels) and the rotational kinetic energy of the wheels (about their center of mass) are, assuming rolling without slipping,

$$K_{trans} = \frac{1}{2}M_{tot}v^2, \qquad K_{rot} = 2\left(\frac{1}{2}I\omega^2\right) = I\left(\frac{v}{R}\right)^2$$

These expressions can be combined to solve for h.

EVALUATE   Substituting the second equation into the first, and using $v = 85$ km/h $= 23.6$ m/s, we find the maximum vertical height reached is

$$h = \frac{v^2}{2g}\left(1 + \frac{2I}{M_{tot}R^2}\right) = \frac{(23.6 \text{ m/s})^2}{2(9.8 \text{ m/s}^2)}\left(1 + \frac{2(2.1 \text{ kg·m}^2)}{(395 \text{ kg})(0.26 \text{ m})^2}\right) = 33 \text{ m}$$

ASSESS   If the rolling motion is ignored, the result would be $h = v^2/2g$, which is what we expect from considering only the linear motion.

69.   **INTERPRET**   In this problem we are given a disk with non-uniform mass density, and asked to find its total mass and rotational inertia. We will therefore need to use the integral expression to calculate the rotational inertia.

DEVELOP   As mass elements, choose thin rings of width dr and radius r (as in Example 10.7) so that

$$dm = \rho(r)\,dV = \left(\frac{\rho_0 r}{R}\right)2\pi rw\,dr = \frac{2\pi\rho_0 w}{R}r^2\,dr$$

The total mass is $M = \int_0^R dm$ and the rotational inertia about the disk axis is $I = \int_0^R r^2\,dm$ (see Equation 10.13).

EVALUATE   (a) The disk's total mass is

$$M = \int_0^R dm = \frac{2\pi\rho_0 w}{R}\int_0^R r^2\,dr = \frac{2\pi\rho_0 wR^2}{3}$$

(b) The disk's rotational inertia about a perpendicular axis through its center is

$$I = \int_0^R r^2\,dm = \frac{2\pi\rho_0 w}{R}\int_0^R r^4\,dr = \frac{2\pi\rho_0 wR^4}{5} = \frac{3}{5}\left(\frac{2\pi\rho_0 wR^2}{3}\right)R^2 = \frac{3}{5}MR^2$$

ASSESS   Our result for I is intermediary between a disk of uniform density and a ring; $\frac{1}{2}MR^2 < I < MR^2$, if expressed in terms of the total mass M, but is less than a disk of uniform density ρ_0; $I < \frac{1}{2}\rho_0\pi R^4 w$, because ρ_0 is the maximum density.

71.   **INTERPRET**   We are asked to show that the rotational inertia of a planar object around an axis perpendicular to the plane of the object is equal to the sum of the rotational inertias around two perpendicular axes within the plane of the object. We will use the integral form of rotational inertia, since the mass is distributed continuously.

DEVELOP   The rotational inertia is $I = \int r^2\,dm$. We will set up our coordinate system such that the two rotational axes within the plane of the object are the x and y coordinate axes.

EVALUATE   The rotational inertia around the x axis is $I_x = \int r^2\,dm = \int y^2\,dm$. The rotational inertia around the y axis is $I_y = \int r^2\,dm = \int x^2\,dm$. The sum of the two is $I_x + I_y = \int x^2\,dm + \int y^2\,dm = \int(x^2 + y^2)\,dm$, but $(x^2 + y^2)$ is just the distance r from the perpendicular z axis, so $I_x + I_y = \int r^2\,dm = I_z$.

ASSESS   We have proven what was requested.

73.   **INTERPRET**   This problem is an exercise in calculating the rotational inertia of an object (in this case, a right-circular cone). Because the mass is distributed continuously throughout the cone, we will apply the integral formula Equation 10.13 to find the rotational inertia.

DEVELOP   Divide the cone into circular slices parallel to the base of the cone, and integrate over all these slices to find the total rotational inertia of the cone. The height of the cone is h and the base radius is R, so the radius of each slice is $r = Rx/h$, where x is the distance from the apex. The volume of the cone is $V = Ah/3$, where A is the area of the base, $A = \pi R^2$, so $V = \pi R^3 h/3$. The volume of each disk-shaped slice is $dV = \pi r^2\,dx$. The cone has uniform mass density M/V, so each disk has mass $dm = MdV/V = 3M(\pi r^2\,dx)/(\pi R^2 h)$. From Table 10.2, the rotational inertia of each disk is $dI = r^2\,dm/2$.

EVALUATE   Evaluating the integral gives

$$I = \int_0^h dI = \frac{1}{2}\int_0^h r^2\,dm = \frac{1}{2}\int_0^h\left(\frac{Rx}{h}\right)^2\frac{3M}{\pi R^2 h}\pi R^2\,d = \frac{3M}{2h^3}\int_0^h x^2\left(\frac{Rx}{h}\right)^2 dx$$

$$I = \frac{3MR^2}{2h^5}\int_0^h x^4\,dx = \frac{3MR^2}{2h^5}\left[\frac{1}{5}h^5\right] = \frac{3}{10}MR^2$$

ASSESS   The units are correct. The value of I is less than that of a cylinder, since a greater proportion of the mass is concentrated along the axis of the cone.

75. **INTERPRET** We are asked to find the torque due to gravity on a rod hanging from one end.

DEVELOP We can break the rod into infinitesimal mass elements, $dm = \mu dr$, where the mass per unit length is $\mu = M / L$. The gravitational force on each element is $F_g = g\,dm$, which means the torque (Equation 10.10) on each mass element is $d\tau = r\left(gdm\right)\sin\theta$, see the figure below.

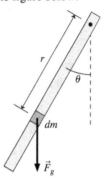

EVALUATE The torque on the full rod is just the integral of the infinitesimal torques over the length of the rod:

$$\tau = \int d\tau = \int_0^L r\mu g \sin\theta\,dr = \mu g \sin\theta \left[\tfrac{1}{2} r^2\right]_0^L = \tfrac{1}{2}MGL\sin\theta$$

ASSESS We could have gotten the same answer by considering just the torque on the center of mass. The rod's center of mass is located at a distance of $r = L/2$ from the pivot point. The gravitational force, Mg, applied at the center of mass, creates a torque of $\tau = \left(L/2\right)\left(MG\right)\sin\theta$, which is what we obtained from integrating over the individual mass elements.

77. **INTERPRET** You want to know if a rotating flywheel has as much energy as its manufacturer claims.

DEVELOP The flywheel can be modeled as a ring with rotational inertia $I = MR^2$. Its rotational kinetic energy is $\tfrac{1}{2}I\omega^2$, from Equation 10.18.

EVALUATE We have to divide the given diameter by 2 to get the radius, and we have to convert the non-SI unit of rpm to rad/s. Following that, the flywheel's kinetic energy is:

$$K_{\text{rot}} = \tfrac{1}{2}MR^2\omega^2 = \tfrac{1}{2}\left(48\text{ kg}\right)\left(\tfrac{1}{2}0.39\text{ cm}\right)^2\left(30{,}000\text{ rpm}\right)^2\left(\frac{\frac{2\pi}{60}\text{ rad/s}}{1\text{ rpm}}\right)^2 = 9.0\text{ MJ}$$

The specs are incorrect. The flywheel's storage capacity is 3 MJ below what the manufacturer claims.

ASSESS A flywheel is like a battery that stores energy as kinetic rotational energy. It has a high rotational inertia and presumably very little friction, so it will spin freely for a long time without slowing down appreciably. When the need arises, the flywheel can be connected to an electric generator, where its rotational energy is converted to electricity.

79. **INTERPRET** We must compare two centrifuges with slightly different designs.

DEVELOP We're told that the two centrifuges have the same mass and radius. But design A looks like a thin ring, while design B looks like a flat disk.

EVALUATE Design A should have approximately a rotational inertia of $I_A \approx MR^2$, compared to the design B with $I_B \approx \tfrac{1}{2}MR^2$.

The answer is (a).

ASSESS The rotational kinetic energy is proportional to rotational inertia $\left(K_{\text{rot}} = \tfrac{1}{2}I\omega^2\right)$. Therefore, it will take twice the work $\left(W = \Delta K\right)$ to spin up centrifuge A to the same rotational speed as centrifuge B.

81. **INTERPRET** We must compare two centrifuges with slightly different designs.

DEVELOP The sample tubes do not rest vertically, but instead tilt outwards. The bottom of the tubes are therefore at a radius greater than the radius of the centrifuges themselves.

EVALUATE If the tubes are made longer, the bottom of the tubes will extend to a greater radius, so the rotational inertia will increase.

The answer is (b).

ASSESS Some centrifuges have a fixed angle (e.g. 45°), at which the tubes are placed. Others have a hinge that lets the tubes swing out when the device starts to turn.

83.   **INTERPRET**   We must compare two centrifuges with slightly different designs.

DEVELOP   For both designs, the rotational inertia is proportional to the mass times the radius squared: $I \propto MR^2$.

EVALUATE   Doubling both the mass and the radius will change the rotational inertia by

$$\frac{I'}{I} = \frac{(2M)(2R)^2}{MR^2} = 8$$

The answer is (c).

ASSESS   For a given rotational speed, the centripetal force is proportional to the radius: $F = m\omega^2 r$. So making the centrifuge bigger will presumably improve its ability to separate materials by their density.

11

ROTATIONAL VECTORS AND ANGULAR MOMENTUM

EXERCISES

Section 11.1 Angular Velocity and Acceleration Vectors

13. **INTERPRET** This problem is an exercise in determining the direction and magnitude of the angular velocity vector. From the direction and speed at which the car is traveling, we are to deduce the angular velocity of its wheels.

    DEVELOP From Chapter 10 (Equation 10.3), we know that the magnitude of the angular velocity (i.e., the angular speed) is given by $\omega = v_{cm} r$. For this problem, we have $v_{cm} = (70 \text{ km/h})(10^3 \text{ m/km})(1 \text{ h}/3600 \text{ s}) = 19.44$ m/s and $r = d/2 = (0.62 \text{ m})/2 = 0.31$ m. The direction of the angular velocity vector can be determined using the right-hand rule (see Figure 11.1).

    EVALUATE Inserting the given quantities into Equation 10.3 gives an angular speed of
    $$\omega = v_{cm}/r = (19.44 \text{ m/s})/(0.31 \text{ m}) = 63 \text{ s}^{-1}$$
    to two significant figures. If the car is rolling noth, the right-hand rule determines that the direction of the anular velocity vector is to the left, which is west. Therefore $\vec{\omega} = 63 \text{ s}^{-1}$ west .

    ASSESS Notice that the angular speed may be reported in units of rad/s, but since radians are a dimensionless quantity, they are often left out, leaving s^{-1}, which is a frequency (Hz).

    EVALUATE Using Equation 11.1, we find the angular acceleration to be
    $$\vec{\alpha}_{ave} = \frac{\Delta \vec{\omega}}{\Delta t} = \frac{\vec{\omega}_f - \vec{\omega}_i}{\Delta t} = \frac{-\omega \hat{j} - (-\omega \hat{i})}{\Delta t} = \frac{\omega}{\Delta t}(\hat{i} - \hat{j})$$
    $$= \frac{62.7 \text{ s}^{-1}}{25 \text{ s}}(\hat{i} - \hat{j}) = (2.5 \text{ s}^{-2})(\hat{i} - \hat{j})$$

    The magnitude of $\vec{\alpha}_{ave}$ is
    $$|\vec{\alpha}_{ave}| = \frac{\sqrt{2}\omega}{\Delta t} = \frac{\sqrt{2}(62.7 \text{ rad/s})}{25 \text{ s}} = 3.6 \text{ rad/s}^2$$

    and $\vec{\alpha}_{ave}$ points in the south-east direction (in the direction of the vector $\hat{i} - \hat{j}$).

    ASSESS Angular acceleration $\vec{\alpha}_{ave}$ points in the same direction as $\Delta \vec{\omega}$. The units can be reported as either rad/s^2 or s^{-2}.

15. **INTERPRET** This problem involves calculating the magnitude of the average acceleration given the initial and final angular velocities, and the time interval between the two. We are also asked to find the angle that the average angular acceleration vector makes with the horizontal.

    DEVELOP Let the x axis be the horizontal direction (positive to the right), and the upward direction be the y axis. The the average angular acceleration vector is simply the difference between the final and initial angular velocities divided by the time interval between these two speeds (i.e., Equation 11.1). The initial angular velocity is $\omega_i = (45 \text{ rpm})\hat{j}$, the final angular speed is $\omega_f = (60 \text{ rpm})\hat{i}$, and the time interval is $t = 15$ s. To find the angle θ the average angular acceleration vector makes with the horizontal, use the fact that that $\tan\theta = \bar{\alpha}_y/\bar{\alpha}_x$.

EVALUATE   (a) Inserting the given quantities into Equation 11.1, we find

$$\bar{\alpha} = \frac{\omega_f - \omega_i}{\Delta t} = \frac{(60 \text{ rpm})\hat{i} - (45 \text{ rpm})\hat{j}}{0.25 \text{ min}^{-1}} = (240 \text{ min}^{-2})\hat{i} - (180 \text{ min}^{-2})\hat{j} = (86,400 \text{ s}^{-2})\hat{i} - (64,800 \text{ s}^{-2})\hat{j}$$

The magnitude of the average acceleration is thus $\bar{\alpha} = \sqrt{(86,400 \text{ s}^{-2})^2 + (64,800 \text{ s}^{-2})^2} = 1.1 \times 10^6 \text{ s}^{-2}$ to two significant figures.

(b) The angle of the average angular acceleration vector with respect to the horiztonal is

$$\theta = \text{atan}\left(\frac{\bar{\alpha}_y}{\bar{\alpha}_x}\right) = \text{atan}\left(\frac{-64,800 \text{ s}^{-2}}{86,400 \text{ s}^{-2}}\right) = -37°$$

ASSESS   Note that the quantities used to calculate part (b) were intermediate quantities, so more significant figures are retained. The final result, however, is reported to two significant figures, which reflects the precision of the data.

Section 11.2 Torque and the Vector Cross Product

17.   INTERPRET   This problem involves finding the torque about the origin given a force and the position vector that indicates where the force is applied.

DEVELOP   Use Equation 11.2 to find the torque. The position vector r is $\vec{r} = (3 \text{ m})\hat{i} + (1 \text{ m})\hat{j}$.

EVALUATE   (a) For a force $\vec{F} = (12 \text{ N})\hat{i}$, the torque is

$$\vec{\tau} = \vec{r} \times \vec{F} = \left[(3 \text{ m})\hat{i} + (1 \text{ m})\hat{j}\right] \times (12 \text{ N} \cdot \text{m})\hat{i} = \begin{vmatrix} \hat{i} & \hat{j} & \hat{k} \\ 3 \text{ m} & 1 \text{ m} & 0 \text{ m} \\ 12 \text{ N} & 0 \text{ N} & 0 \text{ N} \end{vmatrix} = (-12 \text{ N} \cdot \text{m})\hat{k}$$

(b) For a force $\vec{F} = (12 \text{ N})\hat{j}$

$$\vec{\tau} = \vec{r} \times \vec{F} = \left[(3 \text{ m})\hat{i} + (1 \text{ m})\hat{j}\right] \times (12 \text{ N} \cdot \text{m})\hat{j} = \begin{vmatrix} \hat{i} & \hat{j} & \hat{k} \\ 3 \text{ m} & 1 \text{ m} & 0 \text{ m} \\ 0 \text{ N} & 12 \text{ N} & 0 \text{ N} \end{vmatrix} = (36 \text{ N} \cdot \text{m})\hat{k}$$

(c) For a force $\vec{F} = (12 \text{ N})\hat{k}$

$$\vec{\tau} = \vec{r} \times \vec{F} = \left[(3 \text{ m})\hat{i} + (1 \text{ m})\hat{j}\right] \times (12 \text{ N} \cdot \text{m})\hat{k} = \begin{vmatrix} \hat{i} & \hat{j} & \hat{k} \\ 3 \text{ m} & 1 \text{ m} & 0 \text{ m} \\ 0 \text{ N} & 0 \text{ N} & 12 \text{ N} \end{vmatrix} = (12 \text{ N} \cdot \text{m})\hat{i} + (36 \text{ N} \cdot \text{m})\hat{j}$$

ASSESS   For part (c), the magnitude is $\tau = \sqrt{(12 \text{ N} \cdot \text{m})^2 + (36 \text{ N} \cdot \text{m})^2} = 38 \text{ N} \cdot \text{m}$ and the direction is $\theta = \text{atan}(36 \text{ N} \cdot \text{m}/12 \text{ N} \cdot \text{m}) = 72°$ counter clockwise from the x axis and in the x-y plane.

19.   INTERPRET   You want to know what torque is supplied by the deltoid muscle about the shoulder joint when your arm is outstretched.

DEVELOP   From Equation 11.2, the torque is $\vec{\tau} = \vec{r} \times \vec{F}$, with the magnitude equaling $rF \sin \theta$.

EVALUATE   The distance between the shoulder joint (i.e., where the arm pivots) and where the deltoid force is applied is given as $r = 18$ cm. The angle between the corresponding radial vector and the muscle force is $\theta = 180° - 15° = 165°$. The magnitude of the torque is then

$$\tau = rF \sin \theta = (0.18 \text{ m})(67 \text{ N}) \sin 165° = 3.1 \text{ N} \cdot \text{m}$$

By the right-hand rule, we start with our fingers pointing to the right in the direction of $\vec{r}$, and then rotate them upwards in the direction of $\vec{F}$. Our thumb points up, so the torque of 3.1 N·m points out of the page.

ASSESS   Is this enough torque to keep the arm outstretched? Let's assume the arm has a mass of about 3 kg (corresponding to a weight of about 30 N), and its center of mass is 30 cm from the shoulder joint. The gravitational force will pull the arm down at 90° to the horizontal arm direction, thus generating a torque in the opposite direction with a magnitude of $\tau \simeq (30 \text{ N})(0.3 \text{ m}) = 9 \text{ N} \cdot \text{m}$. Therefore, the deltoid muscle would need help from other muscles to keep the arm horizontal.

Section 11.3 Angular Momentum

21. **INTERPRET**   This problem asks us to find the angular momentum of a ball given its linear velocity, its mass, and the distance from its axis of rotatoin.

DEVELOP   The angular momentum of an object about a point is defined as (see Equation 11.3)

$$L = \vec{r} \times \vec{p}$$

where $\vec{p}$ is the linear momentum and $\vec{r}$ is the position vector of the object relative to that point. We may also express $\vec{L}$ as

$$\vec{L} = \vec{r} \times \vec{p} = (rp \sin \theta) \hat{n}$$

where θ is the angle between $\vec{r}$ and $\vec{p}$ and $\hat{n}$ is a unit vector perpendicular to both $\vec{r}$ and $\vec{p}$. For this problem, we can assume that the ball is traveling in a circle of radius r and speed v. Since the velocity of the ball, $\vec{v}$, is perpendicular to $\vec{r}$, the magnitude of the angular momentum about the center is $L = |\vec{r} \times \vec{p}| = rp = rmv$.

EVALUATE   From the problem statement, we have $r = 1.2\text{m} + 0.9 \text{ m} = 2.1 \text{ m}$ and $v = 27$ m/s. Therefore,

$$L = rmv = (2.1 \text{ m})(7.3 \text{ kg})(27 \text{ m/s}) = 4.1 \ \text{kg} \cdot \text{m}^2 \cdot \text{s}^{-1}$$

ASSESS   The direction of $\vec{L}$ is parallel to the axis of rotation. It is perpendicular to both $\vec{v}$ and $\vec{r}$.

23. **INTERPRET**   We are given the elements of rotational inertia and the angular velocity of the hoop and are to find the corresponding angular momentum. We will need to use Table 10.2 to find the rotational inertia.

DEVELOP   For an object rotating about a fixed axis, its angular momentum can be expressed as (see Equation 11.4) $\vec{L} = I\vec{\omega}$, where I is the moment of inertia of the object, and $\vec{\omega}$ is its angular velocity about its axis. From Table 10.2, we find that the rotational inertia of a hoop rotating about its axis is $I = mr^2$.

EVALUATE   With $\omega = 170$ rpm $= 17.89$ rad/s, the magnitude of $\vec{L}$ is

$$L = I\omega = mr^2 \omega = (0.64 \text{ kg})(0.45 \text{ m})^2 (17.8 \text{ rad/s}) = 2.3 \text{ J} \cdot \text{s}$$

The direction of $\vec{L}$ is along the axis of rotation according to the right-hand rule.

ASSESS   The angular momentum vector $\vec{L}$ points in the same direction as $\vec{\omega}$.

Section 11.4 Conservation of Angular Momentum

25. **INTERPRET**   This problem involves conservation of angular momentum, which we can use to find the angular speed of a spinning wheel after a piece of clay is dropped onto it and sticks to its surface.

DEVELOP   If the clay is dropped vertically onto a horizontally spinning wheel, the angular momentum about the vertical spin axis is conserved. Conservation of angular momentum is expressed as

$$\vec{L}_i = \vec{L}_f \quad \Rightarrow \quad I_i \vec{\omega}_i = I_f \vec{\omega}_f$$

For this problem, the direction of the angular velocity does not change, so this expression for conservation of angular momentum reduces to its scalar form, $I_i \omega_i = I_f \omega_f$. The intial rotational inertia is $I_i = I_{\text{wheel}} = 6.40$ kg $\cdot$ m$^2 \cdot$ s^{-1}, and the final rotational inertia is $I_f = I_{\text{wheel}} + m_{\text{clay}} r^2$.

EVALUATE   Inserting the given quantities into the expression from conservation of angular momentum, the final angular velocity is

$$\omega_f = \frac{I_i}{I_f} \omega_i = \left(\frac{I_{\text{wheel}}}{I_{\text{wheel}} + m_{\text{clay}} r^2} \right) \omega_i = \frac{6.40 \text{ kg} \cdot \text{m}^2}{6.40 \text{ kg} \cdot \text{m}^2 + (2.70 \text{ kg})(0.460 \text{ m})^2} (19.0 \text{ rpm}) = 17.4 \text{ rpm}$$

ASSESS   The clay increases the total rotational inertia of the system, so the angular speed decreases, as required by conservation of angular momentum.

27.   **INTERPRET**   In this problem we are asked about the period of a star formed by a collapsing cloud. We can use conservation of angular momentum to find the answer.

DEVELOP   If we assume there are no external torques and no mass loss during the collapse of the star-forming cloud, its angular momentum is conserved, so $I_i \omega_i = I_f \omega_f$ The initial and final rotational inertias may be found from Table 10.2, which gives $I_i = 2mr_i^2/5$ and $I_f = 2mr_f^2/5$. From this, we can solve for the final period of the star using $T_f = 2\pi/\omega_f$.

EVALUATE   Given that the mass involved does not change, conservation of angular momentum gives

$$I_i \omega_i = I_f \omega_f$$

$$\frac{2}{5} MR_i^2 \omega_i = \frac{2}{5} MR_f^2 \omega_f$$

$$\frac{\omega_i}{\omega_f} = \left(\frac{R_f}{R_i} \right)^2$$

Thus, the final period is

$$T_f = T_i \left(\frac{R_f}{R_i} \right)^2 = \left(1.4 \times 10^6 \text{ y} \right) \left(\frac{7.0 \times 10^8 \text{ m}}{1.0 \times 10^{13} \text{ m}} \right)^2 = 6.86 \times 10^{-3} \text{ y} = 2.5 \text{ days}$$

ASSESS   In current models of star formation, the collapsing cloud does not maintain a spherical shape, forming a flattened disk instead, and the central star retains just a fraction of the original cloud's mass.

PROBLEMS

29.   **INTERPRET**   This problem is an exercise in calculating torque, given the force and the position relative to an axis at which the force is applied.

DEVELOP   Use Equation 11.2, $\vec{\tau} = \vec{r} \times \vec{F}$ to calculate the torque, given that $\vec{r} = (18 \text{ cm})\hat{i} + (5.5 \text{ cm})\hat{j}$ and. $\vec{F}(88 \text{ N})\hat{i} - (23)\hat{j}$.

EVALUATE   Evaluating the cross product gives

$$\vec{\tau} = \left[(18 \text{ cm})\hat{i} + (5.5 \text{ cm})\hat{j} \right] \times \left[\vec{F}(88 \text{ N})\hat{i} - (23)\hat{j} \right] = \det \begin{pmatrix} \hat{i} & \hat{j} & \hat{k} \\ 18 \text{ cm} & 5.5 \text{ cm} & 0 \text{ cm} \\ 88 \text{ N} & -23 \text{ N} & 0 \text{ N} \end{pmatrix} = (-9.0 \text{ N·m})\hat{k}$$

ASSESS   Thus the torque is the direction defined by the bolt.

31.   **INTERPRET**   We're asked to calculate the torque exerted by the ball player in order to bring the baseball to rest.

DEVELOP   The player exerts a torque around his shoulder, which results in a stopping force on the ball. From Equation 11.2, the average torque is $\bar{\tau} = r\bar{F}_{stop}$, where we have taken into account that the vertically-held arm and the horizontally-directed force are at right angles, so $\sin \theta = 1$. We won't worry about the direction of the torque, just the magnitude. The average stopping force is equal to $\bar{F}_{stop} = m\bar{a}$, where the average acceleration can be found through Equation 2.11: $\bar{a} = v_0^2/2\Delta x$. Here, v_0 is the initial speed, and Δx is the stopping distance. We have neglected the negative sign because we're only looking for magnitudes.

EVALUATE   The average torque exerted by the player on the ball is:

$$\bar{\tau} = \frac{rmv_0^2}{2\Delta x} = \frac{(63 \text{ cm})(0.145 \text{ kg})(42 \text{ m/s})^2}{2(5.00 \text{ cm})} = 1600 \text{ N·m}$$

ASSESS   One can arrive at the answer by using Equation 10.11: $\bar{\tau} = I\bar{\alpha}$. In this case, the rotational inertia is that of the ball rotating around the shoulder joint: $I = mr^2$. The average angular acceleration relates to the ball's average linear acceleration through Equation 10.5: $\bar{\alpha} = \bar{a}/r$, so the final expression is the same: $\bar{\tau} = rm\bar{a}$.

33. **INTERPRET**   This problem involves calculating the angular momentum of an object. We are given the mass distribution of the object, so we can find its rotational inertia, and we also know its angular velocity.

DEVELOP   Use Equation 11.4, $\vec{L} = I\vec{\omega}$, to compute the angular momentum. The rotational inertia of the weights and bar about the specified axis is (see Table 10.2)

$$I = 2m_{wt}\left(\frac{L}{2}\right)^2 + \frac{1}{12}m_{bar}L^2$$

EVALUATE   With $\omega = 10.0$ rpm $= 1.05$ rad/s, the angular momentum about this axis is

$$L = I\omega = \left[2(25\text{ kg})(0.8\text{ m})^2 + \frac{1}{12}(15\text{ kg})(1.6\text{ m})^2\right](1.05\text{ rad/s}) = 37\text{ J}\cdot\text{s}$$

ASSESS   The greater the angular speed, the larger the angular momentum.

35. **INTERPRET**   We need to find the angular momentum of a disk-shaped rotor that is part of a micromechanical device that measures blood flow.

DEVELOP   The angular momentum of the rotor is $L = I\omega$, where the rotational inertia is that of a disk: $I = \frac{1}{2}MR^2$. We don't explicitly know the rotor's mass, but the material is silicon, which has a density of $\rho = 2.33$ g/cm^3.

EVALUATE   The mass of the rotor is the density times the volume: $M = \rho(d\cdot\pi R^2)$, where d is the rotor's thickness. The radius is half the diameter: $R = 150\mu m$, and the 800-rpm rotational speed converted to SI units is: $\omega = 83.8$ rad/s. So the angular momentum of the rotor during the tests is

$$L = I\omega = \frac{\pi}{2}\rho dR^4\omega$$

$$= \frac{\pi}{2}(2.33\times10^3\text{ kg/m}^3)(2.0\times10^{-6}\text{ m})(150\times10^{-6}\text{ m})^4(83.8\text{ rad/s}) = 3.1\times10^{-16}\text{ J}\cdot\text{s}$$

ASSESS   This is a very small angular momentum, but we expect it to be. Otherwise, the device would significantly disturb the blood flow it is designed to measure.

37. **INTERPRET**   This problem asks us to calculate the rotational inertia of a tire if the design reduces the angular momentum by a certain percentage, while keeping the linear speed fixed.

DEVELOP   The linear speed of the car is related to its angular speed as $v = \omega r$ (see Equation 10.3). Keeping v fixed implies

$$\omega_1 r_1 = \omega_2 r_2$$

From Equation 11.4, $L = I\omega$, the new rotational inertia can be computed.

EVALUATE   The new specifications require that

$$\frac{L_2}{L_1} = \frac{I_2\omega_2}{I_1\omega_1} = 0.7 \quad\Rightarrow\quad \frac{I_2}{I_1} = 0.7\frac{\omega_1}{\omega_2}$$

Using $\omega_1 = \omega_2 r_2/r_1$, we obtain

$$I_2 = (0.70)I_1\frac{\omega_1}{\omega_2} = (0.70)I_1\frac{R_2}{R_1} = (0.70)(0.32\text{ kg}\cdot\text{m}^2)\left(\frac{35\text{ cm}}{38\text{ cm}}\right) = 0.21\text{ kg}\cdot\text{m}^2$$

ASSESS   The general condition is

$$\frac{L_2}{L_1} = \frac{I_2\omega_2}{I_1\omega_1} = \frac{I_2R_1}{I_1R_2} \quad\Rightarrow\quad L_2 = \left(\frac{I_2}{I_1}\right)\left(\frac{R_1}{R_2}\right)L_1$$

A decrease in angular momentum ($L_2 < L_1$) can be achieved by either decreasing r_1/r_2 or I_2/I_1. In our problem, the ratio $r_1/r_2 = (38\text{ cm})/(35\text{ cm}) = 1.09$ actually is increased. However, this change is accompanied by a greater decrease in rotational inertia $I_2/I_1 = (0.206\text{ kg}\cdot\text{m}^2)/(0.32\text{ kg}\cdot\text{m}^2) = 0.64$.

39. **INTERPRET**   This problem involves conservation of angular momentum, which we can use to calculate the motion of the dog relative to the ground.

DEVELOP   Walking once around relative to the turntable, the dog describes an angular displacement of $\Delta\theta_D$ relative to the ground, and the turntable one of $\Delta\theta_T$ in the opposite direction, such that $\Delta\theta_D - \Delta\theta_T = 2\pi$. The vertical component of the angular momentum of the dog-and-turntable system is conserved (which was zero initially), so

$$L_i = L_f \quad \Rightarrow \quad 0 = I_D\omega_D + I_T\omega_T = I_D\left(\frac{\Delta\theta_D}{\Delta t}\right) - I_T\frac{\Delta\theta_T}{\Delta t}$$

where the angular velocities (which are in opposite directions) have been rewritten in terms of the angular displacements and the common time interval. The rotational inertias about the axis of rotation are

$$I_D = mR^2 = (17 \text{ kg})(1.81 \text{ m})^2 = 55.7 \text{ kg}\cdot\text{m}^2$$

and $I_T = 95 \text{ kg}\cdot\text{m}^2$. These results allow us to solve for $\Delta\theta_D$.

EVALUATE   Eliminating $\Delta\theta_T$, we find

$$0 = I_D\Delta\theta_D - I_T(2\pi - \Delta\theta_D) = (I_D + I_T)\Delta\theta_D - 2\pi I_T$$

or

$$\frac{\Delta\theta_D}{2\pi} = \frac{I_T}{I_D + I_T} = \frac{95 \text{ kg}\cdot\text{m}^2}{55.7 \text{ kg}\cdot\text{m}^2 + 95 \text{ kg}\cdot\text{m}^2} = 0.63$$

In other words, $\Delta\theta_D$ is 63% of a full circle relative to the ground.

ASSESS   We find that $\Delta\theta_D$, the angular displacement relative to the ground, decreases with I_D. This is what we expect from conservation of angular momentum.

41. **INTERPRET**   This problem is about the rotational motion of the skaters, given their initial linear speed and radius of the circle they traverse. The aim is to keep the final linear speed and centripetal force below the stated maximums. The key concept here is conservation of angular momentum.

DEVELOP   If the ice is frictionless, the only external force on the skaters is the force that brings the end-skater to a sudden stop at a point we'll call P. (Note: The forces they exert on each other through their hands are internal forces.) The stopping force exerts no torque about point P, so the total angular momentum about a vertical axis through P is conserved. Initially, the other seven skaters are each moving with the same linear momentum $(p = mv_0)$ in a direction perpendicular to the line that connects them $(\sin\theta = 1)$. So from Equation 11.3, the angular momentum of each skater about P is

$$L_{0n} = |\vec{r}_n \times \vec{p}_n| = r_n(mv_0)\sin\theta = mv_0 r_n$$

where r_n is the distance between the n-th skater and the point P: $r_n = n(\ell/7)$ for $n = 1, 2, \ldots, 7$ and $\ell = 12 \text{ m}$. The total initial angular momentum is the sum $L_0 = \sum_{n=1}^{7} L_{0n}$, which will be conserved when the group starts rotating and has an angular momentum of $L_f = I\omega$. Here, the rotational inertia is $I = \sum_{n=1}^{7} mr_n^2$. From all this we can determine the rotational speed, which will give us the linear speed and centripetal force on the outside skater $(n = 7)$.

EVALUATE   The total initial angular momentum is

$$L_0 = \sum_{n=1}^{7} mv_0 r_n = \frac{mv_0\ell}{7}\sum_{n=1}^{7} n = \frac{mv_0\ell}{7}\left[\frac{7\times8}{2}\right] = 4mv_0\ell$$

where we have used $\sum_{n=1}^{N} n = N(N+1)/2$. Similarly, the rotational inertia of the 7 skaters around point P is

$$I = \sum_{n=1}^{7} mr_n^2 = \frac{m\ell^2}{49}\sum_{n=1}^{7} n^2 = \frac{m\ell^2}{49}\left[\frac{7\times8\times15}{6}\right] = \frac{20m\ell^2}{7}$$

where we have used $\sum_{n=1}^{N} n^2 = N(N+1)(2N+1)/6$. Since angular momentum is conserved $\left(L_0 = L_f\right)$, we can solve for the angular speed:

$$\omega = \frac{L_0}{I} = \frac{4mv_0\ell}{\frac{20}{7}m\ell^2} = \frac{7v_0}{5\ell}$$

The outside skater will have a tangential speed of $v = \omega\ell$, so in order to keep this below 8.0 m/s, the initial speed can't exceed:

$$v_0 = \tfrac{5}{7}v < \tfrac{5}{7}\left(8.0 \text{ m/s}\right) = 5.7 \text{ m/s}$$

The force on the outside skater's hand is the centripetal force: $F = ma_c = m\ell\omega^2$. To keep This below 300 N, the initial speed can't exceed:

$$v_0 = \frac{5}{7}\sqrt{\frac{F\ell}{m}} < \frac{5}{7}\sqrt{\frac{\left(300 \text{ N}\right)\left(12 \text{ m}\right)}{\left(60 \text{ kg}\right)}} = 5.5 \text{ m/s}$$

This limit is stricter than the one above. The greatest speed that the skaters can go before the rotational maneuver is 5.5 m/s.

ASSESS Notice that the outside skater will be going 1.4 times faster following the maneuver. By contrast, the skaters closer to the point P will slow down after the maneuver $\left(v_n = \omega r_n\right)$. This makes sense: to keep the total angular momentum constant, some skaters will gain angular momentum, while others will lose it.

43. **INTERPRET** This problem involves conservation of angular momentum, which we can use to find the angular speed of the bird feeder after the bird lands on it. We will need to consider the inertia of the bird feeder, which is given, and that of the bird, which we will take as a point particle of mass $m_b = 140$ g rotating at 19 cm from the axis. Since the bird and the feeder initially have opposite angular momenta with respect to the bird-feeder axis, it is possible that the direction of the feeder's angular momentum will change; so we will keep track of the direction by the sign of ω, which is the angular speed of the bird-feeder.
DEVELOP Apply conservation of angular momentum. The initial angular momentum is the sum of that due to the bird feeder and that due to the bird. Mathematically, this is expressed as

$$L_i = L_{bf} + L_b = I_{bf}\omega_{bf} + I_b\omega_b$$

If we define the initial angular velocity of the bird feeder as the positive direction, the $\omega_{bf} = 5.6$ rpm, $\omega_b = -v_b/r_{bf}$, where we have introduced the negative sign because the bird's initial angular velocity is opposite to that of the bird feeder. The rotational inertia of the bird can be taken as $m_b r_{bf}^2$. The final angular momentum is

$$L_f = L_{bfb} = \left(I_{bf} + I_b\right)\omega_{bfb}$$

where the subscript bfb indicates the bird-feeder-bird combination. By conservation of angular momentum, we can equate L_i and L_f and solve for the final angular speed, $\omega_{\beta\phi\beta}$.

EVALUATE Equating the initial and final angular momenta, we find

$$I_{bf}\omega_{bf} + \left(m_b r_{bf}^2\right)\left(-\frac{v_b}{r_{bf}}\right) = \omega_{bfb}\left(I_{bf} + m_b r_{bf}^2\right)$$

$$\omega_{bfb} = \frac{I_{bf}\omega_{bf} - m_b v_b r_{bf}}{I_{bf} + m_b r_{bf}^2}$$

$$= \frac{\left(0.12 \text{ kg·m}^2\right)\left(5.6 \text{ rpm}\right)\left(2\pi \text{ rad/rev}\right)\left(1 \text{ min}/60 \text{ s}\right) - \left(0.14 \text{ kg}\right)\left(1.1 \text{ m/s}\right)\left(0.19 \text{ m}\right)}{\left(0.12 \text{ kg·m}^2\right) + \left(0.14 \text{ kg}\right)\left(0.19 \text{ m}\right)^2}$$

$$= 0.329 \text{ rad/s} = 3.1 \text{ rpm}$$

ASSESS The sign of the final angular speed is the same as the sign of the initial angular speed, so the bird feeder continues to rotate in the same direction, albeit at a slower speed. Thus, the angular momentum of the bird feeder decreases, because it has absorbed the oppositely directed angular momentum of the bird.

45.   INTERPRET   The problem is about the rotational motion of the turntable. Tossing a piece of clay onto its surface is like a totally inelastic collision from Section 9.5. In this case, the total angular momentum is conserved.

DEVELOP   The forces that cause the clay to stick to the turntable are internal forces (i.e. between clay and turntable). There are no external forces that can generate a torque around the turntable's axis, so the angular momentum of the turntable/clay system in the vertical direction is conserved. If we take the sense of rotation of the turntable to define the positive direction of vertical angular momentum, then the system's initial angular momentum is

$$L_i = I\omega + mvd$$

where we assume here that the clay hits the turntable with the same direction that the table is turning. After the collision, the clay turns at the same speed as the table, so the final angular momentum is

$$L_f = I\omega_f + md^2\omega_f = (I + md^2)\omega_f$$

By conservation of angular momentum, the clay's initial velocity is equal to

$$v = d\omega_f + \frac{I(\omega_f - \omega)}{md}$$

EVALUATE   (a) If $\omega_f = \frac{1}{2}\omega$, then the clay hits the table with speed:

$$v = \frac{d\omega}{2} + \frac{-I\omega}{2md} = d\omega\left(\tfrac{1}{2} - I/2md^2\right)$$

(b) If $\omega_f = \omega$, then the clay hits the table with speed: $v = d\omega$.

(c) If $\omega_f = 2\omega$, then the clay hits the table with speed:

$$v = 2d\omega + \frac{I\omega}{md} = d\omega\left(2 + I/md^2\right)$$

ASSESS   We have written the clay velocity in terms of $d\omega$, which is the initial linear speed of the turntable at the radius where the clay hits. If the clay hits with $v < d\omega$, then the collision will slow down the turntable, but if $v > d\omega$, the turntable will speed up.

47.   INTERPRET   This problem asks us to calculate angular momenta given the mass distribution and rotational speeds (Appendix E) of the various planets of our solar system and the Sun. In particular, we are asked to estimate how much of the solar system's angular momentum about its center is associated with the Sun.

DEVELOP   The planets orbit the Sun in planes approximately perpendicular to the Sun's rotation axis, so most of the angular momentum in the solar system is in this direction. We can estimate the orbital angular momentum of a planet by mvr, where m is its mass, v its average orbital speed, and r its mean distance from the Sun. Compared to the orbital angular momentum of the four giant planets, everything else is negligible, except for the rotational angular momentum of the Sun itself, which can be estimated by assuming the Sun to be a uniform sphere rotating with an average period of $\frac{1}{2}(27 + 36)$ days. (The Sun's period of rotation at the surface varies from approximately 27 days at the equator to 36 days at the poles.)

EVALUATE   The numerical data in Appendix E results in the following estimates:

Orbital Angular Momentum (mvr)		%
Jupiter	19.2×10^{42} J·s	59.7
Saturn	7.85×10^{42} J·s	24.4
Uranus	1.69×10^{42} J·s	5.2
Neptune	2.52×10^{42} J·s	7.8
Rotational Angular Momentum ($\frac{2}{5}MR^2\omega$)		
Sun	0.89×10^{42} J·s	2.8
Total	32.2×10^{42} J·s	99.9

ASSESS   With $L_{orb} \gg L_{rot}$, we find that more than 97% of the total angular momentum of the solar system comes from the orbital angular momentum. In particular, the orbital motion of Jupiter alone accounts for roughly 60% of the total angular momentum.

49.   **INTERPRET**   This problem looks just like an inelastic collision, but instead of using conservation of linear momentum, we will use conservation of *angular* momentum. The angular momentum of each disk is in a single direction, so we can treat this as a one-dimensional problem.

DEVELOP   The masses of disk 1 and 2 are $m_1 = 440$ g and $m_2 = 270$ g, respectively. The radii are $r_1 = 0.035$ m and $r_2 = 0.23$ m. The initial angular speed of disk 1 is $\omega_1 = 180$ rpm . Use conservation of angular momentum, $L_i = L_f$, to find the final angular speed of both disks stuck together, and $\eta = 1 - K_f / K_i$, where $K = \frac{1}{2} I \omega^2$, to find the fraction of energy lost.

EVALUATE

(a) The initial angular momentum is

$$L_1 = I_1 \omega_{1i} + I_2 \overset{=0}{\omega_{2i}} = \frac{1}{2} m_1 r_1^2 \omega_{1i}.$$

The final angular momentum is

$$L_2 = \left(I_1 + I_2 \right) \omega_f = \left(\frac{1}{2} m_1 r_1^2 + \frac{1}{2} m_2 r_2^2 \right) \omega_f.$$

Conservation of angular momentum tells us that

$$L_1 = L_2$$

$$\frac{1}{2} m_1 r_1^2 \omega_{1i} = \left(\frac{1}{2} m_1 r_1^2 + \frac{1}{2} m_2 r_2^2 \right) \omega_f$$

$$\omega_f = \omega_{1i} \left(\frac{m_1 r_1^2}{m_1 r_1^2 + m_2 r_2^2} \right) = (180 \text{ rpm}) \left(\frac{(440 \text{ g})(3.5 \text{ cm})^2}{(440 \text{ g})(3.5 \text{ cm})^2 + (270 \text{ g})(2.3 \text{ cm})^2} \right) = 140 \text{ rpm}$$

(b) The initial kinetic energy is

$$K_i = \frac{1}{2} I_1 \omega_{1i}^2 + \frac{1}{2} I_2 \overset{=0}{\omega_{2i}^2} = \frac{1}{4} m_1 r_1^2 \omega_{1i}^2.$$

The final kinetic energy is

$$K_f = \frac{1}{2} \left(I_1 + I_2 \right) \omega_f^2 = \frac{1}{4} \left(m_1 r_1^2 + m_2 r_2^2 \right) \omega_f^2.$$

so the fraction of the initial kinetic energy lost to friction is

$$\eta = 1 - \frac{K_f}{K_i} = 1 - \frac{m_1 r_1^2 \omega_{1i}^2}{\left(m_1 r_1^2 + m_2 r_2^2 \right) \omega_f^2} = 1 - \frac{m_1 r_1^2 \, \omega_{1i}^2}{\left(m_1 r_1^2 + m_2 r_2^2 \right) \left[\omega_{1i} \left(\frac{m_1 r_1^2}{m_1 r_1^2 + m_2 r_2^2} \right) \right]^2}$$

$$= 1 - \frac{m_1 r_1^2 + m_2 r_2^2}{m_1 r_1^2} = \frac{m_2 r_2^2}{m_1 r_1^2} = \frac{(270 \text{ g})(2.3 \text{ cm})^2}{(440 \text{ g})(3.5 \text{ cm})^2} = 0.265 = 27 \%$$

to two significant figures.

ASSESS   Note that the fractional energy loss doesn't depend on the initial energy. For this particular set of disks, 27% of the initial energy will be lost in the collision regardless of how fast the bottom disk is spinning!

51.   **INTERPRET**   A solid spinning ball drops onto a frictional surface. At first it slides, but due to friction its spin will slow down and its linear speed will increase until it is purely rolling without sliding. We want to find the ball's angular speed when it begins purely rolling, and how long it takes.

DEVELOP   From the problem statement, we see that the ball's mass is M, its radius is R, and its initial angular velocity around the horizontal axis is ω_0. The coefficient of kinetic friction between the ball and the surface is μ_k, so the frictional force is $F_f = \mu_s F_n = \mu Mg$. A torque acts on the ball due to the frictional force, which acts on the edge of the ball. This torque $\tau = -\mu_s MgR$ serves to slow the ball's rotation. Use $\tau = I\alpha$ to find the angular acceleration α and then use $\omega = \omega_0 + \alpha t$ to find the resulting angular speed. The frictional force on the ball also accelerates the ball, so we can use $F = Ma$ and $v = v_0 + at$ to find the speed of the ball. Combining this with the fact that the ball is no longer sliding when $R\omega = v$ allows us to find the time it takes to achieve rolling motion.

EVALUATE   (a) The angular acceleration is

$$\alpha = \frac{\tau}{I} = \frac{-\mu_k MgR}{2MR^2/5} = -\frac{5\mu_k g}{2R}$$

where the negative sign comes from the fact that the frictional force always acts to counter the motion. Inserting this into the kinematic equation $\omega = \omega_0 + \alpha t$ gives

$$\omega = \omega_0 - \frac{5\mu_k g}{2R}t$$

Using the result from part (b) that $t = R\omega/\mu_k g$, we find that

$$\omega = \omega_0 - \frac{5\mu_k g}{2R}\left(\frac{R\omega}{\mu_k g}\right) = \omega_0 - \frac{5\omega}{2}$$

$$= \frac{2}{7}\omega_0$$

(b) The time it takes to achieve rolling motion is found from

$$a = \frac{F}{M} = \frac{\mu_k Mg}{M} = \mu_k g$$

so

$$v = v_0 + \mu_s gt$$

Inserting the condition $R\omega = v$ for rolling motion gives

$$R\omega = \mu_k gt$$

$$t = \frac{R\omega}{\mu_k g}$$

Using the result from part (a) that $\omega = 2\omega_0/7$, we find that

$$t = \frac{2R\omega_0}{\mu_k g}$$

ASSESS   The answer to part (a) is surprising—it says that no matter what the size or speed of the ball, or the coefficient of friction, the angular speed of the ball when it stops sliding is 2/7 of its original value! However, the time it takes the ball to achieve rolling motion depends on the radius of the ball, its initial angular speed, and the coefficient of kinetic friction. Notice that the more slippery is the surface (i.e., for smaller μ_k), the longer it will take for the ball to achieve rolling motion, which is reasonable.

INTERPRET   This problem

DEVELOP   The speed

EVALUATE   Using Equation

ASSESS   Angular

53.   **INTERPRET**   We're asked to derive the precession rate for a spinning gyroscope.
   DEVELOP   The torque is due to gravity. From Equation 11.2, it has a magnitude of $\tau = rF_g \sin\theta$, where θ is the angle between $\vec{L}$ and the vertical line extending up from the point where the gyroscope touches the bottom

support. By the right-hand rule, the torque points in the direction perpendicular to the plane defined by $\vec{r}$ and the vertical.

EVALUATE Over a short time interval, Δt, the angular momentum changes in the direction given by the torque: $\Delta \vec{L} = \vec{\tau} \cdot \Delta t$, as shown in Figure 11.9. This change in $\vec{L}$ corresponds to a change in the rotational axis, since $\vec{L} = I \vec{\omega}$. We can characterize how the axis moves with a small angle $\Delta \phi = \Delta L / L \sin \theta$, as defined in the figure below. The view here is from above looking down at the gyroscope.

After the axis moves, the torque points in a new direction, but always in the direction perpendicular to the plane defined by $\vec{r}$ and the vertical. This leads to circular motion with a rotational speed of

$$\omega_p = \frac{\Delta \phi}{\Delta t} = \frac{1}{\Delta t} \left(\frac{\Delta L}{L \sin \theta} \right) = \frac{\tau}{L \sin \theta} = \frac{mgr \sin \theta}{L \sin \theta} = \frac{mgr}{L}$$

ASSESS This says the precession speed will be faster if the gyroscope has a larger mass and/or a longer radial length. It also says that the rate is inversely proportional to the angular momentum. Since $L = I \omega$, we have $\omega_p \propto 1 / \omega$, which means that as the gyroscope gradually spins slower around its axis (due to friction forces), it will precess faster around the vertical. You may have observed this behavior in a gyroscope or a spinning top.

55. **INTERPRET** We are asked to determine what happens to a spinning gyroscope when different torques are applied to it.

DEVELOP Initially, the gyroscope has no torque on it, and the angular velocity and angular momentum both point to the right. By applying a force on the gyroscope between the arrowhead and disk, you exert a torque given by $\vec{\tau} = \vec{r} \times \vec{F}$ (Equation 11.2).

EVALUATE In this case, the force $\vec{F}$ points into the page and is applied at a radius $\vec{r}$ that points to the right. By the right-hand rule, the torque points upward. By Equation 11.5 $\left(d\vec{L} / dt = \vec{\tau} \right)$, the angular momentum will move in the torque's direction. Because the arrowhead points in the direction of the angular momentum, it too will move upward.

The answer is (d).

ASSESS It might seem odd that you push something in one direction, and it moves in a perpendicular direction. But this is just how the rotational analog of Newton's second law works.

57. **INTERPRET** We are asked to determine what happens to a spinning gyroscope when different torques are applied to it.

DEVELOP The added weight means the gyroscope is no longer balanced on the stand. There will be more downward force on the left-side than on the right-side. By the right-hand rule, this generates a torque that points out of the page.

EVALUATE This torque will cause the angular momentum to move slightly in the direction of the torque, i.e. out of the page. Recall Figure 11.9, where $\Delta \vec{L}$ points the same way as $\vec{\tau}$. This shift in the angular momentum will start the gyroscope turning in a clockwise direction as seen from above. As it moves, the torque will change so that the gyroscope continues to precess clockwise about the stand.

The answer is (d).

ASSESS One might have wrongly assumed that since the torque is out of the page it will "push" the left-hand side of the gyroscope (where the weight was added), thus resulting in a counter-clockwise rotation. The torque does not act on a specific point, but instead acts on the whole system through its angular momentum.

STATIC EQUILIBRIUM

EXERCISES

Section 12.1 Conditions for Equilibrium

15. **INTERPRET** We have been told that, for calculating the net torque, the choice of pivot point does not matter if the sum of the forces is zero (i.e., the object is in static equilibrium). For this problem, we will test this claim by calculating the torques from the previous problem about two new points.

DEVELOP The three forces are $\vec{F_1} = 2\hat{i} + 2\hat{j}$ N applied at point $(x, y) = (2\text{ m}, 0\text{ m})$, $\vec{F_2} = -2\hat{i} - 3\hat{j}$ N applied at $(-1\text{ m}, 0\text{ m})$, and $\vec{F_3} = 1\hat{j}$ N applied at $(-7\text{ m}, 1\text{ m})$. Use Equation 11.2 $\vec{\tau} = \vec{r} \times \vec{F}$ to find the torques due to these three forces around the points $(3\text{ m}, 2\text{ m})$ and $(-7\text{ m}, 1\text{ m})$. To find the value of $\vec{r}$ for an arbitray point, take the vector difference between the point where the force is applied and the point used as the pivot:

$$\vec{r} = (\vec{r}_{\text{applied}} - \vec{r}_{\text{pivot}}).$$

EVALUATE If applying the first force about the point $(3\text{ m}, 2\text{ m})$, the position vector is $\vec{r_1} = (2\text{ m} - 3\text{ m})\hat{i} + (0\text{ m} - 2\text{ m})\hat{j} = (-1\text{ m})\hat{i} + (-2\text{ m})\hat{j}$, so the torque due to $\vec{F_1}$ is

$$\vec{\tau_1} = \vec{r_1} \times \vec{F_1} \begin{vmatrix} \hat{i} & \hat{j} & \hat{k} \\ -1\text{ m} & -2\text{ m} & 0\text{ m} \\ 2\text{ N} & 2\text{ N} & 0\text{ N} \end{vmatrix} = \left[(-2+4)\text{ N}\cdot\text{m} \right]\hat{k} = (2\text{ N}\cdot\text{m})\hat{k}$$

For the second force, the position vector is $\vec{r_2} = (-1\text{ m} - 3\text{ m})\hat{i} + (0\text{ m} - 2\text{ m})\hat{j} = (-4\text{ m})\hat{i} + (-2\text{ m})\hat{j}$, so the torque due to $\vec{F_2}$ is

$$\vec{\tau_2} = \vec{r_2} \times \vec{F_2} \begin{vmatrix} \hat{i} & \hat{j} & \hat{k} \\ -4\text{ m} & -2\text{ m} & 0\text{ m} \\ -2\text{ N} & -3\text{ N} & 0\text{ N} \end{vmatrix} = \left[(12-4)\text{ N}\cdot\text{m} \right]\hat{k} = (8\text{ N}\cdot\text{m})\hat{k}$$

For the third force, the position vector is $\vec{r_3} = (-7\text{ m} - 3\text{ m})\hat{i} + (1\text{ m} - 2\text{ m})\hat{j} = (-10\text{ m})\hat{i} + (-1\text{ m})\hat{j}$, so the torque due to $\vec{F_2}$ is

$$\vec{\tau_3} = \vec{r_3} \times \vec{F_3} \begin{vmatrix} \hat{i} & \hat{j} & \hat{k} \\ -10\text{ m} & -1\text{ m} & 0\text{ m} \\ 0\text{ N} & 1\text{ N} & 0\text{ N} \end{vmatrix} = (-10\text{ N}\cdot\text{m})\hat{k}$$

Summing these torques to find the net torque gives $\vec{\tau}_{\text{net}} = (2+8-10)\hat{k}$ N·m = 0.

If applying the first force about the point $(-7\text{ m}, 1\text{ m})$, the position vector is $\vec{r_1} = (2\text{ m} + 7\text{ m})\hat{i} + (0\text{ m} - 1\text{ m})\hat{j} = (9\text{ m})\hat{i} + (-1\text{ m})\hat{j}$, so the torque due to $\vec{F_1}$ is

$$\vec{\tau_1} = \vec{r_1} \times \vec{F_1} \begin{vmatrix} \hat{i} & \hat{j} & \hat{k} \\ 9\text{ m} & -1\text{ m} & 0\text{ m} \\ 2\text{ N} & 2\text{ N} & 0\text{ N} \end{vmatrix} = (20\text{ N}\cdot\text{m})\hat{k}$$

For the second force, the position vector is $\vec{r_2} = (-1\text{ m} + 7\text{ m})\hat{i} + (0\text{ m} - 1\text{ m})\hat{j} = (6\text{ m})\hat{i} + (-1\text{ m})\hat{j}$, so the torque due to $\vec{F_2}$ is

$$\vec{\tau}_2 = \vec{r}_2 \times \vec{F}_2 \begin{vmatrix} \hat{i} & \hat{j} & \hat{k} \\ 6\,\text{m} & -1\,\text{m} & 0\,\text{m} \\ -2\,\text{N} & -3\,\text{N} & 0\,\text{N} \end{vmatrix} = (-20\,\text{N} \cdot \text{m})\,\hat{k}$$

For the third force, the position vector is $\vec{r}_3 = (-7\,\text{m} + 7\,\text{m})\,\hat{i} + (1\,\text{m} - 1\,\text{m})\,\hat{j} = (0\,\text{m})\,\hat{i} + (0\,\text{m})\,\hat{j}$, so the torque due to $\vec{F}_2$ is $\vec{\tau}_3 = 0$. Summing these torques to find the net torque gives $\vec{\tau}_{\text{net}} = (20 - 20 + 0\,\text{N} \cdot \text{m})\,\hat{k} = 0$.

ASSESS Note that the torque due to force 3 around the second pivot is zero because the force acts at the pivot.

Section 12.2 Center of Gravity

17. **INTERPRET** This problem involves finding the center of gravity of an object and calculating the torque that gravity applies on an object about several different points. We can treat gravity as if it acts only at the center of gravity of the object, and so calculate the position vectors from this point.

DEVELOP Make a drawing of the plate that shows the center of gravity and the position vectors from each point to the center of gravity (see figure below). By symmetry, the center of gravity is located at the center of the uniform plate. The force due to gravity acts at the center of gravity in the downward direction. Apply Equation 11.2, or its scalar analog, Equation 10.10, to find the torque due to gravity about the three points given.

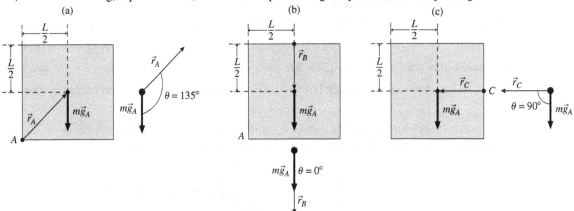

EVALUATE (a) For point A, the angle θ between the position vector and the force vector is 135°, so the magnitude of the torque is

$$\tau = rF\sin\theta = \frac{L}{\sqrt{2}}mg\sin(135°) = \frac{mgL}{2}$$

(b) For point B, the angle θ between the positioin vector and the force vector is 0°, so the torque is zero because $\sin(0°) = 0$.

(c) For point C, the angle θ between the position vector and the force vector is 90°C, so the torque is

$$\tau = rF\sin\theta = \frac{L}{2}mg\sin(90°) = \frac{mgL}{2}$$

and the magnitude of the torque is $mgL/2$.

ASSESS The torque has the same magnitude in (a) and (b), but acts in opposite directions. You can confirm this by the right-hand rule.

19. **INTERPRET** The log is in equilibrium under the torques exerted by the cable, gravity, and the wall.

DEVELOP Because the tree is in equilibrium, we know that the sum of the torques is zero. Since we don't know the force exerted by the wall at the right end of the tree, we choose the contact point with the wall as our pivot point, so that the torque from this force is zero. As for the given forces, gravity acts at the center of gravity, producing a torque of $\tau_g = x_{\text{CG}}F_g$, where x_{CG} is the unknown distance between the wall and the center of gravity. See the figure below. The cable's tension produces a torque in the opposite direction: $\tau_c = -x_cT$, where

$$x_c = 23\text{m} - 4.0\text{m} = 19\text{m}.$$

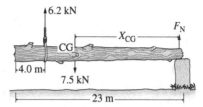

EVALUATE   Setting the sum of the torques to zero gives $\tau_g = -\tau_c$, from which we can find the center of gravity:

$$x_{\text{CG}} = \frac{x_c T}{F_g} = \frac{(19\text{m})(6.2\text{ kN})}{(7.5\text{ kN})} = 16\text{ m}$$

where this is relative to the wall.

ASSESS   This seems reasonable. As expected, the center of gravity is between the cable and the wall, otherwise the tree couldn't possibly be in equilibrium.

Section 12.3 Examples of Static Equilibrium

21.   **INTERPRET**   The interpretation of this problem is the same as for the preceding problem, except that the position of the child must be adjusted so that the scale at the right end of the board will indicate the given quantities.

DEVELOP   The development of this problem is the same as for Problem 20.

EVALUATE   (a) Starting with the same relationship as for Problem 20, we now insert $F_s = 100$ N to find the position of the child. The result is

$$x_c = \frac{m_b g x_{\text{cm}} - F_s x_s}{m_c g} = \frac{(60\text{ kg})(9.8\text{ m/s}^2)(0.40\text{ m}) - (100\text{ N})(1.6\text{ m})}{(40\text{ kg})(9.8\text{ m/s}^2)} = 0.19\text{ m}$$

from the pivot point, or x = 0.80 m − 0.19 m = 0.61 m from the left end of the board.

(b) With $F_s = 300$ N, we have

$$x_c = \frac{(60\text{ kg})(9.8\text{ m/s}^2)(0.40\text{ m}) - (300\text{ N})(1.6\text{ m})}{(40\text{ kg})(9.8\text{ m/s}^2)} = -0.62\text{ m}$$

from the pivot point (i.e., 0.62 m to the right of the pivot point), or x = 0.80 m − (−0.62 m) = 1.42 m from the left end of the board.

ASSESS   The answer to part (b) is reasonable, because if there were no child present, the weight of the board alone could only supply a force

$$F_s = m_b g \left(\frac{x_{\text{cm}}}{x_s}\right) = (60\text{ kg})(9.8\text{ m/s}^2)\left(\frac{0.40\text{ m}}{1.6\text{ m}}\right) = 150\text{ N}$$

to two significant figures. Thus, with the child to the right of the pivot point, we expect the scale reading to be greater than 150 N, which corresponds to what we find in part (b).

23.   **INTERPRET**   This is a problem of static equilibrium, so we know that the sum of the forces must be zero and that the sum of the torques about any point must be zero. We can use these concepts to find the weight of the sumo wrestler standing at the given position.

DEVELOP   Make a sketch of the situation that shows the pivot point, the forces, and where the forces are applied (see figure below). We chose to calculate the torque about the pivot point P, so this will serve as the origin for measuring distances. Applying the condition (Equation 12.2) for equilibrium gives

$$\left(\sum \tau\right)_{\text{P}} = 0 = \overset{=0}{\overbrace{\vec{r}_{\text{P}} \times \vec{F}_{\text{P}}}} + \vec{r}_w \times \vec{F}_w + \vec{r}_{\text{cm}} \times \vec{F}_b + \vec{r}_s \times \vec{F}_s = x_w w_w + x_{\text{cm}} m_b g + x_s F_s$$

where the subscript w refers to the wrestler, b to the board, and s to the scale. We can solve for equation for the upward force F_s exerted by the scale.

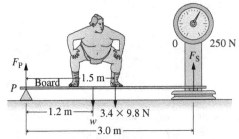

EVALUATE   Inserting the given quantities into the expression for net torque gives

$$x_w w_w = -x_{cm} m_b g + x_s F_s$$

$$w_w = \frac{-x_{cm} m_b g + x_s F}{-x_w} = \frac{-(1.5\text{ m})(3.4\text{ kg})(9.8\text{ m/s}^2) + (3.0\text{ m})(210\text{ N})}{-1.2\text{ m}} = 480\text{ N}$$

to two significant figures.

ASSESS   In pounds, the wrestler weighs about 220 pounds. The maximum scale reading of 250 N is equivalent to about 110 pounds.

Section 12.4 Stability

25.   **INTERPRET**   We are given a potential function, and are asked to find the positions of any stable and unstable equilibria.

DEVELOP   The condition for equilibrium is that $dU/dx = 0$, and the equilibrium point is stable if $d^2U/dx^2 > 0$ and unstable if $d^2U/dx^2 < 0$ The function we are given is $U(x) = 2x^3 - 2x^2 - 7x + 10$.

EVALUATE   We take the derivative and equate it to zero to find the locations of any equilibrium points.

$$\frac{dU}{dx} = \frac{d}{dx}(2x^3 - 2x^2 - 7x + 10) = 6x^2 - 4x - 7 = 0$$

$$x = \frac{-(-4) \pm \sqrt{(-4)^2 - 4(6)(-7)}}{2(6)} = \left\{ \frac{2 - \sqrt{46}}{6}, \quad \frac{2 + \sqrt{46}}{6} \right\}$$

There are equilibrium points at $x_1 = (2 - \sqrt{46})/6 \approx -0.797\text{ m}$ and $x_2 = (2 + \sqrt{46})/6 \approx 1.46\text{ m}$.

Next, we find the second derivative of U, and evaluate it at these two points to ascertain their stability.

$$\frac{d^2U}{dx^2} = \frac{d^2}{dx^2}(2x^3 - 2x^2 - 7x + 10) = \frac{d}{dx}(6x^2 - 4x - 7) = 12x - 4$$

$$\text{for } x = \frac{2 - \sqrt{46}}{6} \quad \Rightarrow \quad \frac{d^2U}{dx^2} = -13.6 < 0$$

$$\text{for } x = \frac{2 + \sqrt{46}}{6} \quad \Rightarrow \quad \frac{d^2U}{dx^2} = 13.6 > 0$$

Thus, the first solution is unstable, and the second is stable.

ASSESS   The potential-energy function is plotted in the figure below, where it can be seen that the points we found are indeed equilibria. The point at –0.797 is unstable, and the point at 1.46 is metastable.

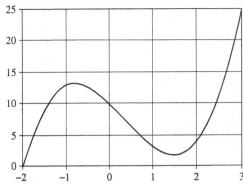

PROBLEMS

27.   **INTERPRET**   This problem involves finding the torque about a pivot point given the forces and the positions at which they are applied. This is also a problem concerning equilibrium, in which the net force and torque on an object must be zero. We can use this to find the force required of the deltoid muscle to keep the arm in equilibrium.

DEVELOP   The figure below shows the forces involved and where they are applied. To find the torque about the shoulder due to the arm and the 6-kg mass, we sum the torques. This gives

$$\tau_a = r_{cm} m_{arm} g \sin\theta + r_{arm} m g \sin\theta$$

where $r_{cm} = 21$ cm, $r_{arm} = 56$ cm , $m_{arm} = 4.2$ kg, $m = 6.0$ kg, and $\theta = 105°$. To find the tensile force required of the deltoid muscle, we require that all the torques (including now the torque due to the deltoid muscle) sum to zero (this is the condition for equilibrium, see Equation 12.2). This gives

$$\left(\sum \tau\right)_{shoulder} = 0 = r_d T \sin\theta_d + r_{cm} m_{arm} g \sin\theta + r_{arm} m g \sin\theta = r_d T \sin\theta_d + \tau_a$$

where $\theta_d = 170°$ and $r_d = 18$ cm. We can solve this for the tensile force T of the deltoid muscle.

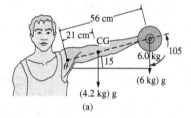

(a)

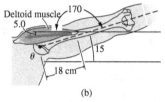

(b)

EVALUATE   (a) The torque due to the arm and the mass is

$$\tau_a = r_{cm} m_{arm} g \sin\theta + r_{arm} m g \sin\theta = \left[(0.21 \text{ m/s})(4.2 \text{ kg}) + (0.56 \text{ m})(6.0 \text{ kg})\right](9.8 \text{ m/s}^2)\sin(255°) = 4.0 \times 10^1 \text{ N·m}$$

to two significant figures.

(b) The magnitude of the tensile force supplied by the deltoid is thus

$$T = \frac{\tau_a}{r_d \sin\theta_d} = \frac{40.2 \text{ N·m}}{(0.18 \text{ m})\sin(170°)} = 1.3 \text{ kN}$$

ASSESS   By the right-hand rule, we see that the torque supplied by the deltoid muscle is out of the page, whereas the torque due to the arm and the mass is into the page. Thus, the deltoid muscle must supply a force that is (1.3 kN)/[(9.8 m/s2)(6.0 kg)] = 22 times weight in order to lift it. This is not very efficient, and it underscores the comment at the end of Example 12.3 that the skeleto-muscular structure of the human extremities evolved for speed and range of motion, not mechanical advantage.

29.   **INTERPRET**   You need to specify the minimum horizontal force needed to push the cart over the step. The cart will be in static equilibrium up until the force is sufficient to overcome the obstacle.

DEVELOP   The forces that you need to account for are: the force, F, exerted by the person; the weight of the cart, mg; the normal force, n, between the wheel and the ground, and the force from the step, F_s. You can assume that the person is only pushing and not lifting the cart in any way. In any case, the person's force and cart's weight are both exerted on the wheel at the wheel's axle. See the figure below.

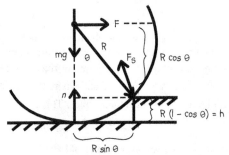

Since you don't know the magnitude or the direction of the step force, you should choose as the pivot point where the wheel meets the step. Before the cart starts moving, the wheel is in static equilibrium, so the sum of the torques around this pivot point should be zero:

$$\left(\sum \tau\right)_{step} = mgR\sin\theta - nR\sin\theta - FR\cos\theta = 0$$

where we have used the trig identity: $\sin(90° - \theta) = \cos\theta$. As the person pushes harder, more of the cart's weight shifts from the ground (supported by n) to the step (supported by F_s). Eventually, the normal force will go to zero, and the wheel will rotate around the point where it meets the step. Therefore, the minimum force needed to push the cart over the step is that which makes $n = 0$:

$$mgR\sin\theta - FR\cos\theta = 0 \quad \rightarrow \quad F = mg\tan\theta$$

EVALUATE   To find the minimum force, we need to find the angle θ. We know that the step height is equal to: $h = R(1 - \cos\theta)$. So

$$\theta = \cos^{-1}\left[1 - \frac{h}{R}\right] = \cos^{-1}\left[1 - \frac{8\text{ cm}}{\frac{1}{2} \cdot 60\text{ cm}}\right] = 42.83°$$

Plugging this into the force equation from above:

$$F = mg\tan\theta = (55\text{ kg})(9.8\text{ m/s}^2)\tan(42.83°) = 500\text{ N}$$

ASSESS   The person is essentially using the wheel as a lever arm to lift the loaded cart up onto the step. The wider the wheel, the longer the lever arm. This is reflected in the fact that a larger R results in a smaller θ and a smaller F.

31.   **INTERPRET**   This problem involves calculating the force required to maintain a leaning ladder in equilibrium. The ladder experiences forces due to gravity (acting on the ladder and on the person climbing the ladder), a normal force due to the frictionless wall against which it is leaning, and forces due to the floor (friction and normal force). Using the conditions for equilibrium (i.e., sum of the forces and torques must be zero), we are asked to find how high a person of a given mass may climb the ladder before the ladder slips.

DEVELOP   Draw a sketch of the ladder that includes all the forces acting on it (see figure below). For the ladder to remain in equilibrium, the forces on it must sum to zero. Summing the forces in the x and y directions, this condition leads to

$$\sum F_x = 0 = f_s - F_{wall} = \mu_s n - F_{wall}$$
$$\sum F_y = 0 = n - m_L g - mg$$

where m is the mass of the person climbing the ladder, and we have expressed the force due to static friction as $f_s = \mu_s n$. To remain in equilibrium, the torques on the ladder must also sum to zero, which gives

$$\left(\sum \tau\right)_A = m_L g \frac{L}{2} \overbrace{\sin(\pi - \theta)}^{\sin(\theta)} + mgr \overbrace{\sin(\pi - \theta)}^{=\sin(\theta)} - F_{wall} L \overbrace{\sin\left(\frac{\pi}{2} + \theta\right)}^{=\cos(\theta)}$$

$$= \frac{m_L g}{2}\sin(\theta) + mg\alpha\sin(\theta) - F_{wall}\cos(\theta)$$

where $\alpha = L/r$ is the position of the person on the ladder, expressed in units of ladder lengths. For the ladder to not slip, the force applied by the wall must be less than or equal to the force due to friction:

$$F_{wall} \leq f_s$$

Using the equations we obtained by imposing the conditions for equilibrium, we obtain

$$\left(\frac{m_L g}{2} + mg\alpha\right)\tan(\theta) \le \mu_s n = \mu_s\left(m_L g + mg\right)$$

$$\alpha \le \frac{\mu_s}{m}\left(m_L + m\right)\cot\theta - \frac{m_L}{2m}$$

The maximum value of α is 1 (this corresponds to the person climbing to the top of the ladder, so $\alpha = L/L = 1$), so if the right-hand side of the expression above is less than unity, the person can climb to the top of the ladder without it slipping. Explicitly, the condition for the person to be able to climb to the top of the ladder is

$$1 \le \frac{\mu_s}{m}\left(m_L + m\right)\cot\theta - \frac{m_L}{2m}$$

The most massive person that can climb to the top of the ladder may be found by setting $\alpha = 1$ and solving for the mass m. This gives

$$1 = \frac{\mu_s}{m}\left(m_L + m\right)\cot\theta - \frac{m_L}{2m}$$

$$m = \frac{\mu_s m_L\left(\cot\theta - 1/2\right)}{1 - \mu_s\cot\theta}$$

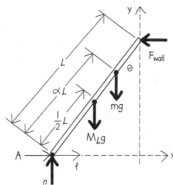

EVALUATE Inserting the values given, we find

$$1 \le \frac{\mu_s}{m}\left(m_L + m\right)\cot\theta - \frac{m_L}{2m} = \frac{0.26}{65\text{ kg}}\left(5.0\text{ kg} + 65\text{ kg}\right)\cot\left(15°\right) - \frac{5.0\text{ kg}}{2\left(65\text{ kg}\right)} = 1.0$$

to two significant figures. Thus, the inequality is satisfied, so the 65-kg person can climb to the top of the ladder without it slipping. The most massive person that can climb the ladder is

$$m = \frac{m_L\left(\mu_s\cot\theta - 1/2\right)}{1 - \mu_s\cot\theta} = \frac{\left(5.0\text{ kg}\right)\left[\left(0.26\right)\cot\left(15°\right) - 0.50\right]}{1 - \left(0.26\right)\cot\left(15°\right)} = 79\text{ kg}$$

ASSESS We see that if the ladder is more massive, a more massive person may climb it. This is because the more massive ladder will generate a greater force due to friction on the floor. If $\mu_s \to 0$, then $m \to 0$, as expected.

33. **INTERPRET** The problem involves static equilibrium, so we can apply the condition that the sum of the forces and torques on the board must be zero.

DEVELOP Sketch the board, showing all the forces acting on it and the position at which the forces are applied (see figure below). The conditions for equilibrium (about the origin drawn on the figure) are:

$$\sum F_x = 0 = T_2\sin\left(60°\right) - T_1\sin\left(35°\right)$$

$$\sum F_y = 0 = T_2\cos\left(60°\right) + T_1\cos\left(35°\right) - w - W$$

$$\left(\sum\tau\right)_0 = 0 = T_2 L\sin\left(20.8°\right) - WL\sin\left(99.2°\right)/2$$

which we can solve for the weight w of the box.

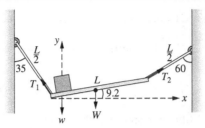

EVALUATE   From the condition of zero torque, we find that

$$T_2 = \frac{W \sin(99.2°)}{2 \sin(20.8°)} = (1.39)W$$

Inserting this result into the x component of the zero-force condition gives

$$T_1 = \frac{T_2 \sin(60°)}{\sin(35°)} = \frac{(1.39)W \sin(60°)}{\sin 35°} = (2.10)W$$

Inserting both these results into the y component of the zero-force condition gives

$$w = T_2 \cos(60°) + T_1 \cos(35°) - W = W\left[(2.1)\cos(60°) + 1.39 \cos(35°) - 1\right] = 1.4\ W$$

ASSESS   Drawing an accurate sketch is helpful for this problem.

35.   **INTERPRET**   This problem is about maintaining a ladder in static equilibrium. The ladder leans against a wall, as in Example 12.2, but the wall now has friction. Thus, in addition to the forces listed in Example 12.2, we must add the force due to static friction at the top of the ladder. This force will resist motion, so it is directed upwards.

DEVELOP   Make a sketch of the situation that shows all the forces acting on the ladder (see figure below). The force f_2 due to static friction at the top of the ladder is given by Equation 5.2, $f_2 \le \mu_{2,s} n_2$. As per Example 12.2, we apply the two conditions for static equilibrium (i.e., zero net force and zero net torque on the object). The net force in the x direction has not changed, but the net force in the y direction has changed due to the addition of f_2. Thus, the net force equilibrium equations read

$$\text{Force, } x \quad \mu_1 n_1 - n_2 = 0$$
$$\text{Force, } y \quad n_1 - mg + f_2 = 0$$

Similarly, the equilibrium equation now reads

$$Ln_2 \sin\phi - \frac{L}{2}mg\cos\phi + Lf_2 \overbrace{\sin(\pi/2 - \phi)}^{=\cos\phi} = 0$$

$$Ln_2 \sin\phi - \frac{L}{2}mg\cos\phi + Lf_2 \cos\phi = 0$$

where the last term is due to the torque by the new friction force. Using the inequalities $f_2 \le \mu_2 n_2$ and $f_1 \le \mu_1 n_1$ we can now find the minimum angle ϕ at which the ladder can be positioned without slipping.

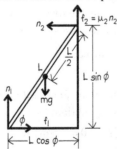

EVALUATE   We will take the maximum values for the friction forces, which will give the maximum angle ϕ at which the ladder can lean. Solving for ϕ from the torque equation, we find

$$\tan\phi = \frac{mg - 2f_2}{2n_2} = \frac{mg - 2\mu_2 n_2}{2n_2} = \frac{mg}{2n_2} - \mu_2$$

From the horizontal force equation, we have $\mu_1 n_1 = n_2$. Inserting this into the horizontal equation leads to

$$mg = n_1 + f_2 = n_1 + \mu_2 n_2 = n_1 + \mu_2 \mu_1 n_1 = n_1 \left(1 + \mu_2 \mu_1\right)$$

Using these two results in the expression for $\tan\phi$ gives

$$\tan\phi = \frac{n_1\left(1+\mu_2\mu_1\right)}{2\mu_1 n_1} - \mu_2 = \frac{\left(1+\mu_1\mu_2\right)}{2\mu_1} - \frac{2\mu_1\mu_2}{2\mu_1} = \frac{1-\mu_1\mu_2}{2\mu_1}$$

$$\phi = \text{atan}\left[\frac{1-\mu_1\mu_2}{2\mu_1}\right]$$

ASSESS   Does this make sense? The units of the argument for atan are dimensionless, as they should be. If we let $\mu_1 \to 0$, then $\phi \to 90°$, meaning the ladder can only remain in equilibrium if it is vertical. Letting $\mu_2 \to 0$, we find the result of Example 12.2, as expected.

37.   **INTERPRET**   This problem is an equilibrium problem, as long as the log does not slip! We use the conditions for equilibrium to find the maximum allowable mass for a climber standing at the upper end of the log.

DEVELOP   We start by drawing a free-body diagram showing all of the forces on the log, as shown in the figure below. The maximum frictional force on the lower end of the log is $f = \mu_s n_1$. Because the log is in equilibrium, this frictional force must be balanced by the normal force n_w from the wall (see Equation 12.1), so if the force from the wall is *greater* than the maximum frictional force, then the log will slip. We can find the force due to the wall by using Equation 12.2, $\sum \tau = 0$ with the lower end of the log as the pivot point. Because the vertical forces must sum to zero (Equation 12.1), we see from the figure that $n_1 = (M + m)g$, where $M = 340$ kg is the mass of the log, and m is the unknown mass of the climber. The length of the log is $L = 6.3$ m and the log forms an angle $\theta = 27°$ with the horizontal. The center of mass of the log is located $L/3$ from the left end, and the coefficient of friction between the left end of the log and the ground is $\mu_s = 0.92$.

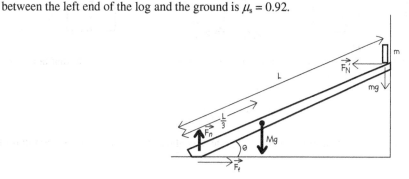

EVALUATE   From the torque condition for equilibrium, we find

$$\sum \vec{\tau} = 0 = mgL\cos\theta + \frac{MgL}{3}\cos\theta - n_w L\sin\theta$$

$$n_w = \left(m + M/3\right)g\cot\theta$$

From the equilibrium equation for the horizontal forces, we find

$$f = n_w$$

$$\mu n_1 = \mu\left(m+M\right)g = \left(m+M/3\right)g\cot\theta$$

$$\mu m + \mu M = \frac{m}{\tan\theta} + \frac{M}{3\tan\theta}$$

$$m\left(\mu - \frac{1}{\tan\theta}\right) = M\left(\frac{1}{3\tan\theta} - \mu\right)$$

$$m = M\left(\frac{\mu - \cot\theta}{\frac{1}{3}\cot\theta - \mu}\right) = M\left(\frac{3(0.92)\tan\left(27°\right)-3}{1-3\tan\left(27°\right)}\right) = 3.0M = 1.0\times10^3\,\text{kg}$$

ASSESS   This mass limit is large enough that any climber can safely cross.

39.   **INTERPRET**   In this problem we are given a block that is balancing on its corner so that its long side makes an angle θ with the horizontal, and we want to find the values of θ for which the block is in equilibrium. We are to comment on whether these are stable or unstable equilibria.

DEVELOP   Let the block be tilted in a plane perpendicular to its thickness, as shown in the figure below. The two conditions for equilibrium (see Equations 12.1 and 12.2) are that (1) there is zero net force on the block and (2) there is zero net torque on the block:

$$\sum F = 0$$
$$(\sum \tau)_P = 0$$

If the block is lying on its long side we can consider the normal force from the floor to act directly under the center of gravity, as shown in the sketch (because the block is uniform, so the normal force over the entire surface may be replaced by a sum of the normal forces acting at the geometric center of the block). This force cancels the force due to gravity. Furthermore, there is no torque on the block since the two forces (normal and gravity) are antiparallel. If the block is lying on its short side, the situation is the same, except that it would be easier to destabilize the block because the pivot point is closer to the line along which gravity acts. If the block is balanced on its corner, as in the middle sketch, then the zero-torque condition demands that the force due to gravity act through the pivot point, so $\alpha + \theta = 90°$.

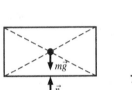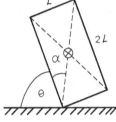

EVALUATE $\alpha + \theta = 90°$ is an unstable equilibrium. The angle θ may be found by noting that $\tan\alpha = L/(2L) = 1/2$, so

$$\theta = 90° - \operatorname{atan}\left(\frac{1}{2}\right) = 63.4°$$

ASSESS   For a rectangular block of length L and width W, one may show that the general expression for the torque about the pivot point is given by

$$\tau = r_\perp F = \frac{mg}{2}\left(L\cos\theta - W\sin\theta\right)$$

Thus, the condition that $\tau = 0$ implies that

$$\tan\theta = \frac{L}{W}$$

In our case, we have $L/W = 1/2$, which gives $\theta = 63.4°$.

41.   **INTERPRET**   In this problem a cubical block is placed on an incline. Given the coefficient of friction between the block and the incline, we'd like to find out whether the block first slides or tips when the angle of the incline is increased. The condition for tipping forward is that the sum of the counter-clockwise torques about the leading corner becomes nonzero. The condition for sliding is that the sum of the forces parallel to the incline becomes nonzero.

DEVELOP   We suppose that the block is oriented with two sides parallel to the direction of the incline, and that its center of mass is at the center. Make a sketch of the situation that shows all the forces and the positions at which they act (see figure below). The condition for sliding is that the sum of the forces parallel to the incline becomes nonzero. Taking the direction down the incline to be positive, this condition gives

$$mg\sin\theta_{\text{slide}} - f_s^{\text{max}} > 0$$

$$mg\sin\theta_{\text{slide}} > \mu_s n = \mu_s mg\cos\theta$$

$$\tan\theta_{\text{slide}} > \mu_s$$

where we have used Equation 5. 2 for static friction ($f_s \le \mu_s n$) and we have applied Newton's second law ($F = ma$) in the direction normal to the incline to find $mg\cos\theta_{\text{slide}} = n$. The condition for tipping over is that the center of mass lies to the left of the lower corner of the block (see sketch), or $\theta_{\text{tip}} > 45°$. Compare θ_{tip} to θ_{slide} to see if the cube will slide first or tip.

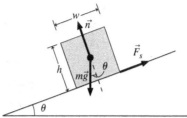

EVALUATE   The block will slide at the angle

$$\theta_{\text{slide}} = \text{atan}\left(\mu_s\right) = \text{atan}\left(0.95\right) = 43.5° < \theta_{\text{tip}}$$

Thus, the cube slides before tipping.

ASSESS   For a general rectangular solid, the angle at which the block will tip is $\theta = \text{atan}\,(h/w)$, where w is the width and h is the height of the object. We find that sliding happens first if $\mu_s < w/h$. This makes sense because when the coefficient of friction is small, the block has a greater tendency to slide. On the other hand, when the coefficient of friction is large ($\mu_s > w/h$), we'd expect tipping to take place first. Another way to understand this is to consider a given coefficient of static friction, and let the width w decrease. This will make the block tip because the ratio w/h will decrease to less than μ_s, as expected for the geometry of a tall, thin block placed on an incline.

43.   **INTERPRET**   In this problem a ladder is leaning against the wall and we want to find the mass of the heaviest person who can climb to the top of the ladder while keeping it in static equilibrium. We will therefore apply the conditions for static equilibrium; namely, that the sums of the forces and torques on the ladder are zero.

DEVELOP   The forces on the uniform ladder are shown in the sketch below, with the force exerted by the (frictionless) wall being horizontal. Consider a person who has climbed up the ladder a fraction α of its length. Equilibrium conditions (Equations 12.1 and 12.2) require that

$$0 = \sum F_x = f - F_{\text{wall}}$$

$$0 = \sum F_y = n - \left(m_L + m\right)g$$

$$0 = \left(\sum \tau\right)_A = F_{\text{wall}}L\overbrace{\sin\left(\pi - \theta\right)}^{=\cos\theta} + \left(\frac{L}{2}\right)m_L g\overbrace{\sin\left(\theta + \pi/2\right)}^{=-\sin\theta} + mg\alpha L\overbrace{\sin\left(\theta + \pi/2\right)}^{=-\sin\theta}$$

$$= F_{\text{wall}}L\cos\theta - \left(\frac{L}{2}\right)m_L g\sin\theta - mg\alpha L\sin\theta$$

The ladder will not slip if $f \le \mu_s n$. Using the equations above, this condition can be rewritten as

$$f = F_{\text{wall}} = \left(\frac{1}{2}m_L + \alpha m\right)g\tan\theta \le \mu_s n = \mu_s\left(m_L + m\right)g$$

or

$$\alpha \le \frac{\mu_s\left(m_L + m\right)\cot\theta - m_L/2}{m} = \mu_s\cot\theta + \frac{m_L}{m}\left(\mu_s\cot\theta - \frac{1}{2}\right)$$

Here, we used the horizontal force equation to find f, the torque equation to find F_{wall}, and the vertical force equation to find n.

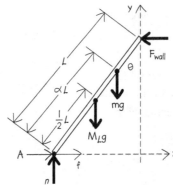

EVALUATE   For a person at the top of the ladder, $\alpha = 1$, and the condition for no slipping becomes

$$m \le m_{\rm L}\left(\frac{\mu_{\rm s} \cot\theta - 1/2}{1 - \mu_{\rm s} \cot\theta} \right)$$

With the data given for the ladder [note that $\cot\theta = \cot\left(\pi/2 - 66°\right) = \tan\left(66°\right)$], we obtain

$$m \le (9.5 \text{ kg})\frac{(0.42)\tan\left(66°\right) - 1/2}{1 - (0.42)\tan\left(66°\right)} = 74 \text{ kg}$$

ASSESS   The above equation shows that when the coefficient of friction becomes too small, $\mu_{\rm s} \cot\theta < 1/2$, or $\mu_{\rm s} < \tan\theta/2$ (see Example 12.2), slipping will occur and it's no longer possible for the ladder to remain in static equilibrium. In this situation, nobody can climb up to the top of the ladder without making the ladder slip, regardless of his or her mass.

45. **INTERPRET**   We want to find the energy required to bring a cube of side s from a stable to an unstable equilibrium. The problem is equivalent to finding the increase in potential energy of the system. We can consider that the entire mass of the cube is concentrated at its center of gravity, which is its geometric center for a uniform cube in a uniform gravitational field.

DEVELOP   When resting in a stable equilibrium position, the center of mass of a uniform cube of side s is at a distance $y_0 = s/2$ above the tabletop. When balancing on a corner, the center of mass is now a distance

$$y = \sqrt{\left(s/2\right)^2 + \left(s/2\right)^2 + \left(s/2\right)^2} = \frac{\sqrt{3}}{2}s$$

above the corner resting on the tabletop.

EVALUATE   From the above, the potential energy difference is

$$\Delta U = mg\,\Delta y_{\rm cm} = mgs\left(\frac{\sqrt{3}}{2} - \frac{1}{2} \right) = 0.366\ mgs$$

This is the energy required to bring the cube to an unstable equilibrium.

ASSESS   Raising the vertical distance of the center of mass increases the potential energy of the cube. In general, the stability of a system decreases as its potential energy is increased.

47. **INTERPRET**   This problem deals with a ladder leaning against a frictionless wall, and we want to verify the condition under which any person (with any mass) can climb to the top, and also the condition in which nobody can climb to the top. The conditions of equilibrium (Equations 12.1 and 12.2) will apply.

DEVELOP   The forces on the uniform ladder are shown in the sketch below. The person of mass m is positioned on the ladder a fraction α of its total length L from the bottom. Equilibrium conditions of zero net force and zero net torque require that

$$0 = \sum F_x = f - F_{\rm wall}$$

$$0 = \sum F_y = n - \left(m_{\rm L} + m\right)g$$

$$0 = \left(\sum \tau\right)_A = n_{\rm w} L \overbrace{\sin\left(\theta + \pi/2\right)}^{=\cos\theta} - \frac{m_{\rm L}gL}{2}\overbrace{\sin\left(\pi - \theta\right)}^{=\sin\theta} - mg\alpha L \overbrace{\sin\left(\pi - \theta\right)}^{=\sin\theta} = n_{\rm w} L \cos\theta - \frac{m_{\rm L}gL}{2}\sin\theta - mg\alpha L \sin\infty$$

The ladder will not slip if $f \le \mu_s n$. Using the equations above, this condition can be rewritten as

$$f = n_w = \left(\frac{1}{2}m_L + \alpha m\right) g \tan\theta \le \mu_s n = \mu_s(m_L + m)g$$

or

$$\alpha \le \frac{\mu_s (m_L + m)\cot\theta - m_L/2}{m} = \mu_s \cot\theta + \frac{m_L}{m}\left(\mu_s \cot\theta - \frac{1}{2}\right)$$

Here, we used the horizontal force equation to find f, the torque equation to find n_w, and the vertical force equation to find n.

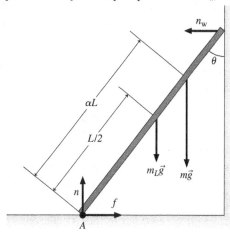

EVALUATE   For a person at the top of the ladder, $\alpha = 1$, and the condition for no slipping becomes

$$m \le m_L \left(\frac{\mu_s \cot\theta - 1/2}{1 - \mu_s \cot\theta}\right) = m_L \left(\frac{\mu_s - \tan\theta/2}{\tan\theta - \mu_s}\right)$$

Since m is positive, this condition cannot be fulfilled if $\mu_s \le \frac{1}{2}\tan\theta$, i.e., no one can climb to the top without causing the ladder to slip. However, if $\mu_s = \tan\theta$, the limit is ∞ so anyone can climb to the top.

ASSESS   When the coefficient of friction becomes too small, $\mu_s \cot\theta < 1/2$, or $\mu_s < \tan(\theta/2)$ (see Example 12.2), slipping will occur and it is no longer possible for the ladder to remain in static equilibrium. In this situation, nobody can climb up to the top of the ladder without making the ladder slip, regardless of his or her mass.

49.   **INTERPRET**   In this problem a wheel has been placed on a slope. We want to apply a horizontal force at its highest point to keep it from rolling down. We will apply the conditions for static equilibrium to solve this problem; namely that the sum of the forces on the wheel must be zero and the sum of the torque on the wheel must be zero.

DEVELOP   Consider the conditions for static equilibrium of the wheel, under the action of the forces shown in the sketch below. Here F_{app} is the applied horizontal force, F_c is the contact force of the incline (normal plus friction), and we assume that the center of mass is at the geometric center of the wheel. For the wheel to remain in static equilibrium, the forces must satisfy Equation 12.1. Our plan is to compute the torques about P using Equation 12.2. Note that the contact force, F_c does not create a torque because it acts at the pivot point so it has no lever arm.

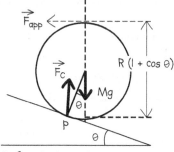

EVALUATE   The torques about the point of contact sum to zero, or

$$0 = \left(\sum \tau\right)_P = F_{app} R(1 + \cos\theta) - MgR\sin\theta$$

Therefore, the applied force is

$$F_{app} = Mg \frac{\sin \theta}{1 + \cos \theta} = Mg \tan\left(\frac{\theta}{2}\right)$$

ASSESS   The applied force vanishes when $\theta = 0$ (flat surface), and becomes maximum when $\theta = 90°$. In this limit, $F_{app} = Mg$ and points vertically upward.

51.   **INTERPRET**   In this problem a rectangular block is placed on an incline. Given the coefficient of friction between the block and the incline, we'd like to find out under what condition the block would slide before tipping.

DEVELOP   We suppose that the block is oriented with two sides parallel to the direction of the incline, and that its center of mass is at the geometric center of the block. The condition for sliding is

$$mg \sin \theta > f_s^{max} = \mu_s n = \mu_s mg \cos \theta$$

or $\tan \theta > \mu_s$. The condition for tipping over is that the center of mass lie to the left of the lower corner of the block (see sketch for Problem 12.50). Thus $\theta > \alpha$, where

$$\alpha = \text{atan}\left(\frac{w}{h}\right)$$

is the diagonal angle of the block.

EVALUATE   For the rectangular block with $w = h/2$, it tips over when $\theta > \alpha = \text{atan}(w/h) = \text{atan}(1/2)$ but will slide when $\theta > \text{atan}(\mu_s)$. Thus, if $\mu_s < \tan \alpha = 1/2$, the block in Problem 50 will slide before tipping.

ASSESS   We find that sliding happens first if $\mu_s < w/h$. This makes sense because when the coefficient of friction is small, the block has a greater tendency to slide. On the other hand, when the coefficient of friction is large $(\mu_s > w/h)$, we would expect tipping to take place first.

53.   **INTERPRET**   In this problem we want to verify the statement that the choice of pivot point does not matter when applying the conditions for static equilibrium.

DEVELOP   With reference to Fig. 12.29, we follow the hint given in the problem statement and write

$$\vec{\tau}_P = \sum_i \vec{r}_{Pi} \times \vec{F}_i = \sum_i \left[(\vec{r}_{Oi} + \vec{R}) \times \vec{F}_i\right] = \sum_i \vec{r}_{Oi} \times \vec{F}_i + \vec{R} \times \sum_i \vec{F}_i = \vec{\tau}_O + \vec{R} \times \vec{F}_{net}$$

EVALUATE   When the system is in static equilibrium, the total force and torque acting on the system vanish; that is $\vec{F}_{net} = 0$ and $\vec{\tau}_P = \vec{0}$. Therefore, we have $\vec{\tau}_P = \vec{\tau}_O = \vec{0}$ so, the total torque about any two points is the same.

ASSESS   If the angle θ is increased, then the corresponding coefficient of friction must also be increased to keep the pole from slipping.

55.   **INTERPRET**   This problem is about static equilibrium. The forces acting on the pole are the tension in the rope, gravity acting at the center of mass of the pole, and the contact force of the incline (perpendicular component n and parallel component f). We want to find the minimum coefficient of friction that will keep the pole from slipping.

DEVELOP   Make a sketch of the situation that shows the forces and the positions at which they are applied (see figure below). Applying the equilibrium condition of zero net torque (Equation 12.2) about the center of mass shows that a frictional force f must act up the plane if the rod is to remain in static equilibrium. Since the weight mg of the rod, and the normal force n contribute no torques about the center of mass, there must be a force to oppose the torque due to the

tension T. The equations for static equilibrium (parallel and perpendicular components of Equation 12.1, and CCW-positive component of Equation 12.2) are:

$$0 = \sum F_{\parallel} = f + T \cos \theta - mg \sin \theta$$

$$0 = \sum F_{\perp} = n - T \sin \theta - mg \cos \theta$$

$$0 = (\sum \tau)_{cm} = T(L/2) \overbrace{\sin(\pi/2 + \theta)}^{= \cos \theta} - f L/2 = T(L/2) \cos \theta - f L/2$$

The solutions for the forces are $f = \frac{1}{2} mg \sin \theta$, $T = \frac{1}{2} mg \tan \theta$, and

$$n = \frac{1}{2} mg \left(2 \cos \theta + \frac{\sin^2 \theta}{\cos \theta}\right)$$

subject to the condition that $f \leq \mu n$.

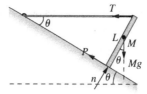

EVALUATE   Therefore,

$$\sin \theta \le \mu \left(2\cos \theta + \frac{\sin^2 \theta}{\cos \theta} \right)$$

$$\mu \ge \frac{\tan \theta}{2 + \tan^2 \theta}$$

ASSESS   If the angle θ is increased, then the corresponding coefficient of friction must also be increased in order to keep the pole from slipping. By use of the identities $\sin 2\theta = 2\sin \theta \cos \theta$, $\cos 2\theta = \cos^2 \theta - \sin^2 \theta$, and $\sin^2 \theta = 1 - \cos^2 \theta$, this may be rewritten as

$$\mu \ge \frac{\sin(2\theta)}{3 + \cos(2\theta)}$$

57.   **INTERPRET**   We use equilibrium methods to find the horizontal component of force on a bookshelf bracket tab. The bookshelf is in equilibrium, so the sum of forces and the sum of torques are both zero. Since the sum of forces is zero, we may use any point as the pivot for calculating the torques. We would expect that the horizontal force is much larger than the weight of the books, since the books have more leverage than the bracket.

DEVELOP   We start by drawing a diagram showing the forces and their approximate locations, as shown in the figure below. The mass of books is $m = 32$ kg, the distance $y = 4.5$ cm, and the distance $x = 12$ cm. Since we know nothing about the force $\vec{F}_b$ acting on the bottom corner of the bracket, we will use that point as our pivot point. The sum of torques around this point must be zero, which gives

$$\sum \tau = 0 = F_g x - F_h y.$$

Use this to find F_h.

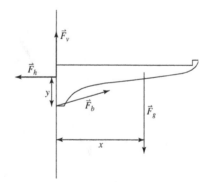

EVALUATE   Solving for F_h and inserting the know values gives

$$F_h = F_g \frac{x}{y} = mg \frac{x}{y} = \frac{(32 \text{ kg})(9.8 \text{ m/s}^2)(12 \text{ cm})}{4.5 \text{ cm}} = 840 \text{ N}$$

to two significant figures.

ASSESS   The force is much larger than the weight of the books, as we expected.

59.   **INTERPRET**   We use equilibrium methods to find the horizontal component of force on a bracket mounting screw. The bracket is in equilibrium, so the sums of forces and the sum of torques are both zero. Since the sum of forces is zero, we may use any point as the pivot for calculating the torques. We would expect that the horizontal force is much larger than the weight of the plant, since the plant has more leverage than the bracket.

DEVELOP   We start by drawing a diagram showing the forces and their approximate locations, as shown in the figure below. The mass of the plant is $m_2 = 4.2$ kg, the mass of the bracket is $m_1 = 0.85$ kg, the distance $y = 7.2$ cm, the distance $x_1 = 9.0$ cm, and the distance $x_2 = 28$ cm. Since we know nothing about the force $\vec{F}_b$ acting on the bottom corner of the bracket, we will use that point as our pivot point. The sum of torques around this point must be zero, which gives

$$\left(\sum \tau\right)_b = 0 = F_{g1}x_1 + F_{g2}x_2 - F_{sh}y$$

Use this to find F_{sh}.

EVALUATE   Solving the expression above for F_{sh} and inserting the given quantities gives

$$F_{sh} = \frac{F_{g1}x_1 + F_{g2}x_2}{y} = \frac{\left(m_1x_1 + m_2x_2\right)g}{y} = \frac{\left(0.85 \text{ kg}\right)\left(9.0 \text{ cm}\right) + \left(4.2 \text{ kg}\right)\left(28 \text{ cm}\right)}{7.2 \text{ cm}}\left(9.8 \text{ m/s}^2\right) = 170 \text{ N}.$$

ASSESS   The force is much larger than the weight of the plant, as we expected.

61.   **INTERPRET**   This problem involves two masses, separated by a massless rod, hovering far above the Earth. We need to find the net gravitational force on these masses, the net torque around the center of mass, and the center of gravity. We will use the equation for gravitational force, as well as the equation for torque. The center of gravity is not the same as the center of mass, in this case, since there is a change in gravitational field between the two masses.

DEVELOP   We start with a diagram, as shown in the figure below. We use Newton's law of universal gravity $\vec{F}_g = GM_E m/r^2$ (Equation 8.1) to find the force on each end of the spacecraft, then find the vector sum of the forces to obtain the net gravitational force. Use the forces on each end, and the angle found from the figure, to find the torque on each end and thus the net torque around the center of mass. The center of gravity is the point on the spacecraft at which the torques will be zero. We have labeled the distance from the left end of the spacecraft to the center of gravity x.

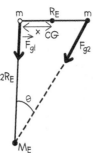

EVALUATE
(a) The gravitational force on the left mass is $\vec{F}_1 = -GM_E m/\left(2R_E\right)^2 \hat{j}$. The gravitational force on the right mass is $F_2 = -GM_E m\left(4R_E^2 + R_E^2\right)^{-2}$, directed at an angle $\theta = \text{atan}\left[R_E/\left(2R_E\right)\right] = \text{atan}\left(1/2\right) = 26.7°$. We note in passing that $\sin \theta = 1/\sqrt{5}$ and $\cos \theta = 2/\sqrt{5}$. We break F_2 into x and y components:

$$\vec{F}_2 = -G\frac{M_E m}{4R_E^2 + R_E^2}\left(\sin \theta \hat{i} + \cos \theta \hat{j}\right) = -G\frac{M_E m}{5R_E^2}\left(\frac{1}{\sqrt{5}}\hat{i} + \frac{2}{\sqrt{5}}\hat{j}\right).$$

Now we add the vector components to find the total force: $\vec{F} = -G\frac{M_E m}{R_E^2}\left[\left(\frac{1}{4} + \frac{2}{\sqrt{5}}\right)\hat{j} + \left(\frac{1}{\sqrt{5}}\right)\hat{i}\right]$. The magnitude of this force is $F = G\frac{M_E m}{R_E^2}\left(1.229\right)$, and the direction of the force is $\text{atan}\left[\left(1/\sqrt{5}\right)/\left(1/4 + 2/\sqrt{5}\right)\right] = 21.3°$ left of the negative y axis.

(b) The net torque around the center of mass is

$$\tau = -F_1 \frac{R_E}{2} + F_{2j} \frac{R_E}{2} = -G \frac{M_E m}{4 R_E^{\cancel{2}}} \left(\frac{\cancel{R_E}}{2} \right) + G \frac{M_E m}{5 R_E^{\cancel{2}}} \frac{\cancel{2}}{\sqrt{5}} \left(\frac{\cancel{R_E}}{\cancel{2}} \right)$$

$$= G \frac{M_E m}{R_E} \left(-\frac{1}{8} + \frac{1}{5\sqrt{5}} \right) = G \frac{M_E m}{R_E} (-0.0356)$$

(c) To find the center of gravity, repeat the calculation for (b) but use x for the left-hand distance and $(R_E - x)$ for the right-hand distance, setting $\tau = 0$. Solving this for x will give the distance of the center of gravity from the left end, which we can compare to $R_E/2$.

$$0 = -F_1 x + F_{2j} (R_E - x) = -G \frac{M_E m}{4 R_E^2} (x) + G \frac{M_E m}{5 R_E^2} \frac{2}{\sqrt{5}} (R_E - x)$$

$$= -\frac{1}{4} x + \frac{2}{5\sqrt{5}} (R_E - x)$$

$$x \left(\frac{1}{4} + \frac{2}{5\sqrt{5}} \right) = R_E \frac{2}{5\sqrt{5}}$$

$$x = 0.417 R_E$$

$$\frac{R_E}{2} - x = 0.083$$

ASSESS   The torque is negative, which in this case means counterclockwise as we would expect. The center of gravity in this case is not at the center of mass, due to the decrease in gravitational force with altitude.

63.   **INTERPRET**   You need to determine the forces on a roof in the event of a snowfall. You use equilibrium techniques, with the sum of torques equaling zero, to determine whether the tie beam will hold.

DEVELOP   The figure below shows the relevant forces acting on the rafter located on the right-side of the roof (the forces will be the same on the left-side due to the roof's symmetry). You are told to assume that the contact force, F_{snow}, coming from the weight of the snow and building material is concentrated at the peak. This downward force is countered by the upward force coming from the wall, F_{wall}. The only horizontal forces are from the tie beam, F_{tie}, and from the contact force between the rafters at the peak, F_{peak}. Because the sum of the horizontal forces is zero, F_{tie} will be equal and opposite to F_{peak}.

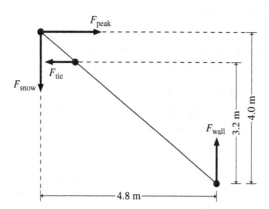

To find the value of F_{tie}, you'll need to consider the torques on the rafter. The most convenient choice for a pivot point is the joint where the rafter meets the wall. The sum of the torques around this point is

$$\sum \tau = F_{peak} (4.0 \text{ m}) - F_{tie} (3.2 \text{ m}) - F_{snow} (4.8 \text{ m}) = 0$$

EVALUATE   Using the fact that $F_{tie} = F_{peak}$, the torque equation gives:

$$F_{tie} = F_{snow} \frac{4.8 \text{ m}}{0.8 \text{ m}} = (170 \text{ kg})(9.8 \text{ m/s}^2)(6) = 10 \text{ kN}$$

Since this is greater than the limit of 7.5 kN, the tie beam will not hold. As for what forces act on the tie beam, recall that the inward-pointing force, F_{tie}, in the figure is the force *from* the tie beam on the rafter. By Newton's third law, the force *on* the tie beam from the rafter will be in the opposite direction. As such, the tie beam will be pulled outward by each rafter, resulting in tension rather than compression.

ASSESS   One way to improve the design would be to choose a longer tie beam that could be placed farther down from the peak. In fact, if the tie beam were 1.1 m, rather than 0.8 m, below the peak in the roof, the tie beam would be able to hold under the weight of the snow.

65.   **INTERPRET**   We're asked to analyze a boom and pulley system.

DEVELOP   To find the maximum tension, we need to parameterize the tension by a singular variable. We can simplify the problem by combining the downward forces from the boom's weight and the sample rope into one variable, F_d, which we assume acts on the end of the boom. Therefore, the sum of the torques around the pivot is:

$$\sum \tau = rT \sin\phi - rF_d \sin\theta = 0$$

where r is the length of the boom and the two angles are defined in the figure below.

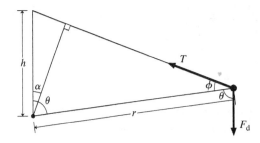

We have defined in the figure another angle, α, as well as the height, h, of the vertical support. The parameters h, r, and F_d are constant, whereas the 3 angles change as the boom rotates. The angles are related to each other by: $\theta + \phi - \alpha = 90°$, and $h\cos\alpha = r\sin\phi$. We plug these 2 relations into the torque equation and use some of the trig identities in Appendix A to arrive at the tension as a function of the angle θ between the boom and the vertical:

$$T = F_d \sqrt{1 + \frac{r^2}{h^2} - 2\frac{r}{h}\cos\theta}$$

EVALUATE   As θ increases from zero (straight up), the tension increases as well, reaching a maximum at $\theta = 90°$ when the boom is horizontal.

The answer is (a).

ASSESS   The tension keeps increasing as θ continues to 180°, but we assume that the boom cannot rotate any lower than a horizontal position, otherwise it would drop into the river.

67.   **INTERPRET**   We're asked to analyze a boom and pulley system.

DEVELOP   Let y be the length of boom rope between the end of the boom and the pulley on the vertical support. From the figure in Problem 12.65 and the law of cosines in Appendix A:
$$y^2 = r^2 + h^2 - 2rh\cos\theta$$

As the rope is pulled at constant speed v_0, the length y will be getting shorter: $dy/dt = \dot{y} = -v_0$, where we use dots to indicate time derivatives. We want to find how the boom's angle changes with respect to the horizontal. Since the angle θ is with respect to the vertical, we switch to its complement: $\beta = 90° - \theta$, such that
$$\sin\beta = \frac{r^2 + h^2 - y^2}{2rh}$$

We'll take the derivative of this equation with respect to time.

EVALUATE   Taking the first derivative with respect to time and using the chain rule (see Appendix A) gives:

$$\left(\cos \beta\right)\dot{\beta} = \frac{-2y\dot{y}}{2rh} \quad \rightarrow \quad \dot{\beta} = \frac{v_0}{rh}\frac{y}{\cos \beta} > 0$$

The derivative $\dot{\beta}$ is positive because all of the other terms are positive (as long as $\beta \geq 0$). This implies that the angle is increasing, as we would expect. To find the rate at which it is increasing, we need to take the second derivative with respect to time:

$$\frac{d}{dt}\dot{\beta} = \ddot{\beta} = \frac{v_0}{rh}\left[\frac{\dot{y}}{\cos \beta} - \frac{y}{\cos^2 \beta}(-\sin \beta)\dot{\beta}\right] = \frac{v_0^2}{rh\cos \beta}\left[1 + \frac{y^2}{rh}\frac{\tan \beta}{\cos \beta}\right]$$

All of the terms are again positive, so the angle increases at an increasing rate.

The answer is (b).

ASSESS   The result implies that the boom's angle is changing faster as it points more and more vertical. This makes sense since the fastest rate at which the angle could be changing is $\dot{\beta} = v_0 / r$, i.e. when the velocity is perpendicular to the radius. This occurs at the top, when the boom nears the vertical.

OSCILLATORY MOTION

13

EXERCISES

Section 13.1 Describing Oscillatory Motion

17. **INTERPRET** The question here is about the oscillatory behavior of the violin string. Given the frequency of oscillation, we are asked to find the period.

DEVELOP The relationship between period and frequency is given by Equation 13.1, $T = 1/f$.

EVALUATE Using Equation 13.1, we obtain

$$T = \frac{1}{f} = \frac{1}{440 \text{ Hz}} = 2.27 \times 10^{-3} \text{ s}$$

ASSESS The period is the oscillation is the inverse of the frequency. Note that the unit of frequency is the hertz; $(1 \text{ Hz} = 1 \text{ s}^{-1})$.

19. **INTERPRET** The problem involves simple harmonic motion. We are to create expressions that characterize oscillations, given their amplitude, frequency, and the magnitude at $t = 0$.

DEVELOP The general expression of the position of an object undergoing simple harmonic motion is given by Equation 13.8:

$$x(t) = A\cos(\omega t + \phi)$$

where A is the amplitude, ω is the angular frequency, and ϕ is the phase. By taking the time derivative of $x(t)$, we obtain the corresponding velocity as a function of time (see Equation 13.9):

$$v(t) = \frac{d}{dt}x(t) = -A\omega\sin(\omega t + \phi)$$

EVALUATE (a) Since the displacement is a maximum at $t = 0$, the phase constant is zero: $\phi = 0$. Using Equation 13.8 with $A = 10$ cm, $\omega = 2\pi(5 \text{ Hz}) = 10\pi \text{ s}^{-1}$, the displacement is found to be

$$x(t) = A\cos(\omega t + \phi) = (10 \text{ cm})\cos\left[(10\pi\text{s}^{-1})t\right]$$

(b) Equation 13.9 shows that the maximum (positive) velocity occurs at $t = 0$ if $\sin\phi = -1$, or $\phi = \pi/2$. Therefore, with $A = 2.5$ cm, $\omega = 5 \text{ s}^{-1}$, we have

$$x(t) = A\cos(\omega t + \phi) = (2.5 \text{ cm})\cos\left[(5 \text{ s}^{-1})t - \pi/2\right] = (2.5 \text{ cm})\sin\left[(5 \text{ s}^{-1})t\right]$$

where we have used $\cos(\omega t - \pi/2) = \sin \omega t$.

ASSESS For a system undergoing simple harmonic motion, once the position as a function of time is given in the form of Equation 13.8, physical quantities such as velocity, acceleration, angular frequency and period can all be readily determined.

21. **INTERPRET** We want to find the period of the hummingbird's wing flap.

DEVELOP From Equation 13.1, the period is the inverse of the frequency: $T = 1/f$.

EVALUATE   The period of a hummingbird wing is:

$$T = \frac{1}{f} = \frac{1}{(45 \text{ Hz})} = 22 \text{ ms}$$

ASSESS   Our eyes cannot follow the beating of a hummingbird's wings. That's because the human eye essentially processes information at a rate of about 30 times per second, or equivalently once every 33 ms.

Section 13.2 Simple Harmonic Motion

23.   INTERPRET   We can consider the car's suspension as a simple spring-mass system.

DEVELOP   Equations 13.7b and 13.7c give the expressions for frequency and period in the case of simple harmonic motion:

$$f = \frac{1}{2\pi}\sqrt{\frac{k}{m}} \quad \text{and} \quad T = 2\pi\sqrt{\frac{m}{k}}$$

EVALUATE   Using the expressions above and the mass and spring constant given, the frequency is:

$$f = \frac{1}{2\pi}\sqrt{\frac{26 \text{ kN/m}}{1900 \text{ kg}}} = 0.59 \text{ Hz}$$

And the period is:

$$T = 2\pi\sqrt{\frac{1900 \text{ kg}}{26 \text{ kN/m}}} = 1.7 \text{ s}$$

ASSESS   If the car hits a bump, it will start to vibrate at this nice slow frequency. However, the shock absorbers will damp this motion, so the vibrations should die away after a few cycles.

25.   INTERPRET   In this problem, a mass attached to a spring undergoes simple harmonic motion. Given the maximum speed and the maximum acceleration of the mass, we want to find the angular frequency, the spring constant, and the amplitude of the oscillation.

DEVELOP   The maximum speed and the maximum acceleration of the mass can be obtained from Equations 13.9 and 13.10 by taking the maximum value of the trigonometric functions (i.e., 1). Taking the absolute value, the result is

$$v_{max} = \omega A \qquad a_{max} = \omega^2 A$$

From a_{max} and v_{max}, the angular frequency may be obtained from $\omega = a_{max}/v_{max}$. Once ω is known, the spring constant and the amplitude of the oscillation can be calculated using:

$$k = m\omega^2 \qquad A = \frac{a_{max}}{\omega^2} = a_{max}\left(\frac{v_{max}}{a_{max}}\right)^2 = \frac{v_{max}^2}{a_{max}}$$

EVALUATE   (a) The angular frequency is

$$\omega = \frac{a_{max}}{v_{max}} = \frac{15 \text{ m/s}^2}{3.5 \text{ m/s}} = 4.286 \text{ s}^{-1} \approx 4.3 \text{ s}^{-1}$$

(b) The spring constant is $k = m\omega^2 = (0.05 \text{ kg})(4.29 \text{ s}^{-1})^2 = 0.92 \text{ N/m}$.
(c) Similarly, the amplitude of the motion is

$$A = \frac{v_{max}^2}{a_{max}} = \frac{(3.5 \text{ m/s})^2}{15 \text{ m/s}^2} = 0.82 \text{ m}$$

ASSESS   To check that our results are correct, let's compute the amplitude A using the value of ω found in (a). In either way, we have

$$A = \frac{v_{max}}{\omega} = \frac{3.5 \text{ m/s}}{4.286 \text{ s}^{-1}} = 0.82 \text{ m}$$

$$A = \frac{a_{max}}{\omega^2} = \frac{15 \text{ m/s}^2}{(4.286 \text{ s}^{-1})^2} = 0.82 \text{ m}$$

The results indeed agree.

27. **INTERPRET** The problem is about simple harmonic motion of a particle. Given its maximum speed and maximum acceleration, we want to find the angular frequency, the period, and the amplitude of the oscillation.
DEVELOP The maximum speed and the maximum acceleration of the mass are derived from Equations 13.9 and 13.10 by setting the trigonometric functions to unity (their maximum value):

$$v_{max} = \omega A \qquad a_{max} = \omega^2 A$$

From a_{max} and v_{max}, the angular frequency may be obtained as $\omega = a_{max}/v_{max}$. Once ω is known, the period and the amplitude of the oscillation can be computed using (see Equations 3.1, 3.9, and 3.10)

$$T = \frac{1}{f} = \frac{2\pi}{\omega} \quad A = \frac{a_{max}}{\omega^2} = a_{max} \left(\frac{v_{max}}{a_{max}}\right)^2 = \frac{v_{max}^2}{a_{max}}$$

EVALUATE (a) The angular frequency is

$$\omega = \frac{a_{max}}{v_{max}} = \frac{3.1 \text{ m/s}^2}{1.4 \text{ m/s}} = 2.2 \text{ rad/s}$$

(b) The period of the motion is $T = 2\pi/\omega = 2\pi/(2.21 \text{ rad/s}) = 2.8 \text{ s}$.
(c) Similarly, the amplitude of the motion is

$$A = \frac{v_{max}^2}{a_{max}} = \frac{(1.4 \text{ m/s})^2}{3.1 \text{ m/s}^2} = 0.63 \text{ m}$$

ASSESS To check that our results are correct, let's compute the amplitude A using the value of ω found in (a):

$$A = \frac{v_{max}}{\omega} = \frac{1.4 \text{ m/s}}{2.21 \text{ s}^{-1}} = 0.63 \text{ m}$$

$$A = \frac{a_{max}}{\omega^2} = \frac{3.1 \text{ m/s}^2}{(2.21 \text{ s}^{-1})^2} = 0.63 \text{ m}$$

The results indeed agree.

Section 13.3 Applications of Simple Harmonic Motion

29. **INTERPRET** This problem is about the simple harmonic motion of the pendulum in a grandfather clock. We want to find the time interval between successive ticks.
DEVELOP The period of a simple pendulum is given by Equation 13.15:

$$T = 2\pi \sqrt{\frac{L}{g}}$$

We note that the clock ticks twice each period of oscillation, so the time between clicks is a half-period.
EVALUATE Using Equation 13.15, the time between ticks is

$$\Delta t = \frac{T}{2} = \pi \sqrt{\frac{L}{g}} = \pi \sqrt{\frac{1.45 \text{ m}}{9.81 \text{ m/s}^2}} = 1.21 \text{ s}$$

ASSESS One tick every 1.21 s seems reasonable. Note that if we increase the length of the pendulum, then the period and the time between ticks will increase as well.

31. **INTERPRET** The problem is about a physical pendulum—a meter stick suspended from one end. We want to know its period of oscillation.
DEVELOP The period of a physical pendulum can be obtained from Equation 13.13.

$$T = \frac{2\pi}{\omega} = 2\pi \sqrt{\frac{I}{mgL}}$$

The meter stick is a physical pendulum whose center of mass is $L = L_0/2 = 0.50$-m below the point of suspension. From Table 10.2, we see that the rotational inertia of a stick about one end is $I = mL_0^2/3$.

EVALUATE Using Equation 13.13, the period of the meter stick is

$$T = 2\pi\sqrt{\frac{I}{mgL}} = 2\pi\sqrt{\frac{mL_0^2/3}{mgL_0/2}} = 2\pi\sqrt{\frac{2L_0}{3g}} = 2\pi\sqrt{\frac{2(1.0 \text{ m})}{3(9.8 \text{ m/s}^2)}} = 1.6 \text{ s}$$

ASSESS   A simple experiment can be carried out to verify that a period of 1.6 s (or roughly 5 complete oscillations in 8 seconds) for the meter stick is reasonable.

Section 13.4 Circular and Harmonic Motion

33.   **INTERPRET**   This problem involves a body undergoing simple harmonic motion in two dimensions, with a different angular frequency in each dimension. We are to find how many oscillations it makes before returning to its initial position.

DEVELOP   Let the angular frequencies be ω_x and ω_y, such that

$$\frac{\omega_x}{\omega_y} = \frac{1.75}{1} = \frac{7}{4}$$

Since $T = 2\pi/\omega$, the ratio of the periods for the x and y components of the motion is then equal to

$$\frac{T_x}{T_y} = \frac{2\pi/\omega_x}{2\pi/\omega_y} = \frac{\omega_y}{\omega_x} = \frac{4}{7}$$

EVALUATE   The above equation gives $7T_x = 4T_y$, which means that seven oscillations in the x direction are completed in the x direction and four are completed in the y direction before the object returns to its intial position.
ASSESS   Since $\omega_x = 1.75\,\omega_y > \omega_y$, we have $T_x = T_y/1.75 < T_y$. The smaller the period, the greater the number of oscillations completed in a given time interval.

Section 13.5 Energy in Simple Harmonic Motion

35.   **INTERPRET**   This problem involves finding the maximum angular displacement and speed given the torsional constant, rotational inertia, and total energy.
DEVELOP   Because only conservative forces (i.e., the torsional force of the wire) act on the torsional oscillator, we can apply conservation of energy to relate its total energy to its oscillation parameters. Specifically, we chose to evaluate the total energy when the rotational speed is maximum, at which point the torsional potential energy is zero). Thus

$$E_{tot} = \frac{1}{2}I\left(\frac{d\theta}{dt}\right)^2_{max}$$

where $\left(d\theta/dt\right)^2_{max}$ is the maximum angular speed (not to be confused with ω, the constant natural frequency of the oscillator). We can determine $\left(d\theta/dt\right)^2_{max}$ by inspecting the equation of motion for a torsional oscillator (Equation 13.11). Because it is the same as for a linear oscillator, with k replaced by κ, we can immediately write

$$\left(\frac{d\theta}{dt}\right)_{max} = \omega A$$

where ω is the natural angular frequency of the oscillator and A is the maximum angular displacement (in radians). Inserting this into the expression for total energy, and using Equation 13.12 $\omega = \sqrt{\kappa/I}$, we find

$$E_{tot} = \frac{1}{2}I\left(\frac{d\theta}{dt}\right)^2_{max} = \frac{1}{2}I\omega^2 A^2 = \frac{1}{2}\kappa A^2$$

From these expressions, we can both find the maximum displacement A and the maximum angular speed $\left(d\theta/dt\right)^2_{max}$.
EVALUATE   The maximum angular displacement is

$$A = \pm\sqrt{\frac{2E_{tot}}{\kappa}} = \pm\sqrt{\frac{2(4.7 \text{ J})}{3.4 \text{ N}\cdot\text{m/rad}}} = \pm 1.7 \text{ rad} = \pm 95°$$

where the two signs indicate that the maximum angular displacement occurs in both the counter-clockwise direction and in the clockwise direction. The maximum angular speed is

$$\left(\frac{d\theta}{dt}\right)_{max} = \pm\sqrt{\frac{2E_{tot}}{I}} = \pm\sqrt{\frac{2(4.7 \text{ J})}{1.6 \text{ kg}\cdot\text{m}^2}} = \pm2.4 \text{ s}^{-1} = \pm15 \text{ rad/s}$$

where the two signs indicate that this speed is attained in both the counter-clockwise direction and the clockwise direction.

ASSESS   Notice how conservation of total mechanical energy allowed us to evaluate the total mechanical energy where it was simplest to do so.

Section 13.6 Damped Harmonic Motion and Section 13.7 and Resonance

37.   **INTERPRET**   The problem is about damped harmonic motion. We are interested in finding out how long it takes for the vibration amplitude of a piano string to drop to half its initial value.

DEVELOP   The solution to the damped harmonic motion is given by Equation 13.17:

$$x(t) = Ae^{-bt/2m}\cos(\omega t + \phi)$$

Thus, the amplitude is half the initial value when $e^{-bt/2m} = 0.5$.

EVALUATE   Taking the natural logarithms of both sides of the equation above gives $bt/2m = \ln 2$. Thus, the time for the amplitude to be halved is

$$t = \frac{2m}{b}\ln 2 = \frac{\ln 2}{2.8 \text{ s}^{-1}} = 0.25 \text{ s}$$

ASSESS   Since the amplitude drops by half in just 0.25 s, we conclude that the damping must be rather strong.

39.   **INTERPRET**   This problem involves nondamped driven harmonic motion. We want to find the speed of the car that leads to a maximum vibration amplitude, which will occur when the frequency with which the car drives over the bumps equals the natural frequency of its suspension system.

DEVELOP   The peak amplitude occurs at resonance, when $\omega_d = \omega_0$, so the car receives an impulse from the bumps once each period. The condition for resonance is therefore that the car travel the distance between bumps in one period:

$$L_0 = \frac{40 \text{ m}}{vT_0}$$

EVALUATE   Solving the resonance condition for the speed v gives

$$v = \frac{L_0}{T_0} = L_0 f_0 = (40 \text{ m})(0.45 \text{ Hz}) = 18 \text{ m/s} = 65 \text{ km/h}$$

where we have used Equation 13.1, $T = 1/f$.

ASSESS   If the spacing between bumps increases, then the car speed must also go up to meet the rather unpleasant resonance condition!

PROBLEMS

41.   **INTERPRET**   The problem involves two identical mass-spring systems undergoing simple harmonic motion. We want to find the time elapsed between releasing the masses so that the two oscillations differ in phase by $\pi/2$.

DEVELOP   Suppose that both masses are released from their maximum positive displacements, the first at $t = 0$ and the second at $t = t_0$. Because the phase (the argument of the cosine in Equation 13.8) for each is zero at release, $\phi_{10} = 0$ and $\omega_2 t_0 + \phi_{20} = 0$. The difference in phase is

$$\Delta\phi = \phi_1 - \phi_2 = (\omega_1 t + \phi_{10}) - (\omega_2 t + \phi_{20}) = (\omega_1 - \omega_2)t + (\phi_{10} - \phi_{20}) = -\phi_{20}$$

where $\omega_1 = \omega_2 = \sqrt{k/m}$ for identical mass-spring systems (see Equation 13.7a). This allows us to solve for t_0.

EVALUATE   With $\Delta\phi = \pi/2$, we have

$$t_0 = \frac{-\phi_{20}}{\omega_2} = \frac{\pi/2}{\sqrt{k/m}} = \frac{\pi/2}{\sqrt{(2.2 \text{ N/m})/(0.43 \text{ kg})}} = 0.70 \text{ s}$$

ASSESS In terms of the period, the time is

$$t_0 = \frac{\phi_{20}}{2\pi}T = \frac{\pi/2}{2\pi}T = \frac{1}{4}T.$$

This makes sense since the phase change in one period is 2π.

43. **INTERPRET** This problem involves simple harmonic motion. We are asked to find the time for which a mass is in contact with a spring given the initial speed of the mass and the parameters of the spring.

DEVELOP The spring is initially in its equilibrium position. The mass will compress the spring, then the spring will extend again to its equilibrium position, at which point the mass will depart to the left because the spring will begin to slow down (because it enters its extension phase). Thus, the mass is in contact with the spring for a half period, so the time duration will be $T/2$, which we can find using Equation 13.7c. From Equation 13.9, which describes the motion of a mass-spring system, we see that the maximum velocity is related to the amplitue by $v_{max} = \omega A$, from which we can find the maximum compression A of the spring, given that $v_{max} = v_0$.

EVALUATE (a) The mass is in contact with the spring for a time t given by

$$t = \frac{T}{2} = \pi\sqrt{\frac{m}{k}}$$

(b) Using the fact that $\omega = 2\pi/T$, the maximum compression is

$$A = \frac{v_0}{\omega} = \frac{v_0 T}{2\pi} = v_0\sqrt{\frac{m}{k}}$$

ASSESS We could use conservation of total mechanical energy to find the maximum compression as well. The initial energy of the mass-spring system is $K = mv_0^2/2$ and the energy when the spring is at maximum compression is completely potential energy (because the velocity of the mass is zero at this point) and is given by $U = kA^2/2$. Equating the two gives

$$\frac{1}{2}mv_0^2 = \frac{1}{2}kA^2$$

$$A = \pm v_0\sqrt{\frac{m}{k}}$$

where the two signs indicate the that maximum displacement from equilibrium of the spring may be an extension or a compression.

45. **INTERPRET** This problem involves the simple harmonic motion of a pendulum. We are asked to find the period of one cycle of the pendulum, and from that find the duration of the physics class.

DEVELOP From Equation 13.13, we know that the period is related to the pendulum parameters by

$$T = 2\pi\sqrt{\frac{I}{mgL}}$$

From Table 10.2, we find that the rotational inertia I for a thin rod rotating abou its end is $I = ML^2/3$, so

$$T = 2\pi\sqrt{\frac{mL^2}{3mgL}} = 2\pi\sqrt{\frac{L}{3g}}$$

From this, we can find the duration of the class, assuming the student starts his pendulum the instant class starts.

EVALUATE We are told the pendulum completes 6279 cycles, so the class duration is

$$t = (6279)(2\pi)\sqrt{\frac{0.17 \text{ m}}{3(9.8 \text{ m/s}^2)}} = 3.0 \times 10^3 \text{ s} = 50 \text{ min}$$

ASSESS Adding the 10-minute break between classes to walk to the next class, the 50-min class time allows the classes to start on the hour, which seems quite reasonable.

47. **INTERPRET** We will analyze a bacteria protein that undergoes simple harmonic motion.

DEVELOP The peak force occurs when the mass-spring system is maximally displaced from equilibrium: $F_{peak} = k \left| x_{max} \right|$. The maximum displacement for simple harmonic motion is by definition the amplitude: $\left| x_{max} \right| = A$. From this we can find the spring constant, and from Equation 13.7b, the effective mass is $m = k / \left(2\pi f \right)^2$.

EVALUATE (a) The spring constant of the dynein-microtubule system is

$$k = \frac{F_{peak}}{A} = \frac{1.0 \text{ pN}}{15 \text{ nm}} = 0.0667 \text{ pN/nm} = 67 \ \mu\text{N/m}$$

(b) Given the frequency of oscillation, the effective mass being oscillated is

$$m = \frac{k}{\left(2\pi f \right)^2} = \frac{66.7 \ \mu\text{N/m}}{\left(2\pi \cdot 70 \text{ Hz} \right)^2} = 3.4 \times 10^{-10} \text{ kg}$$

ASSESS The effective mass is about 100 times that of a typical human cell. This seemingly large value may be because the system is moving through a viscous fluid that acts like additional mass.

49. **INTERPRET** This problem involves the simple harmonic motion of a pendulum. We are to derive the equations of motion by considering the linear forces and displacements of the bob, rather than the angular torques and displacements. This derivation will involve applying Newton's second law.

DEVELOP Draw a diagram of the situation (see figure below). The tangential component of Newton's second law, for a simple pendulum of mass m and length L, is $m \left(d^2 s / dt^2 \right) = -mg \sin \theta$. The radial component guarantees that the motion follows a circular arc. The horizontal displacement is $x = L \sin \theta$. For small displacements, $x \approx L\theta = s$, so the equation of motion is approximately

$$m \frac{d^2 s}{dt^2} \approx m \frac{d^2 x}{dt^2} = -mg \frac{x}{L}$$

$$\frac{d^2 x}{dt^2} = -\left(\frac{g}{L} \right) x$$

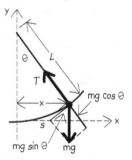

EVALUATE This equation is the same as Equation 13.3, with $k/m = g/L$. Thus, from Equation 13.7c, the period is $T = 2\pi/\omega = 2\pi \sqrt{m/k} = 2\pi \sqrt{L/g}$.

ASSESS Notice that this derivation assumes small displacements from equilibrium, where $\sin \theta \approx \theta$. For larger displacements, the period derived here is not valid.

51.   **INTERPRET**   This problem involves simple harmonic motion of the torsional variety. We are to find the mass of the oscillating beam given the change in its oscillating frequency when two masses are attached to its extremities.

DEVELOP   The frequencies before and after the addition of the steel workers are related by $f_2 = 0.8 f_1$. These frequencies can be related to the rotational inertias of the beam, with and without the steel workers, through Equation 13.12, which gives

$$f_1 = \frac{\omega_1}{2\pi} = \frac{1}{2\pi}\sqrt{\frac{\kappa}{I_1}}$$

$$f_2 = \frac{\omega_2}{2\pi} = \frac{1}{2\pi}\sqrt{\frac{\kappa}{I_2}}$$

since the torsional constant κ does not change, the ratio of these expressions gives $f_2/f_1 = \sqrt{I_1/I_2}$. From Table 10.2, we know that the rotational inertia of a beam about its center is $I_1 = ML^2/12$, and adding two masses at the extremities of the beam changes the rotational inertiat to $I_2 = I_1 + 2m(L/2)^2$, so we can solve for the mass M of the beam.

EVALUATE   Solving for the mass M gives

$$0.80 = \frac{f_2}{f_1} = \sqrt{\frac{I_1}{I_2}} = \sqrt{\frac{ML^2/12}{ML^2/12 + 2m(L/2)^2}}$$

$$0.80 = \left(1 + \frac{6m}{M}\right)^{-1/2}$$

$$M = \frac{6m}{(0.80)^{-2} - 1} = \frac{6(75\ \text{kg})}{(0.80)^{-2} - 1} = 800\ \text{kg}$$

ASSESS   If the steelworkers both have the same mass as the beam, the change in oscillating frequency would be $f_2/f_1 = 34\%$.

53.   **INTERPRET**   The problem asks us to find several characteristics of a object's trajectory given its equation of motion. The object undergoes simple harmonic motion in both the x and y directions, with the motion in each direction being 90° out of phase (i.e. when the displacement is maximum in one direction it is at zero in the other direction).

DEVELOP   The position of the object as a function of time is $\vec{r} = A\sin \omega t\, \hat{i} + A\cos \omega t\, \hat{j}$. Its position can therefore be found by calculating the magnitude $|\vec{r}|$. The velocity may be found by differentiating the position with respect to time, and the speed is the magnitude of the velocity, $|\vec{v}|$. Finally, the angular speed may be found from by dividing the speed by the radius, since this object executes circular motion (i.e., the velocity is always perpendicular to the radial position vector).

EVALUATE   (a) The object's distance from the origin is $|\vec{r}| = \sqrt{(A\sin \omega t)^2 + (A\cos \omega t)^2} = A$ a constant, so its path is a circle with radius A. (b) Differentiating $r(t)$, we find $\vec{v} = d\vec{r}/dt = \omega A\cos \omega t\, \hat{i} - \omega A\sin \omega t\, \hat{j}$. (c) The object's speed is $|\vec{v}| = \sqrt{(\omega A\cos \omega t)^2 + (-\omega A\sin \omega t)^2} = \omega A$, also a constant. (d) From Equation 10.3, the angular speed is $v/r = \omega A/A = \omega$.

ASSESS   Note that $\vec{v} \cdot \vec{r} = 0$, as required for circular motion.

55.   **INTERPRET**   This problem is about simple harmonic motion of a pendulum. We want to know what its period is when it is treated as a simple pendulum or as a physical pendulum.

DEVELOP   Draw a diagram of the situation (see figure below). The periods of a simple pendulum and a physical pendulum are given by

$$T_{\text{simple}} = 2\pi\sqrt{\frac{L}{g}}$$

$$T_{\text{phys}} = 2\pi\sqrt{\frac{I}{mgL}}$$

The rotational inertia I of the physical pendulum can be found using the parallel-axis theorem (see Equation 10.17 with $d = L$ and $I_{cm} = 2MR^2/5$), which gives

$$I = \frac{2}{5}MR^2 + ML^2$$

Use these two expressions to compare the periods of the pendulum in the two cases.

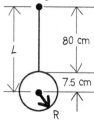

EVALUATE   When treated as a simple pendulum, the period is

$$T_{simple} = 2\pi\sqrt{\frac{L}{g}} = 2\pi\sqrt{\frac{0.875 \text{ m}}{9.81 \text{ m/s}^2}} = 1.877 \text{ s}$$

On the other hand, if we regard the pendulum as a physical pendulum with rotational inertia $I = \frac{2}{5}mR^2 + mL^2$, then its period becomes

$$T_{phys} = 2\pi\sqrt{\frac{I}{mgL}} = 2\pi\sqrt{\frac{\frac{2}{5}mR^2 + mL^2}{mgL}} = 2\pi\sqrt{\frac{L}{g}}\sqrt{1 + \frac{2R^2}{5L^2}} = T_{simple}\sqrt{1 + \frac{2R^2}{5L^2}}$$

$$= T_{simple}\sqrt{1 + \frac{2}{5}\left(\frac{0.075 \text{ m}}{0.875 \text{ m}}\right)^2} = (1.00147)T_{simple} = 1.880 \text{ s}$$

The fractional error is

$$\frac{T_{phys} - T_{simple}}{T_{phys}} = 0.147\%.$$

ASSESS   The period found by assuming the pendulum to be simple is slightly less than if we treat it as a physical pendulum.

57.   **INTERPRET**   We are told how the period of the oscillating device varies with the mass of the animal placed in its cage. From this calibration curve, we can find the spring constant and mass of the cage.

DEVELOP   The square of the period is linear with the mass: $T^2 = (2\pi/k)m$, from Equation 13.7c. Therefore, the slope of the given equation should be equal to $(2\pi/k)$. The value of the equation at $m = 0$ should correspond to the mass of the cage:

$$T^2\big|_{m=0} = (2\pi/k)m_{cage}$$

EVALUATE   (a) The slope in the given equation is 5.0 s²/kg, so the spring constant is

$$k = \frac{2\pi}{5.0 \text{ s}^2/\text{kg}} = 1.26 \text{ kg/s}^2 \approx 1.3 \text{ N/m}$$

(b) The given equation equals 4.0 s² for $m = 0$, so the mass of the cage is

$$m_{cage} = \left(\frac{k}{2\pi}\right)T^2\big|_{m=0} = \left(\frac{1.26 \text{ kg/s}^2}{2\pi}\right)4.0 \text{ s}^2 = 0.80 \text{ kg}$$

ASSESS   When the cage is empty, the small mass measuring device oscillates with a period of 2.0 s. To lengthen the period by one second (i.e. $T^2 = 9.0 \text{ s}^2$), it would take an animal of mass 1 kg. This gives us a sense of how sensitive the SMMD is.

59.   **INTERPRET**   This problem involves the simple harmonic motion of a mass-spring system, with the caveat that the spring is composed of two springs with different spring constants and that are acting in opposite directions.

DEVELOP   When the mass is at its equilibrium position, the springs are either 1) both extended, 2) both compressed, or 3) both in equilibrium because the forces applied by the springs must act opposite to each other (for the first two scenarios), or they must apply zero force (for the last scenario). If the mass is moved by an amount Dx to the right of the equilibrium position, the force of the first spring increases by $k_1 \Delta x$ to the left, and the force of the second spring decreases by $k_2 \Delta x$ to the right (which is also an increase to the left). Thus, the net force is $(k_1 + k_2)\Delta x$ to the left, which represents a restoring force (opposite to the displacement Δx to the right). The effective spring constant is therefore $k_{\text{eff}} = k_1 + k_2$.

EVALUATE   Inserting the effective spring constant into Equation 13.7a, we find the frequency of oscillation to be

$$\omega = \sqrt{(k_1 + k_2)/m}$$

ASSESS   If one spring constant is much larger than the other, than it will dominate, as we would expect.

61.   **INTERPRET**   This problem involves simple harmonic motion and potential energy. For the potential energy, we will take the lowest point of the pendulum's trajectory as the zero of potential energy. We are asked to express the potential energy of a simple pendulum in the small-amplitude limit.

DEVELOP   The potential energy of a simple pendulum or equation is $U = mgh = mgL(1 - \cos\theta)$. Use the small angle formula to approximate the cosine.

EVALUATE   For small angles, $\cos\theta \approx 1 - \theta^2/2$, so the potential energy becomes

$$U \approx mgL\left[1 - \left(1 - \frac{\theta^2}{2}\right)\right] = \frac{1}{2}mgL\theta^2$$

which is proportional to the angular displacement squared, as indicated in the problem statement.

ASSESS   You can do a Taylor-series expansion of $\cos\theta$ to verify the small-angle formula for $\cos\theta$.

63.   **INTERPRET**   This problem involves simple harmonic motion, conservation of energy, and rotational inertia. We are asked to derive the equation of motion from energy considerations for the spring-cylinders system shown in Figure 13.33.

DEVELOP   With reference to Equation 10.20 (and the condition $v = \omega R$ for rolling without slipping), the kinetic energy of the system is

$$K = \frac{1}{2}Mv^2 + \frac{1}{2}I_{cm}\omega^2 = \frac{1}{2}Mv^2 + \frac{1}{2}\left(\frac{1}{2}MR^2\right)\left(\frac{v}{R}\right)^2 = \frac{3}{4}Mv^2$$

The potential energy of the spring is $U = kx^2/2$. Combining these two, the total mechanical energy is

$$E = K + U = \frac{3}{4}Mv^2 + \frac{1}{2}kx^2$$

$$= \frac{3}{4}M\left(\frac{dx}{dt}\right)^2 + \frac{1}{2}kx^2$$

$U = \frac{3}{4}M(dx/dt)^2 + \frac{1}{2}kx^2.$

where we have used $v = dx/dt$. Differentiate this expression to find an expression involving acceleration ($a = d^2x/dt^2$).

EVALUATE   Differentiating, we find:

$$\frac{dE}{dt} = 0 = \frac{3}{4}M \cdot 2\left(\frac{dx}{dt}\right)\left(\frac{d^2x}{dt^2}\right) + \frac{1}{2}k \cdot 2x\left(\frac{dx}{dt}\right) \quad \text{or} \quad \frac{d^2x}{dt^2} = -\frac{2k}{3M}x \equiv -\omega^2 x$$

where we have recognized the prefactor in the last term to be the angular frequency ω. Thus, we have

$$\frac{d^2x}{dt^2} = -\omega^2 x$$

which is the same as the expression found in Problem 62, except that for this problem the angular freqency contains different factors, which reflects the difference in the geometry of the two problems.

ASSESS   The energy method is particularly convenient for analyzing small oscillations, since complicated details of the forces can be avoided.

65. **INTERPRET** This problem involves underdamped harmonic oscillation. We are asked to find how many oscillations a mass-spring system will execute before the damping stops the motion.

DEVELOP From Equation 13.17, the time for the amplitude to decay to 1/e of its original value is $t = 2m/b$ while the period is $T = 2\pi\sqrt{m/k}$. Dividing the former by the latter gives the number of oscillations that occur before the amplitude is reduced to 1/e of its initial value.

EVALUATE Inserting the given quantities, we find

$$N = \frac{t}{T} = \frac{2m}{2\pi b}\sqrt{\frac{k}{m}} = \frac{\sqrt{km}}{\pi b} = \frac{\sqrt{(3.3 \text{ N/m})(0.25 \text{ kg})}}{\pi(8.4 \times 10^{-3} \text{ kg/s})} = 34$$

complete oscillations occur.

ASSESS More oscillations will occur for a larger mass, or a stronger spring constant.

67. **INTERPRET** This problem involves the simple harmonic motion of a vertical mass-spring system. Given the spring parameters, we are to find the amplitude and period of the resulting motion.

DEVELOP The distance from the initial position of the mass on the unstretched spring, to the equilibrium position, where the net force is zero, is just the amplitude of the oscillations, since the initial velocity for a dropped mass is zero. Because the net force at the equilibrium position is zero (by Newton's second law, $F_{net} = ma$ the force must be zero because $a = 0$ at equilibrium) we have

$$F_{net} = 0 = kA - mg$$
$$kA = mg$$

from which we can find the amplitude A. To find the period, recall that the force due to gravity does not affect the period, only the equilibrium position. Thus, the period is given by Equation 13.7 c.

EVALUATE (a) Solving the expression above for A gives $A = mg/k = (0.49 \text{ kg})(9.8 \text{ m/s}^2)/(74 \text{ N/m}) = 6.5$ cm.
(b) The period is $T = 2\pi\sqrt{m/k} = 2\pi\sqrt{(0.49 \text{ kg})/(74 \text{ N/m})} = 0.51$ s.

ASSESS For part (a), we can also consider dropping the mass from the unstretched position (zero spring force), so the initial acceleration is at its maximum magnitude, which is just g, so $a_{max} = g = \omega^2 A = k/(mA)$, or $A = mg/k$ as before.

69. **INTERPRET** We're given a general equation for the potential energy of a particle and asked to find the frequency of its simple harmonic motion.

DEVELOP Recall that the force on a particle is related to the potential energy through Equation 7.8:
$F = -dU/dx$.

EVALUATE The force on the particle as a function of distance is

$$F = -\frac{d}{dx}\left[ax^2\right] = -2ax$$

This has the form of a restoring force, as in Equation 13.2 with $k = 2a$. So the particle undergoes simple harmonic motion with a frequency given by Equation 13.7b:

$$f = \frac{1}{2\pi}\sqrt{\frac{2a}{m}}$$

ASSESS What this tells us is that we have simple harmonic motion whenever the potential energy is proportional to the distance squared.

71. **INTERPRET** In this problem we are given two mass-spring systems with same mass but different frequencies, and we want to compare their energies and maximum accelerations.

DEVELOP As shown in Section 13.5, the energy of a mass-spring system is $E = m\omega^2 A^2/2$. Also, from Equation 13.10, the maximum acceleration is $a_{max} = \omega^2 A$. Use the fact that $\omega_1 = 2\omega_2$ to find the ratio of the energies and accelerations.

EVALUATE   (a) If m and A are the same but $\omega_1 = 2\omega_2$, we have

$$\frac{E_1}{E_2} = \frac{m\omega_1^2 A/2}{m\omega_2^2 A/2} = \left(\frac{\omega_1}{\omega_2}\right)^2 = (2)^2 = 4$$

$$E_1 = 4E_2$$

(b) Comparing the maximum accelerations of the two systems, we find

$$\frac{a_{max,1}}{a_{max,2}} = \frac{\omega_1^2 A}{\omega_2^2 A} = \left(\frac{\omega_1}{\omega_2}\right)^2 = (2)^2 = 4$$

$$a_{max,1} = 4a_{max,2}$$

ASSESS   Both the energy and the maximum acceleration of the mass-spring system increase with ω^2. Doubling the frequency quadruples E and a_{max}.

73.   INTERPRET   This problem involves simple harmonic motion. Given the maximum tension that could be sustained by a thread, we want to find the maximum-allowable amplitude for the pendulum motion.

DEVELOP   It is shown in Example 13.3 that the greatest tension in a simple pendulum occurs at the bottom of its swing, where $T_{max} = mg(1 + A^2)$ and A is the angular amplitude (i.e., $A = \theta_{max}$).

EVALUATE   For the thread in this problem, the maximum tension is $T_{max} = 6.0$ N. Therefore, the maximum allowable amplitude is

$$A_{max} = \sqrt{\frac{T_{max}}{mg} - 1} = \sqrt{\frac{6.0 \text{ N}}{(0.50 \text{ kg})(9.8 \text{ m/s}^2)} - 1} = 0.474 \text{ rad} = 27°$$

ASSESS   If the maximum tension the thread can sustain is only equal to the weight of the object, $T_{max} = mg$, then $A = 0$, which implies that any pendulum motion will break the thread.

75.   INTERPRET   The rolling ball on the track executes simple harmonic motion. Given the vertical height y as a function of the position, we can determine its period of oscillation by applying the principle of conservation of total mechanical energy. To do so, we choose the bottom of the track to have zero potential energy.

DEVELOP   The potential energy of the ball, relative to the bottom of the track, is

$$U(x) = mgy = mgax^2$$

and its kinetic energy from rolling (without slipping) is

$$K = \frac{1}{2}mv^2 + \frac{1}{2}I_{cm}\omega^2 = \frac{1}{2}mv^2 + \frac{1}{2}\left(\frac{2}{5}mR^2\right)\left(\frac{v}{R}\right)^2 = \frac{7}{10}mv^2$$

where we have used $I_{cm} = 2mR^2/5$ from Table 10.2. Since the total mechanical energy, $E = U + K$ is constant, we have

$$0 = \frac{dE}{dt} = \frac{d}{dt}\left(mgax^2 + \frac{7}{10}mv^2\right) = 2mgax\frac{dx}{dt} + \frac{7}{5}mv\frac{dv}{dt}$$

If we assume that $v = \sqrt{(dx/dt)^2 + (dy/dt)^2} \approx dx/dt$ (i.e., the vertical displacement is small), then the above expression can be simplified to

$$0 = 2mgax\frac{dx}{dt} + \frac{7}{5}mv\frac{dv}{dt} = 2mgaxv + \frac{7}{5}mv\frac{d^2x}{dt^2} = mv\left(2gax + \frac{7}{5}\frac{d^2x}{dt^2}\right)$$

or

$$\frac{d^2x}{dt^2} = -\frac{10}{7}gax$$

This expression allows us to solve for the angular frequency, and hence the period.

EVALUATE   Comparing the above expression with Equation 13.3 and with the help of Equation 13.7a, we see that the angular frequency of the oscillation is

$$\omega = \sqrt{\frac{10ga}{7}}$$

Therefore, the period is

$$T = \frac{2\pi}{\omega} = 2\pi\sqrt{\frac{7}{10ga}}$$

ASSESS   In Problem 64, the period of the sliding point mass is found to be $2\pi/\sqrt{2ga}$. Therefore, we see that the period for the rolling ball is longer, which is expected because increasing rotational inertia increases the period. Checking the units of the expression for T gives

$$\sqrt{\frac{1}{(m/s^2)(m^{-1})}} = s$$

as expected for a period.

77.   **INTERPRET**   Since the two blocks stick together, the collision is completely inelastic. That means some of the kinetic energy is lost, but momentum will still be conserved. We can use this fact to help find the frequency, amplitude, and phase constant after the two masses stick and oscillate together.

$$f_2 = \frac{\omega_2}{2\pi} = \frac{1}{2\pi}\sqrt{\frac{k}{m_1 + m_2}} = \frac{3.39\ \text{s}^{-1}}{2\pi} = 0.54\ \text{Hz}$$

$$A = \sqrt{x(t_c)^2 + [-v(t_c)/\omega_2]^2} = \sqrt{(10\ \text{cm})^2 + (68\ \text{cm}/3.39)^2} = 22.4\ \text{cm} \approx 22\ \text{cm}$$

$$\phi = \tan^{-1}\left(\frac{-v(t_c)}{\omega_2 x(t_c)}\right) - \omega_2 t_c = \tan^{-1}\left(\frac{68\ \text{cm/s}}{(3.39\ \text{s}^{-1})(10\ \text{cm})}\right) - 69.7° = -6.22° = -0.11\ \text{rad}$$

DEVELOP   The simple harmonic motion with just the first block on the spring can be described by Equation 13.8 and the given amplitude and phase constant;

$$x(t) = (10\ \text{cm})\cos(\omega_1 t - \pi/2) = (10\ \text{cm})\sin\omega_1 t$$

where

$$\omega_1 = \sqrt{\frac{k}{m_1}} = \sqrt{\frac{23\ \text{N/m}}{1.2\ \text{kg}}} = 4.38\ \text{s}^{-1}$$

This equation holds up to the time of the collision, i.e., for $t < t_c$, where $t_c = \frac{\pi}{2\omega_1}$, since for the rightmost point of oscillation, $\sin\omega_1 t_c = 1$ or $\omega_1 t_c = \pi/2$. (This specifies the original zero of time appropriate to the given phase constant of $-\pi/2$.)

Equation 13.8 also describes the simple harmonic motion after the collision;

$$x(t) = A\cos(\omega_2 t + \phi)$$

for $t > t_c$, where $\omega_2 = \sqrt{k/(m_1 + m_2)} = 3.39\ \text{s}^{-1}$ is the angular frequency when both blocks oscillate on the spring. It follows from this that

$$v(t) = -\omega_2 A\sin(\omega_2 t + \phi)$$

The amplitude A and phase constant ϕ can be determined from these two equations evaluated just after the collision, essentially at t_c, if we assume that the collision takes place almost instantaneously; then conservation of momentum during the collision can be applied (see Equation 9.11). Just after the collision, $x(t_c) = 10$ cm (given) and

$$v(t_c) = \frac{m_1 v_1 + m_2 v_2}{m_1 + m_2}$$

where just before the collision, $v_1 = 0$ (given m_1 at rightmost point of its original motion) and $v_2 = -1.7$ m/s (also given).

EVALUATE   The frequency of oscillation after the inelastic collision is

$$f_2 = \frac{\omega_2}{2\pi} = \frac{1}{2\pi}\sqrt{\frac{k}{m_1 + m_2}} = \frac{3.39\ \text{s}^{-1}}{2\pi} = 0.540\ \text{Hz}$$

To solve for the amplitude, we note that numerically,

$$v(t_c) = v(t_c) = \frac{m_1 v_1 + m_2 v_2}{m_1 + m_2} = \frac{0 + (0.8 \text{ kg})(-1.7 \text{ m/s})}{1.2 \text{ kg} + 0.8 \text{ kg}} = -0.68 \text{ m/s}$$

Thus, with

$$x(t_c) = 10 \text{ cm} = A\cos(\omega_2 t_c + \phi),$$
$$v(t_c) = -68 \text{ cm/s} = -\omega_2 A \sin(\omega_2 t_c + \phi)$$

we solve for A (using $\sin^2 + \cos^2 = 1$) and find

$$A = \sqrt{x(t_c)^2 + [-v(t_c)/\omega_2]^2} = \sqrt{(10 \text{ cm})^2 + (68 \text{ cm}/3.39)^2} = 22.4 \text{ cm}$$

To find the phase, we first note that

$$\omega_2 t_c = \omega_2 \left(\frac{\pi}{2\omega_1} \right) = \frac{\pi}{2} \frac{\sqrt{k/(m_1 + m_2)}}{\sqrt{k/m_1}} = \frac{\pi}{2} \sqrt{\frac{m_1}{m_1 + m_2}} = \frac{\pi}{2} \sqrt{\frac{1.2 \text{ kg}}{1.2 \text{ kg} + 0.8 \text{ kg}}} = 1.22 \text{ radians}$$
$$= 69.7°$$

Solving for ϕ (using $\sin\alpha/\cos\alpha = \tan\alpha$), we find $\tan(\omega_2 t_c + \phi) = \frac{-v(t_c)/\omega_2}{x(t_c)}$, or

$$\phi = \tan^{-1}\left(\frac{-v(t_c)}{\omega_2 x(t_c)} \right) - \omega_2 t_c = \tan^{-1}\left(\frac{68 \text{ cm/s}}{(3.39 \text{ s}^{-1})(10 \text{ cm})} \right) - 69.7° = 63.5° - 69.7° = -6.22°$$
$$= -0.109 \text{ rad}$$

ASSESS The solution for A is equivalent to calculating the various energies in the second simple harmonic motion, since just after the collision,

$$K(t_c) = \frac{1}{2}(m_1 + m_2)v(t_c)^2 = \frac{1}{2}(2 \text{ kg})(-0.68 \text{ m/s})^2 = 0.462 \text{ J}$$

$$U(t_c) = \frac{1}{2}kx(t_c)^2 = \frac{1}{2}(23 \text{ N/m})(0.1 \text{ m})^2 = 0.115 \text{ J}$$

$$E = K(t_c) + U(t_c) = 0.577 \text{ J} = \frac{1}{2}kA^2$$

or $A = \sqrt{2(0.577 \text{ J})/(23 \text{ N/m})} = 22.4 \text{ cm}$. Once A is known, ϕ can also be found from either expression for $x(t_c)$ or $v(t_c)$, e.g.,

$$\omega_2 t_c + \phi = \cos^{-1}\left(\frac{10 \text{ cm}}{22.4 \text{ cm}} \right) = \sin^{-1}\left(\frac{68 \text{ cm/s}}{(3.39 \text{ s}^{-1})(22.4 \text{ cm})} \right)$$

79. **INTERPRET** In this problem, we want to show by direct substitution that $x = A\cos(\omega_d t + \phi)$ is a solution to the differential equation given in Equation 13.18, with A given by Equation 13.19.

DEVELOP The algebra involved in this problem is straightforward but somewhat tedious.

EVALUATE When $x = A\cos(\omega_d t + \phi)$ is substituted into Equation 13.18, one obtains

$$m\left[-\omega_d^2 A \cos(\omega_d t + \phi) \right] = -kA\cos(\omega_d t + \phi) - b\left[-\omega_d A \sin(\omega_d t + \phi) \right]$$
$$+ F_0 \left[\cos(\omega_d t + \phi)\cos\phi + \sin(\omega_d t + \phi)\sin\phi \right]$$

where we let $\omega_d t = \omega_d t + \phi - \phi$ in the F_0 term, and used a trigonometric identity. This equation is true if the coefficient of the $\sin(\omega_d t + \phi)$ and $\cos(\omega_d t + \phi)$ terms on each side are equal, respectively, that is,

$$-m\omega_d^2 A = -kA + F_0 \cos\phi$$
$$0 = b\omega_d A + F_0 \sin\phi$$

Let $\omega_0^2 = k/m$ and these equations become

$$F_0 \cos\phi = -m(\omega_d t - \phi)A$$
$$F_0 \sin\phi = -b\omega_d A$$

Squaring and adding, we get Equation 13.19:

$$A = \frac{F_0/m}{\sqrt{\left(\omega_d^2 - \omega_0^2\right)^2 + b^2\omega_d^2/m^2}}$$

ASSESS   By substitution, we have verified that the function $x = A\cos(\omega_d t + \phi)$ indeed is a solution to the differential equation

$$m\frac{d^2x}{dt^2} = -kx - b\frac{dx}{dt} + F_0\cos\omega_d t$$

81. **INTERPRET**   We are to show that $x(t) = a\cos(\omega t) - b\sin(\omega t)$ represents simple harmonic motion, or that it is equivalent to $x(t) = A\cos(\omega t + \phi)$, where $A = \sqrt{a^2 + b^2}$ and $\phi = \operatorname{atan}(b/a)$. To do this, we use trigonometric identities.

DEVELOP   Use the trigonometric identity $\cos(\alpha + \beta) = \cos\alpha\cos\beta - \sin\alpha\sin\beta$, using $x(t) = A\cos(\omega t + \phi)$ as a starting point.

EVALUATE

$$x(t) = A\cos(\omega t + \phi) = A(\cos\omega t\cos\phi - \sin\omega t\sin\phi)$$
$$= (A\cos\phi)\cos\omega t - (A\sin\phi)\sin\omega t$$

Now we draw an arbitrary right triangle, as shown in the figure below. We can see from the figure that $a = A\cos\phi$, $b = A\sin\phi$, $A = \sqrt{a^2 + b^2}$ and $\tan\phi = b/a$. Inserting these results into the expression above for $x(t)$ gives $x(t) = a\cos(\omega t) - b\sin(\omega t)$.

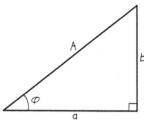

ASSESS   This is as much a trigonometry problem as a physics problem—it allows us to see that simple harmonic motion can be expressed in many ways. You can also demonstrate that $x(t) = a\cos(\omega t) - b\sin(\omega t)$ represents simple harmonic motion by substituting it into Equation 13.3, which gives

$$m\frac{d^2}{dt^2}\left[a\cos(\omega t) - b\sin(\omega t)\right] = -\omega^2 m\left[a\cos(\omega t) - b\sin(\omega t)\right] = -\omega^2 mx(t)$$

which satisfies Equation 13.3 provided $\omega = \sqrt{k/m}$.

83. **INTERPRET**   A pendulum at rest on Earth has a given frequency. The frequency becomes higher when accelerating. We use a vector sum to find the effective value of g in the simple pendulum equation and calculate the acceleration. We will use the simple pendulum approximation.

DEVELOP   Start with a sketch, as shown in the figure below. From the ratio of the initial period

$T = 2\pi\sqrt{L/g} = \frac{60\text{ s}}{90\text{ cycles}} = 0.667$ s and the final period $T' = 2\pi\sqrt{L/g_{eff}} = \frac{60\text{ s}}{91\text{ cycles}} = 0.659$ s, we can calculate g_{eff}. Once we know g_{eff}, we can use $g_{eff}^2 = g^2 + a^2$ to find a.

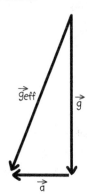

EVALUATE Taking the ratio of the periods gives

$$\frac{T}{T'} = \sqrt{\frac{g_{eff}}{g}}$$

$$g_{eff} = g\left(\frac{T}{T'}\right)^2$$

so

$$a = \sqrt{g_{eff}^2 - g^2} = \sqrt{g^2\left(\frac{T}{T'}\right)^4 - g^2} = g\sqrt{\left(\frac{T}{T'}\right)^4 - 1}$$

$$= \left(9.8 \text{ m/s}^2\right)\sqrt{\left(\frac{0.667 \text{ s}}{0.659 \text{ s}}\right)^4 - 1} = 2.1 \text{ m/s}^2$$

ASSESS It turns out that our use of ratios eliminates the need to know much of anything about the physical setup. In fact, we get the same result even if we use the equation for the physical pendulum, rather than for the simple pendulum.

85. **INTERPRET** We are asked to consider the body mass measuring device used to measure the mass of astronauts in space.

DEVELOP The oscillation period of the BMMD is related to the mass, m_a, of a given astronaut by

$$T = 2\pi\sqrt{\frac{m_c + m_a}{k}}$$

where m_c is the mass of the chair, and k is the spring constant.

EVALUATE The period for a 90-kg astronaut can be related to the period for a 60-kg astronaut:

$$\frac{T_{90}}{T_{60}} = \sqrt{\frac{20 \text{ kg} + 90 \text{ kg}}{20 \text{ kg} + 60 \text{ kg}}} = 1.17$$

The time is therefore 17% longer with the more massive astronaut.

The answer is (c).

ASSESS The 90-kg astronaut is 50% more massive than the 60-kg astronaut, but the period does not scale linearly with the mass, as we have just shown.

87. **INTERPRET** We are asked to consider the body mass measuring device used to measure the mass of astronauts in space.

DEVELOP Let's assume the astronaut's mass declines according to $m_a(t) = m_{a0} - bt$.

EVALUATE Taking the time derivative of the equation relating the period to the mass:

$$\frac{dT}{dt} = \frac{d}{dt}\left[2\pi\sqrt{\frac{m_c + m_{a0} - bt}{k}}\right] = \frac{T_0}{2}\frac{-b/M_0}{\sqrt{1 - bt/M_0}}$$

Where T_0 and M_0 are the period and total mass at the beginning when $t = 0$. Since the derivative is negative, the period is decreasing. We take the second derivative to see the rate at which the period is decreasing:

$$\frac{d^2T}{dt^2} = \frac{d}{dt}\left[\frac{T_0}{2}\frac{-b/M_0}{\sqrt{1 - bt/M_0}}\right] = -\frac{T_0}{4}\left(\frac{b}{M_0}\right)^2\left(1 - bt/M_0\right)^{-3/2}$$

The second derivative is negative, which means the rate is getting more negative with time. Or to say it another way its decreasing at an ever-increasing rate.

The answer is (c).

ASSESS We can check that our result makes sense by putting in some representative values for the time. At $t = 0$, the period is decreasing at $\dot{T} = -\frac{1}{2}\left(T_0 b/M_0\right)$, whereas at $t = 3M_0/4b$, it is decreasing at $\dot{T} = -\left(T_0 b/M_0\right)$, which is twice as fast. So yes, the period is decreasing at an ever-increasing rate.

14

WAVE MOTION

EXERCISES

Section 14.1 Waves and Their Properties

17. **INTERPRET** This problem is about wave propagation. Given the speed and frequency of the ripples, we are asked to compute the period and the wavelength of the wave.

    DEVELOP Equation 14.1 relates the speed of the wave to its period, frequency, and wavelength:

    $$v = \frac{\lambda}{T} = \lambda f$$

    Apply this equation to solve the problem.

    EVALUATE Equation 14.1 gives (a) $T = 1/f = 1/(5.2 \text{ Hz}) = 0.19 \text{ s}$, and (b)
    $\lambda = v/f = (34 \text{ cm/s})/(5.2 \text{ Hz}) = 6.5 \text{ cm}$.

    ASSESS The unit of frequency is Hz, with $1 \text{ Hz} \equiv 1 \text{ s}^{-1}$. If the frequency is kept fixed, then increasing the wavelength will increase the speed of propagation.

19. **INTERPRET** This problem is about wave propagation. Given the speed and frequency of various electromagnetic waves, we are asked to compute their wavelength.

    DEVELOP Equation 14.1 relates the speed of the wave to its period, frequency, and wavelength:

    $$v = \frac{\lambda}{T} = \lambda f \quad \text{or} \quad \lambda = \frac{v}{f}$$

    Use this equation to solve the problem.

    EVALUATE Since the speed of propagation of electromagnetic waves in vacuum is simply the speed of light, $v = c = 3.0 \times 10^8$ m/s, Equation 14.1 gives

    (a) $\lambda = c/f = \frac{3.0 \times 10^8 \text{ m/s}}{1.0 \times 10^6 \text{ Hz}} = 3.0 \times 10^2 \text{ m/s} = 300 \text{ m}$;

    (b) $\lambda = c/f = \frac{3.00 \times 10^8 \text{ m/s}}{190 \times 10^6 \text{ Hz}} = 1.58 \text{ m}$;

    (c) $\lambda = c/f = \frac{3.0 \times 10^8 \text{ m/s}}{10 \times 10^9 \text{ Hz}} = 0.03 \text{ m} = 3.0 \text{ cm}$;

    (d) $\lambda = c/f = \frac{3 \times 10^8 \text{ m/s}}{4 \times 10^{13} \text{ Hz}} = 8 \times 10^{-6} \text{ m} = 8 \text{ μm}$;

    (e) $\lambda = c/f = \frac{3.0 \times 10^8 \text{ m/s}}{6.0 \times 10^{14} \text{ Hz}} = 5.0 \times 10^{-7} \text{ m} = 500 \text{ nm}$;

    (f) $\lambda = c/f = \frac{3.0 \times 10^8 \text{ m/s}}{1.0 \times 10^{18} \text{ Hz}} = 3.0 \times 10^{-10} \text{ m} = 3.0 \text{ Å}$ (See Appendix C for units.)

    ASSESS If the speed of propagation is kept fixed, then a higher frequency means a shorter wavelength.

21. **INTERPRET** We are asked to find the wavelengths of ultrasound waves used in medical imaging.

    DEVELOP The wavelength is related to the speed and frequency through Equation 14.1: $\lambda = v/f$.

    EVALUATE (a) For fetal imaging, the ultrasound wavelength is

    $$\lambda = \frac{v}{f} = \frac{1500 \text{ m/s}}{8.0 \times 10^6 \text{ Hz}} = 0.19 \text{ mm}$$

(b) For adult kidney imaging, the ultrasound wavelength is

$$\lambda = \frac{v}{f} = \frac{1500 \text{ m/s}}{3.5 \times 10^6 \text{ Hz}} = 0.43 \text{ mm}$$

ASSESS   The wavelength gives us a sense of the image resolution. Both techniques should be able to discriminate objects that are at least a millimeter-wide.

Section 14.2 Wave Math

23.   INTERPRET   We are given a function that describes a traveling sinusoidal wave, and asked to compute various physical quantities associated with the wave.

DEVELOP   Consider a traveling wave of the form given in Equation 14.3:

$$y(x,t) = A\cos(kx \pm \omega t)$$

The amplitude of the wave is A; its wavelength is given by Equation 14.2; $\lambda = 2\pi/k$, its period is given by Equation 13.5; $T = 2\pi/\omega$. The speed of propagation is $v = \lambda f = \omega/k$, and the direction of propagation is $+x$ if the argument is $kx - \omega t$ and $-x$ if the argument is $kx + \omega t$.

EVALUATE   (a) Comparing $y = 1.3\cos(0.69x + 31t)$ with Equation 14.3, we find the amplitude to be $A = 1.3$ cm. (b) The wavelength is

$$\lambda = \frac{2\pi}{k} = \frac{2\pi}{0.69 \text{ cm}^{-1}} = 9.1 \text{ cm}$$

(c) Equation 13.5 gives

$$T = \frac{2\pi}{\omega} = \frac{2\pi}{31 \text{ s}^{-1}} = 0.20 \text{ s}$$

(d) The speed of the wave is

$$v = \frac{\omega}{k} = \frac{31 \text{ s}^{-1}}{0.69 \text{ cm}} = 45 \text{ cm/s.}$$

(e) A phase of the form $kx + \omega t$ describes a wave propagating in the negative x direction.

ASSESS   This problem demonstrates that the wave function of the form given in Equation 14.3 contains all the information about the amplitude, the wavelength, the period, the phase and the speed of propagation of a wave. Thus, once the wave function is given, all these quantities can be calculated.

25.   INTERPRET   The problem concerns a simple harmonic wave that we wish to characterize by its vertical displacement as a function of time.

DEVELOP   The wave number is related to the wavelength: $k = 2\pi/\lambda$ (Equation 14.2). The angular frequency is related to the wave speed: $\omega = vk$ (Equation 14.4). A simple harmonic wave will have the form of a sinusoidal function as in Equation 14.3: $y(x,t) = A\sin(kx \pm \omega t)$.

EVALUATE   (a) The angular frequency of this wave is

$$\omega = \frac{2\pi v}{\lambda} = \frac{2\pi(35 \text{ cm/s})}{16 \text{ cm}} = 14 \text{ s}^{-1}$$

(b) The wave number is

$$k = \frac{2\pi}{\lambda} = \frac{2\pi}{16 \text{ cm}} = 0.39 \text{ cm}^{-1}$$

(c) We're told that the wave is propagating in the negative x direction, which means we use the positive sign in Equation 14.3. We're also told that y is at a maximum at $x=0$ when $t=0$. We could include a phase constant, or just as easily change to the cosine function, which has its maximum value at the origin:

$$y(x,t) = (2.5 \text{ cm})\cos\left[\left(0.39 \text{cm}^{-1}\right)x + \left(14\text{s}^{-1}\right)t\right]$$

ASSESS The wave has a 2.5-cm-high crest at $x=0$ when $t=0$. At a later time, t, this crest should be found at $x = -vt$. We can verify this by plugging this x value into the displacement equation:

$$y(-vt, t) = (2.5 \text{ cm}) \cos\left[(0.39 \text{cm}^{-1})(-35 \text{cm/s})t + (14 \text{s}^{-1})t\right] = 2.5 \text{ cm}$$

The displacement remains constant as long as we move with the wave at its wave speed.

Section 14.3 Waves on a String

27. INTERPRET Given the tension in the cable and the linear mass density (mass per unit length) of the cable, we want to find the speed at which a transverse wave propagates along the crystal.

DEVELOP The relationship between the speed of propagation, the tension, and the mass per unit length of a cable is given by Equation 14.5, $v = \sqrt{F/\mu}$. Given F and μ, we can calculate v.

EVALUATE The speed of the transverse wave in the cables is

$$v = \sqrt{\frac{F}{\mu}} = \sqrt{\frac{2.5 \times 10^8 \text{ N}}{4100 \text{ kg/m}}} = 250 \text{ m/s}$$

to two significant figures.

ASSESS Increasing the tension results in a greater acceleration of the disturbed cables, and hence the wave propagates more rapidly.

29. INTERPRET In this problem we are given the mass per unit length of a spring and the speed of propagation of a transverse wave on the spring, and we would like to know the tension in the spring.

DEVELOP The relationship between the speed of propagation, the tension, and the mass per unit length of the medium is given by Equation 14.6, $v = \sqrt{F/\mu}$. Given v and μ, we can calculate F.

EVALUATE Using Equation 14.6, we find the tension in the spring is

$$F = \mu v^2 = (0.17 \text{ kg/m})(6.7 \text{ m/s})^2 = 7.6 \text{ N}$$

ASSESS Our result indicates that, keeping μ fixed, increasing the tension will result in a greater propagation speed.

31. INTERPRET This problem requires us to find the average power carried by a wave propagating along a rope. We are given mass per unit length of the rope and its tension and the frequency and amplitude of the wave, from which we can find the average power.

DEVELOP The average power transmitted by transverse traveling waves in a string is given by Equation 14.7 $\bar{P} = \frac{1}{2}\mu\omega^2 A^2 v$. The speed of propagation can be obtained by using Equation 14.6 $v = \sqrt{F/\mu}$.

EVALUATE Using the values given in the problem statement, we find the average power is

$$\bar{P} = \frac{1}{2}\mu\omega^2 A^2 v = \frac{1}{2}\mu\omega^2 A^2 \sqrt{\frac{F}{\mu}} = \frac{1}{2}(0.28 \text{ kg/m})(2\pi \text{ rad} \times 3.3 \text{ s}^{-1})^2 (0.061 \text{ m})^2 \sqrt{\frac{550 \text{ N}}{0.28 \text{ kg/m}}}$$
$$= 9.9 \text{ W}$$

ASSESS The wave power is proportional to the speed of propagation. It is also proportional to the square of the amplitude and the square of the angular frequency.

Section 14.4 Sound Waves

33. INTERPRET This problem involves finding the wave speed of sound waves traveling through air under standard conditions.

DEVELOP Use Equation 14.9, $v = \sqrt{\gamma P/\rho}$ relates the speed of sound v to pressure P and density ρ. For air, $\gamma = 7/5$.

EVALUATE Inserting the given quantities, we find the speed is

$$v = \sqrt{\gamma P/\rho} = \sqrt{\frac{7(101 \times 10^3 \text{ N/m}^2)}{5(1.20 \text{ kg/m}^3)}} = 343 \text{ m/s}.$$

ASSESS This is the accepted value for the speed of sound at standard temperature and pressure.

35.   **INTERPRET**   For this problem we are to find the speed of sound in a gaseous medium.

DEVELOP   The speed of sound in a medium is given by Equation 14.9:

$$v = \sqrt{\frac{\gamma P}{\rho}}$$

where γ is a constant characteristic of the gas, P is the pressure, and ρ is the density of the gas.

EVALUATE   Using the values given in the problem, the speed of sound in NO_2 gas is

$$v = \sqrt{\frac{\gamma P}{\rho}} = \sqrt{\frac{1.29\left(4.8 \times 10^4 \text{ N/m}^2\right)}{0.35 \text{ kg/m}^3}} = 420 \text{ m/s}$$

ASSESS   The speed of sound at room temperature (20 °C) is about 343 m/s. The speed depends on the thermodynamic properties of the medium.

37.   **INTERPRET**   For this problem, we are to find the frequency of a sound wave in a gaseous medium of the underwater habitat, and compare this frequency to that of sound waves in air under standard conditions.

DEVELOP   To compute the frequency, we first calculate the speed of sound in the underwater habitat using Equation 14.9, $v = \sqrt{\gamma P / \rho}$. Once v is known, we use Equation 14.1, $v = \lambda f$, to find the frequency.

EVALUATE   The speed of sound is

$$v = \sqrt{\frac{\gamma P}{\rho}} = \sqrt{\frac{1.61\left(6.2 \times 10^5 \text{ N/m}^2\right)}{4.5 \text{ kg/m}^3}} = 471 \text{ m/s}$$

Therefore, the frequency of 0.5-m wavelength sound waves is

$$f = \frac{v}{\lambda} = \frac{471 \text{ m/s}}{0.50 \text{ m}} = 940 \text{ Hz}$$

to two significant figures.

ASSESS   In "normal air," the frequency would be about

$$f = \frac{v}{\lambda} = \frac{343 \text{ m/s}}{0.50 \text{ m}} = 686 \text{ Hz}$$

Section 14.5 Interference

39.   **INTERPRET**   This problem is about wave interference. Given the condition for the second calm region where waves interfere destructively, we want to compute the wavelength of the ocean wave.

DEVELOP   The condition for destructive interference is a phase difference of $k_2 \Delta r = \left(2\pi / \lambda_2\right) \Delta r = 3\pi$, or an odd multiple of $\pi = 180°$. The second node occurs when the path difference is three half-wavelengths, or $\Delta r \equiv AP - BP = 3\lambda_2 / 2$.

EVALUATE   From Example 14.5, we have $\Delta r = 8.1$ m, so the wavelength is

$$\lambda_2 = \frac{2}{3} \Delta r = \frac{2}{3}(8.1 \text{ m}) = 5.4 \text{ m}$$

ASSESS   Comparing with Example 14.5, we expect the wavelength in this case to be shorter since at the same distance away from the source the calm region encountered here is the second one.

Section 14.7 Standing Waves

41.   **INTERPRET**   This problem is about standing-wave modes in a string that is either clamped at both ends or clamped at a single end with the opposite end free.

DEVELOP   If the string is clamped at both ends, the amplitudes there must be zero. If L is the length of the string, then the standing waves must satisfy the condition given in Equation 14.13:

$$L = \frac{m\lambda}{2} \quad m = 1, 2, 3, \ldots$$

It follows that the frequencies of the standing-wave modes of a string fixed at both ends are all the (positive) integer multiples of the fundamental frequency,

$$f_m^e = \frac{v}{\lambda_m} = m\left(\frac{v}{2L}\right) = mf_1^e \quad m = 1, 2, 3, \ldots$$

where $f_1^e = v/(2L)$ is the fundamental frequency. However, if only one end of the string is fixed, then from Figure 14.28, the last wavelength will either be a quarter-wavelength or three-quarters of a wavelength. In other words, an odd number of quarter wavelengths must fit within the total length L. Mathematically, this is expressed as

$$L = \frac{(2m+1)\lambda}{4} \quad m = 0, 1, 2, \ldots$$

and the corresponding frequencies are

$$f_m^o = \frac{v}{\lambda_m} = (2m+1)\left(\frac{v}{4L}\right) = mf_1^o, \quad m = 0, 1, 2, \ldots$$

where $f_1^o = v/(4L) = f_1^e/2$ is the fundamental frequency.

EVALUATE (a) With both ends fixed, the next higher frequency above the fundamental frequency is

$$f_2^e = 2f_1^e = 2(140 \text{ Hz}) = 280 \text{ Hz}$$

(b) The fundamental frequency for the string fixed at one end is

$$f_1^o = (2 \cdot 0 + 1)\frac{v}{4L} = \left(\frac{1}{2}\right)\left(\frac{v}{2L}\right) = \frac{1}{2}f_1^e = \frac{1}{2}(140 \text{ Hz}) = 70 \text{ Hz}$$

which is one half the fundamental frequency of the string fixed at both ends.

(c) In this case, the standing-wave frequencies are only the odd multiples of the fundamental frequency, therefore the second standing-wave mode has frequency

$$f_2^e = (2 \cdot 2 + 1)f_1^e = 3f_1^e = 3(70 \text{ Hz}) = 210 \text{ Hz}$$

ASSESS When the string is clamped at both ends, it can accommodate an integer number of half-wavelengths. However, if it's clamped only at one end and the other end is free, then the string can accommodate only an odd number of quarter-wavelengths.

43. **INTERPRET** Using a crude model, we can estimate the length of a person's vocal tract by the lowest pitch they can generate.

DEVELOP Treating the vocal tract as a pipe closed at one end, the air inside the vocal tract will form standing waves that satisfy $|\sin kL| = 1$, or

$$L = \frac{m\lambda}{4}, \quad m = 1, 3, 5, 7\ldots.$$

To write this in terms of frequency $(f = v/\lambda)$, we need the speed of sound waves at body temperature. From Equation 14.9, we know that $v = \sqrt{\gamma P/\rho}$, and in Chapter 17, we will learn that for most gases the pressure and temperature are related such that $P/\rho \propto T$, where the temperature is in Kelvin. The speed of sound at standard temperature $(20°\text{C} = 293 \text{ K})$ is 343 m/s, so at body temperature $(37°\text{C} = 310 \text{ K})$ the speed of sound is

$$v = (343 \text{ m/s})\sqrt{\frac{310 \text{ K}}{293 \text{ K}}} = 353 \text{ m/s}$$

EVALUATE Using the given fundamental mode $(m = 1)$, we can solve for the length of the vocal tract:

$$L = \frac{\lambda}{4} = \frac{v}{4f} = \frac{353 \text{ m/s}}{4(620 \text{ Hz})} = 14 \text{ cm}$$

ASSESS This seems like a reasonable length for the vocal tract. And indeed, an outside reference says the average length of the human vocal tract is about 17 cm in males and 14 cm in females.

Section 14.8 The Doppler Effect and Shock Waves

45.   **INTERPRET**   This problem involves the Doppler effect. We want to find the frequency perceived by an observer that is moving toward the source, so that the source is at rest with respect to the medium through which the wave moves.
DEVELOP   The Doppler-shifted frequency perceived by the firefighter moving toward the siren is given by Equation 14.16:

$$f' = f\left(1 + u/v\right)$$

where we use the positive sign because the observer is approaching the source at a speed $u = 120$ km/h $= 33.3$ m/s. From Problem 14.33, we know that the speed of sound in air under standard conditions is 343 m/s.
EVALUATE   Inserting the given quantities into Equation 14.16, the frequency perceived by the firefighter is

$$f' = f\left(1 + \frac{u}{v}\right) = 85 \text{ Hz}\left[1 + (33.3 \text{ m/s})/(343 \text{ m/s})\right] = 93 \text{ Hz}$$

ASSESS   As expected, because the firefighter is moving toward the sound source, the frequency he perceives is higher than when he is at rest (i.e., $f' > f$). On the other hand, $f' < f$ if he were to move away from the source.

47.   **INTERPRET**   This problem is about using the Doppler effect for light to deduce the galaxy's motion relative to Earth.
DEVELOP   The formula for the Doppler shift for light is different than for sound, but when the relative velocity u of the source with respect to the observer is very small compared to the wave speed c for light, the result is the same as Equations 14.14a and b.
EVALUATE   For the galaxy described in this problem, the observed wavelength is greater (red-shifted) than the laboratory wavelength, so the galaxy is receding with speed

$$\frac{u}{c} = \frac{\lambda'}{\lambda} - 1 = \frac{708 \text{ nm}}{656 \text{ nm}} - 1 = 7.93 \times 10^{-2}$$

$$u = 0.0793c \approx 2.38 \times 10^{7} \text{ m/s}$$

ASSESS   The red shift observed in light from distant galaxies is an indication that the universe is expanding, as suggested by the Big Bang theory.

PROBLEMS

49.   **INTERPRET**   This problem involves transverse waves on a spring. We are given the initial string tension and wave speed, and are asked to find the wave speed if the string tension is increased to a new value.
DEVELOP   The relationship between the wave speed, the tension, and the mass per unit length of a string is given by Equation 14.6, $v = \sqrt{F/\mu}$. Given the initial speed of the transverse wave and the initial tensile force on the string, we can find the mass per unit length μ. Because this quantity remains essentially constant as we increase the tension on the string. we can use this value to find the wave speed that results when the tensile force is increased to a new value. Thus, solving for μ, we find

$$\mu = \frac{F_1}{v_1^2}$$

where $F_1 = 14$ N and $v_1 = 18$ m/s. Insert this mass per unit length back into Equation 14.6 along with the new tensile force $F_2 = 40$ N to find the new wave speed v_2.

EVALUATE   The new wave speed is

$$v_2 = \sqrt{\frac{F_2}{\mu}} = v_1\sqrt{\frac{F_2}{F_1}} = \sqrt{\frac{40 \text{ N}}{14 \text{ N}}}\,(18 \text{ m/s}) = 30 \text{ m/s}$$

ASSESS   Our result indicates that, keeping μ fixed, increasing the tension will result in a greater propagation speed.

51. **INTERPRET** The problem is about the total energy carried by the traveling wave, given the string tension F, the wave amplitude A, and the wavelength λ.

DEVELOP The average wave energy, $d\bar{E}$, in a small element of string of length dx is transmitted in time dt at the wave speed, $v = dx/dt$. From Equation 14.7,

$$d\bar{E} = \bar{P}dt = \frac{1}{2}\mu\omega^2 A^2 v \; dt = \frac{1}{2}\mu\omega^2 A^2 \, dx$$

Thus, the average linear energy density is $d\bar{E}/dx = \mu\omega^2 A^2/2$. The total average energy in a wave train of length $L = 2\lambda$ is

$$\bar{E} = \frac{d\bar{E}}{dx}L = \frac{1}{2}\mu\omega^2 A^2 (2\lambda)$$

Use Equation 14.1 $v = f\lambda = \omega\lambda/(2\pi)$ and Equation 14.6 $v = \sqrt{F/\mu}$ to express this result in terms of F, λ, and A.

EVALUATE In terms of the quantities specified in this problem, the total energy in the wave train is

$$\bar{E} = \frac{1}{2}\mu\omega^2 A^2 (2\lambda) = \frac{1}{2}\left(\frac{F}{v^2}\right)\left(\frac{2\pi v}{\lambda}\right)^2 A^2 (2\lambda) = \frac{4\pi^2 F A^2}{\lambda}$$

ASSESS The relationship derived can be written as $\bar{P} = (d\bar{E}/dx)v$. For a one-dimensional wave, $\bar{P}$ is the intensity, so the average intensity equals the average energy density times the speed of wave-energy propagation. This is a general wave property.

53. **INTERPRET** This problem is about the power emitted by a localized wave source that emits uniformly in all directions. We are given its intensity at a certain distance and are asked to find the power of the source.

DEVELOP Assuming the light bulb emits uniformly in all directions, the intensity at a distance r from the light bulb that has an average power output P is given by Equation 14.8, $I = P/(4\pi r^2)$. Solve this equation for P.

EVALUATE From Equation 14.8, we have

$$P = 4\pi r^2 I = 4\pi(3.3 \text{ m})^2 (0.73 \text{ W/m}^2) = 99.9 \text{ W} \approx 1.0\times10^2 \text{ W}$$

where scientific notation is used to show the significant figures.

ASSESS The light source is a 100-watt light bulb. Note that the intensity decreases with the square of the distance.

55. **INTERPRET** This problem involves the superposition of waves. We want to show that the superposition of two harmonic waves results in a third harmonic wave.

DEVELOP Using the identity

$$\cos\alpha + \cos\beta = 2\cos\left(\frac{\alpha-\beta}{2}\right)\cos\left(\frac{\alpha+\beta}{2}\right)$$

we find that the superposition of the two waves y_1 and y_2 gives

$$y = y_1 + y_2 = A\cos(kx - \omega t) + A\cos(kx - \omega t + \phi) = 2A\cos\left(\frac{\phi}{2}\right)\cos\left(kx - \omega t + \tfrac{1}{2}\phi\right)$$

EVALUATE We see that from the expression above that the superposition of the two harmonic waves results in a third harmonic wave. Writing $y = A_s \cos(kx - \omega t + \phi_s)$ and comparing with the expression above, we find the amplitude and the phase to be

$$A_s = 2A\cos\left(\frac{\phi}{2}\right) \qquad \phi_s = \frac{\phi}{2}$$

ASSESS Let's check our results by considering the following limits: **(i)** $\phi = 0$: In this case, $y_1 = y_2$, and the resultant amplitude is simply $A_s = A + A = 2A$. **(ii)** $\phi = \pi$: In this case, we have [using the identity $\cos(\theta + \pi) = -\cos\theta$]

$$y_2 = A\cos(kx - \omega t + \pi) = -A\cos(kx - \omega t) = -y_1$$

and therefore, $y = y_1 + y_2 = 0$.

57.   **INTERPRET**   For this problem, we are to calculate how the propagation speed of a transverse wave on a spring is affected as the spring is stretched, increasing the tension force in the spring.

DEVELOP   The speed of propagation can be obtained by using Equation 14.6: $v = \sqrt{F/\mu}$. We regard the spring as a stretched string with tension, $F = k\Delta x = k(L - L_0)$. In addition, its linear mass density is $\mu = m/L$.

EVALUATE   Equation 14.5 gives the speed of transverse waves as

$$v = \sqrt{\frac{F}{\mu}} = \sqrt{\frac{k\Delta x}{m/L}} = \sqrt{\frac{k(L - L_0)}{m/L}} = \sqrt{\frac{kL(L - L_0)}{m}}$$

ASSESS   There are two effects here that affect the speed of propagation. The first is the amount of stretching, Δx. This makes the speed of propagation go up as $\sqrt{\Delta x}$. The second effect is the change in mass per unit length, characteristic of the inertia of the spring. The mass density decreases as the spring is stretched. This makes it easier for the wave to propagate on the spring.

59.   **INTERPRET**   This problem involves spherical sound waves emitted from a localized source. Therefore, the sound waves propagate in all directions (4p steradians) from the source. The total power carried by the sound waves is constant (ignoring loss mechanisms), but the intensity (i.e., power per unit area) decrease as the waves get farther from the source, much as an image on a balloon becomes larger as you blow up the balloon.

DEVELOP   As a function of distance, the intensity of spherical waves emitted from a point source is given by Equation 14.8:

$$I = \frac{P}{A} = \frac{P}{4\pi r^2}$$

Ignoring loss mechanisms, the power P carried by the waves remains constant (ignoring loss mechanisms), so we can relate the intensity measured at the two distances as follows:

$$P = I_1(4\pi r_1^2) = I_2(4\pi r_2^2)$$

Solving this for r_2 gives

$$r_2 = \pm r_1 \sqrt{\frac{I_1}{I_2}}$$

We will use the positive square root since we are interested in the distance the person needs to walk directly away from the source, which is $d = r_2 - r_1$.

EVALUATE   Inserting the given values for the intensities and distance, we find

$$r_2 = (15 \text{ m})\sqrt{\frac{750 \text{ mW/m}^2}{270 \text{ mW/m}^2}} = 25 \text{ m}$$

Thus, the person needs to walk a distance $d = r_2 - r_1 = 25 \text{ m} - 15 \text{ m} = 10 \text{ m}$ away from the source.

ASSESS   The intensity falls off as the inverse square of distance. The further you walk away from the source, the weaker the intensity. Note that if we use the negative square root for r2, we find the distance the person needs to walk directly toward the source is $d = -25 \text{ m} - 10 \text{ m} = -35 \text{ m}$.

61.   **INTERPRET**   This problem is about the relationship between the wave speed of a transverse wave on a spring and the amount the spring is stretched. As the spring is stretched, the tensile force increases, which causes the wave speed to increase. Given the relationship between the wave speed and the spring distortion (i.e., the amount by which it is stretched), we are to calculate the spring's equilibrium length.

DEVELOP   The wave speed can be obtained from Equation 14.6: $v = \sqrt{F/\mu}$. We regard the spring as a stretched string with tension $F_1 = k\Delta x = k(L_1 - L_0)$, where L_0 is its equilibrium length. In addition, its linear mass density is $\mu_1 = m/L_1$. When the spring is stretched to $L_2 = 2L_1$, the tension becomes $F_2 = k(L_2 - L_0) = k(2L_1 - L_0)$ and the linear mass density becomes $\mu_2 = m/L_2 = m/(2L_1)$. Therefore, since the speed of the transverse wave on the spring stretched a to a total length L_2 is triple that of the spring stretched to a total length L_1, we can write

$$3v_1 = v_2$$

$$3\sqrt{\frac{F_1}{\mu_1}} = \sqrt{\frac{F_2}{\mu_2}}$$

$$3\sqrt{\frac{k(L_1 - L_0)}{m/L_1}} = \sqrt{\frac{k(2L_1 - L_0)}{m/(2L_1)}}$$

which we can solve for the unstretched length L_0.

EVALUATE Solving the above expression gives

$$3\sqrt{L_1(L_1 - L_0)} = \sqrt{2L_1(2L_1 - L_0)}$$

$$9(L_1 - L_0) = 2(2L_1 - L_0)$$

$$L_0 = L_1\left(\frac{5}{7}\right)$$

ASSESS There are two effects here that affect the speed of propagation. The first one is the amount of stretching, Δx. This makes the speed of propagation go up as $\sqrt{\Delta x}$. The second effect is the change in mass per unit length, which is characteristic of the inertia of the spring. The linear mass density decreases as the spring is stretched, which makes it easier for the wave to propagate along the spring.

63. **INTERPRET** This problem is about finding the speed of the airplane, given its apparent location as the source of sound.

DEVELOP Make a diagram of the situation (see figure below). The travel time for the sound from the airplane, reaching you along a line making an angle of $\theta = 35°$ with the vertical from the airplane's actual location, is $\Delta t = d/v$, where v is the speed of sound in air under standard conditions (~343 m/s). During this time, the airplane moved a horizontal distance $\Delta x = d\sin\theta$. Once Δt and Δx are known, we can calculate the speed of the airplane using $u = \Delta x/\Delta t$.

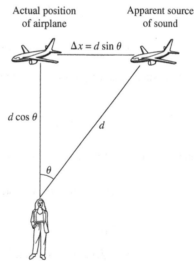

EVALUATE Using the values given in the problem statement, the speed of the airplane is

$$u = \frac{\Delta x}{\Delta t} = \frac{d\sin\theta}{d/v} = v\sin\theta = (343 \text{ m/s})\sin(35°) = 200 \text{ m/s}$$

or, approximately 440 mi/h.

ASSESS The value is reasonable for an airplane's speed. Note that the airplane's altitude, 5.2 km = $d\cos\theta$, is not needed in this calculation.

65. **INTERPRET** This problem involves converting sound intensity from the decibel scale to the linear scale.

DEVELOP The sound intensity level in decibels is given by Equation 14.10:

$$\beta = 10\log\left(\frac{I}{I_0}\right)$$

where I is the intensity (measured in W/m^2), and $I_0 = 10^{-12}$ W/m^2 is the standard threshold of hearing a 1 kHz.

EVALUATE   If the sound intensity is doubled, then $I' = 2I$. Equation 14.10 shows that

$$\beta' = 10\log\left(\frac{I'}{I_0}\right) = 10\log\left(\frac{2I}{I_0}\right) = 10\log\left(\frac{I}{I_0}\right) + 10\log 2$$

$$= \beta + 3.01$$

Thus, the decibel level increases by about 3 dB.

ASSESS   The problem demonstrates that doubling the intensity corresponds to a 3 dB increase. Human ears, however, do not respond linearly to the intensity change. For each 10-dB increase, you perceive an increase in loudness by roughly a factor of 2.

67.   **INTERPRET**   This problem deals with the variation with distance of sound intensity measured in decibels. We want to find the distance one must move away from the sound source for the loudness to drop by a factor of two.

DEVELOP   The sound intensity level in decibels is given by Equation 14.10:

$$\beta = 10\log\left(\frac{I}{I_0}\right)$$

where I is the intensity (measured in W/m^2), and $I_0 = 10^{-12}$ W/m^2 is the standard threshold of hearing a 1 kHz. For the perceived loudness to decrease by a factor of two, the decibels must decrease by 5. Therefore,

$$\beta' - \beta = -10$$

$$10\log\left(\frac{I'}{I_0}\right) - 10\log\left(\frac{I}{I_0}\right) = -10$$

$$\log\left(\frac{I'}{I}\right) = -1$$

$$\frac{I'}{I} = 10^{-1}$$

From Equation 14.8, we see that the sound intensity I drops as the inverse square of the distance from the source, so

$$\frac{I'}{I} = \frac{r^2}{r'^2}$$

where r = 2.0 m. Combining these two expressions allows us to solve for r', which is the distance at which the loudness will decease by a factor of 2.

EVALUATE   Equating the two expressions for the ratio I'/I, we find

$$\frac{r^2}{r'^2} = \frac{1}{10}$$

$$r' = \sqrt{10}r = \sqrt{10}\,(2.0\text{ m}) = 6.3\text{ m}$$

ASSESS   We find that for the loudness perceived to go down by half, the intensity I must decrease by 10 dB, or a factor of 10.

69.   **INTERPRET**   This problem is about the standing-wave condition and the requirement it imposes on the time it takes for the wave to complete one round trip in the medium. We are to show that this time is an integer number of wave periods.

DEVELOP   Because the string is clamped at both ends, the amplitudes there must be zero. If L is the length of the string, then the standing waves must satisfy the condition given in Equation 14.13:

$$L = \frac{m\lambda}{2} \quad m = 1, 2, 3, \ldots$$

On the other hand, the round-trip time for waves on a string of length L and clamped at both ends is

$$t = \frac{2L}{v} = \frac{2L}{\lambda f} = \frac{2LT}{\lambda}$$

using Equation 14.1, $v = \lambda f = \lambda / T$.

EVALUATE Substituting the first equation into the second gives

$$t = \frac{2LT}{\lambda} = \frac{m\lambda T}{\lambda} = mT \quad m = 1,2,3,\ldots$$

Therefore, we see that t is an integer multiple of the wave period.

ASSESS The conclusion can also be drawn by examining Figure 14.28. For example, the wavelength of the fundamental harmonic is $\lambda_1 = 2L$. This gives $t_1 = \lambda_1 / v = T$.

71. **INTERPRET** We are asked to show that the simple harmonic wave with a sinusoidal shape is a solution to the general wave equation.

DEVELOP Equation 14.3 describes a simple harmonic wave:

$$y(x,t) = A\cos(kx - \omega t)$$

where we choose the negative value for a wave traveling in the positive x direction. We will take the partial derivatives of y with respect to x and t, in order to see if it satisfies Equation 14.5:

$$\frac{\partial^2 y}{\partial x^2} = \frac{1}{v^2}\frac{\partial^2 y}{\partial t^2}$$

EVALUATE By the chain rule, the second derivative of y with respect to x is

$$\frac{\partial^2 y}{\partial x^2} = \frac{\partial}{\partial x}\left[-kA\sin(kx - \omega t)\right] = -k^2 A\cos(kx - \omega t) = -k^2 y$$

Similarly, the second derivative of y with respect to time t is

$$\frac{\partial^2 y}{\partial t^2} = \frac{\partial}{\partial t}\left[-(-\omega)A\sin(kx - \omega t)\right] = -\omega^2 A\cos(kx - \omega t) = -\omega^2 y$$

Since both second derivatives are proportional to y, we can equate them to arrive at:

$$\frac{-1}{k^2}\frac{\partial^2 y}{\partial x^2} = \frac{-1}{\omega^2}\frac{\partial^2 y}{\partial t^2} \quad \rightarrow \quad \frac{\partial^2 y}{\partial x^2} = \frac{1}{v^2}\frac{\partial^2 y}{\partial t^2}$$

Where we have used Equation 14.4: $v = \omega / k$. This proves that Equation 14.3 satisfies the wave equation.

ASSESS With a little more effort, one can show that a sum of harmonic waves with different wavelengths will also satisfy the wave equation:

$$y(x,t) = \sum_{i=1}^{N} A_i \cos(k_i x - \omega_i t)$$

The only requirement is that all the individual waves travel at the same speed: $v = \omega_i / k_i$.

73. **INTERPRET** This problem is about the Doppler effect in sound from a moving source, which we are to use to find the speed of a moving truck.

DEVELOP Suppose a wave source approaches you at constant speed, and you measure a wave frequency f_1 and as the source passes and then recedes, you measure frequency f_2. From Equation 14.15, we find that

$$f_1 = \frac{f}{1 - u/v} \qquad f_2 = \frac{f}{1 + u/v}$$

The two equations can be combined to solve for u, which is the speed of the source.

EVALUATE The above expressions for f_1 and f_2 imply

$$f = (1 - u/v)f_1 = (1 + u/v)f_2$$

which can be solved to give

$$\frac{u}{v} = \frac{f_1 - f_2}{f_1 + f_2} = \frac{1100 \text{ Hz} - 950 \text{ Hz}}{1100 \text{ Hz} + 950 \text{ Hz}} = 0.0732$$

For sound waves in "normal" air, this implies a truck speed of

$$u = 0.0732v = 0.0732(343 \text{ m/s}) = 25.1 \text{ m/s} = 90 \text{ km/h}$$

to two significant figures.

ASSESS   A speed of 90.4 km/h is reasonable for a truck on a freeway. Note that an increase in the speed of the truck would result in a greater difference between f_1 and f_2.

75.   **INTERPRET**   In this problem we want to find the altitude of a supersonic plane, given its speed and the time you hear the sonic boom.

DEVELOP   Since the plane is moving at a supersonic speed, shock waves are formed and the Mach angle is given by $\sin\theta = u/v$. The altitude and distance traveled in level flight are related by $h = (u\Delta t)\tan\theta$ (the shock front moves with the same speed as the aircraft).

EVALUATE   The Mach angle is $\theta = \sin^{-1}(1/2.2) = 27.0°$ for the plane. Its altitude is

$$h = (u\Delta t)\tan\theta = (2.2 \times 340 \text{ m/s})(19 \text{ s})\tan(27.0°) = 7.3 \text{ km}$$

ASSESS   The altitude is greater than the typical flying altitude of commercial aircraft (about 10,000 ft, or 3000 m). However, a high-altitude surveillance plane (such as Lockheed U-2) can fly at an altitude greater than 21 km!

77.   **INTERPRET**   This problem is about the Doppler effect, with the source receding from the observer. Given the frequency shift, we are to calculate the speed at which the source is receding.

DEVELOP   The wavelength measured by an observer as the wave source recedes is given by Equation 14.14b:

$$\lambda' = \lambda\left(1 + \frac{u}{v}\right)$$

where v is the speed of the wave, and $u = 8.2$ m/s is the speed of the source.

EVALUATE   Solving the above equation for the wave speed gives

$$v = \frac{u}{(\lambda'/\lambda) - 1} = \frac{8.2 \text{ m/s}}{1.2 - 1} = 41 \text{ m/s}$$

ASSESS   The Doppler-shifted wavelength increases as the source recedes from the observer. However, the Doppler-shifted frequency decreases, and is given by Equation 14.15, $f' = f/(1 + u/v)$.

79.   **INTERPRET**   You are trying to get out of a speeding ticket by arguing that the police radar was defective.

DEVELOP   The radar works by recording the Doppler shift in high-frequency radio waves that bounce off your car. Let's assume the radar was up the road from your car when it measured your speed, as shown in the figure below.

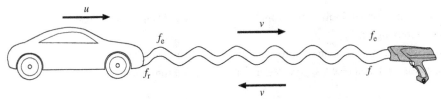

The reflection of the radio waves should be thought of as two separate interactions. First, the radio waves are "received" on the surface of your car, and second, they are "emitted" by your car's surface back to the radar gun. In this way your car is both a moving observer and a moving source.

The Doppler shift equations (14.14 and 14.15) are approximately valid in the case of radio waves, as long as the velocity of the source/observer is much less than the speed of light $(v = 3.00 \times 10^8 \text{ m/s})$. This is certainly the case for your car.

EVALUATE   As explained above, the car first acts as a moving observer, so it "receives" radio waves with frequency

$$f_r = f\left(1 + \frac{u}{v}\right)$$

where we have taken the positive sign because the car is approaching the source in our assumed scenario. This received frequency is immediately "emitted" (i.e. reflected) by the moving car with a further shifted frequency:

$$f_e = \frac{f_r}{1 - u/v}$$

where we have taken the negative sign because the car is an approaching source. The police radar picks up this reflected signal and compares it to the original frequency:

$$\Delta f = f_e - f = f\left[\frac{1 + u/v}{1 - u/v} - 1\right] \simeq 2f\frac{u}{v}$$

Here we have used the fact that $u \ll v$. Solving for the car's velocity gives

$$u = \frac{\Delta f}{2f}v = \frac{15.6 \text{ kHz}}{2(70 \text{ GHz})}\left(3.00 \times 10^8 \text{ m/s}\right) = 120 \text{ km/h}$$

The judge should rule that the radar was working properly.

ASSESS   You wrongly assumed that there is only one Doppler shift, either from the car as a moving observer or a moving source. This caused your calculation of your car's speed to be twice what it actually was.

81. **INTERPRET**   You want to verify that a Doppler radar system can measure the velocity of rain drops to the required accuracy.

    DEVELOP   In Problem 14.79, we explained how police radar work, and this Doppler radar system is the same. The radar waves reflect off the rain drops, which act both as moving observers and moving sources. Since the velocity of the rain drops is much less than the speed of light, the total frequency shift measured by the system will be:

$$\Delta f \simeq 2f\frac{u}{v}$$

    EVALUATE   The vendor's 5.0-GHz radar can detect velocities down to

$$u = \frac{\Delta f}{2f}v = \frac{50 \text{ Hz}}{2(5.0 \text{ GHz})}\left(3.00 \times 10^8 \text{ m/s}\right) = 1.5 \text{ m/s} = 5.4 \text{ km/h}$$

No, apparently the vendor's radar is not sufficient.

    ASSESS   The vendor could try to reduce the minimum frequency shift, Δf, that its radar can measure. Equally, it could increase the frequency that its device emits at, since $u \propto 1/f$.

83. **INTERPRET**   This problem involves the relationship between tension and wave speed for a transverse wave traveling on a string. We are told the mass that provides the tension in a rope and the resulting wave speed. To change the wave speed to the (greater) desired wave speed, we need to increase the mass hanging on the rope. From the ratio of the initial and final speeds, and the equation for speed of a transverse wave in a rope, we can find the mass required to give the desired wave speed.

    DEVELOP   The speed of a wave on a rope is given by Equation 14.6, $v = \sqrt{F/\mu}$, where F is the tension and m is the linear mass density of the string. The tension, in this case, is provided by hanging mass m, so (assuming a frictionless pulley and massless string) $F = mg$. The wave speed is initially $v = 18$ m/s and the desired speed is $v' = 30$ m/s. Take the ratio of v' to v to find the new mass m' needed to give the desired wave speed.

EVALUATE Taking the ratio v'/v and solving for m' gives

$$\frac{v'}{v} = \frac{\sqrt{F'/\mu}}{\sqrt{F/\mu}} = \sqrt{\frac{F'}{F}} = \sqrt{\frac{m'}{m}}$$

$$m' = m\left(\frac{v'}{v}\right)^2 = (1.4 \text{ kg})\left(\frac{30 \text{ m/s}}{18 \text{ m/s}}\right)^2 = 3.9 \text{ kg}$$

ASSESS Note that to solve this problem we never did bother to find what the tensions were, or what was μ. Using ratios to solve problems such as this allows us to avoid that extra work!

85. **INTERPRET** We explore the physics of tsunami waves.

DEVELOP If you were to follow the motion of a single water molecule at the surface of the ocean, you would see that it loops around in a nearly circular pattern as each wave goes by. Water molecules beneath the surface also loop around in generally smaller patterns, as shown in the figure below.

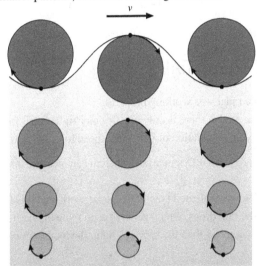

This subsurface motion is driven by pressure differences in the water, which are the result of the uneven ocean surface above. As you go deeper beneath the surface, the wave-driven pressure differences become less important. At a certain depth, water molecules no longer move in response to waves at the surface. This depth depends on the wavelength of the waves.

EVALUATE Shallow water waves are those in which subsurface motion continues all the way to the bottom. That is to say, the whole water column moves. For a tsunami to move the whole water column in deep water, its wavelength must be sufficiently long.

The answer is (a).

ASSESS One can find references that say the depth at which waves begin to behave like shallow water waves is at 1/20th of their wavelength. A typical tsunami might have a wavelength of 200 km, which means it would be a shallow water wave in the deepest parts of the ocean (around 10 km).

87. **INTERPRET** We explore the physics of tsunami waves.

DEVELOP We already know that the speed of the tsunami decreases as it approaches the shore. If the wavelength is also decreasing, then by Equation 14.1, we'd assume that the wave frequency remains roughly constant:

$f = v / \lambda.$

EVALUATE It's hard to imagine how the total energy of the wave, E_{tot}, could increase without some external force acting on it. But it's possible that the *rate* at which the wave carries energy to the shore increases. This power will be the total energy in the wave divided by the time it takes for all the wave to reach the shore, which is just the period:

$$P = \frac{E_{tot}}{T} = E_{tot} \cdot f$$

We've already argued that the total energy and the frequency are constant, so the power must be constant as well. If we were wrong and one of these quantities is increasing, then this equation says that one of the other quantities would have to change as well. So the only answer that makes sense is that none of these quantities changes. The answer is (d).

ASSESS   Ocean wave theory says that the energy per unit horizontal surface depends only on the density, ρ, and the wave amplitude squared:

$$\sigma_E = \tfrac{1}{2}\rho g A^2$$

The horizontal area of a wave is the wavelength multiplied by its width, which will call w. Therefore, the total energy of a tsunami wave would be $E_{tot} = \sigma_E \lambda w$. If indeed the total energy is constant, the wave amplitude will be proportional to

$$A \propto \frac{1}{\sqrt{\lambda}} \propto \frac{1}{\sqrt{v}} \propto \frac{1}{\sqrt[4]{d}}$$

If the depth decreases by a factor of 100, the amplitude will increase by a factor of 3.

15

FLUID MOTION

EXERCISES

Section 15.1 Density and Pressure

15. **INTERPRET** This problem requires us to calculate the total mass of a substance given its density and volume.

DEVELOP The density ρ is defined as the mass per unit volume, or $\rho = m/V$. Given the volume and the density, the mass can be calculated by solving this equation for m: $m = \rho V$. Because we are given the mass in units of kg/m3 and the volume in terms of L, we will convert L to m^3 using the conversion factor

$$(1\,\text{L})\left(\frac{10^3\,\text{cm}^3}{L}\right)\left(\frac{1\,\text{m}}{10^2\,\text{cm}}\right)^3 = 10^{-3}\,\text{m}^3$$

$$1 = 10^{-3}\,\text{m}^3/\text{L}$$

EVALUATE Inserting the given quantities $\rho = 1600$ kg/m^3 and $V = (0.75\,\text{L})(10^{-3}\,\text{m}^3/\text{L}) = 0.75 \times 10^{-3}$ m^3, we find

$$m = \rho V = \left(1600\,\text{kg}/m^3\right)\left(0.75 \times 10^{-3}\,m^3\right) = 1.2\,\text{kg}$$

ASSESS The mass is linearly proportional to the volume and to the density.

17. **INTERPRET** This problem involves calculating the density given the mass and the volume of a substance, and calculating the volume were this same mass to have a different density.

DEVELOP Use the definition of density $\rho = m/V$ to find the density of the air in the cylinder, which we will call ρ_1. For part (b), let ρ_1 go to $\rho_2 = 1.2$ kg/m3 and keep the mass the same ($m = 8.8$ kg) to calculate the new volume V_2 occupied by the air.

EVALUATE (a) The density of the air in the cylinder is

$$\rho_1 = \frac{m}{V_1} = \frac{8.8\,\text{kg}}{0.050\,\text{m}^3} = 180\,\text{kg}/m^3$$

(b) The volume occupied by this mass of air at atmospheric density is

$$V_2 = \frac{m}{\rho_2} = \frac{8.8\,\text{kg}}{1.2\,\text{kg/m}^3} = 7.3\,\text{m}^3$$

ASSESS These volumes are small enough that any variation in the density of the air due to gravity may be ignored.

19. **INTERPRET** This problem is similar to the preceding one, except that we are to convert the pressure needed to support inches of water (instead of inches of Hg) to SI units.

DEVELOP The density of water (by definition) is

$$\rho_w \equiv \left(1\,\text{g/cm}^3\right)\overbrace{\left(\frac{10^2\,\text{cm}}{\text{m}}\right)^3}^{=1}\overbrace{\left(\frac{1\,\text{kg}}{10^3\,\text{g}}\right)}^{=1} = 10^3\,\text{k g/m}^3$$

Because the pressure $p_0 = 0$ at the upper surface of the water (see Figure 15.4), the hydrostatic pressure given by Equation 15.3 reduces to $p = \rho g h$. For this problem, $\rho = \rho_w$ and $h = 1$ in $= 25.4 \times 10^{-3}$ m.

EVALUATE Inserting the given quantities into the expression for pressure gives

$$p = \left(1.00 \times 10^3 \ kg/m^3\right)\left(9.81 \ m/s^2\right)\left(25.4 \times 10^{-3} \ m\right) = 249 \ kPa$$

ASSESS The pressure needed to support 25.4 mm (= 1 in) of water is about twice that needed to support 1 mm of Hg, which is expected because Hg is more dense than water.

21. **INTERPRET** This problem involves calculating the pressure given the force and the area over which the force is applied.

DEVELOP The force exerted on the ground is the elephant's weight $F_g = mg$, which we can insert into Equation 15.1 to find the pressure.

EVALUATE The pressure exerted by the elephant's foot is

$$p = \frac{F_g}{A} = \frac{mg}{A} = \frac{(4300 \ kg)(9.8 \ m/s^2)}{\pi(0.15 \ m)^2} = 600 \ kPa.$$

ASSESS Note that this is the average pressure. An elephant's foot is undoubtedly nonuniform, so the pressure may be greater than this in some areas and less than this in other areas. Note also that this pressure is some 6 times the pressure due to the atmosphere, which seems reasonable.

Section 15.2 Hydrostatic Equilibrium

23. **INTERPRET** This problem involves calculating the density of an incompressible fluid, given the pressure increase as a function of depth.

DEVELOP For an incompressible fluid, the increase in pressure with depth is given by Equation 15.3, $p = p_0 + \rho g h$. Applying this equation to two depths h_1 and h_2, we can write

$$p_1 = p_0 + \rho g h_1$$
$$p_2 = p_0 + \rho g h_2$$
$$p_2 - p_1 = \rho g (h_2 - h_1)$$

Given that $p_2 - p_1 = 100$ kPa for $h_2 - h_1 = 6$m, we can calculate the density ρ.

EVALUATE Solving for the density, we find

$$\rho = \frac{p_2 - p_1}{g(h_2 - h_1)} = \frac{100 \times 10^3 \ Pa}{(9.8 \ m/s^2)(6.0 \ m)} = 1.7 \times 10^3 \ kg/m^3$$

ASSESS Checking the units of this expression, we find

$$\frac{Pa}{(m/s^2)(m)} = \frac{N \cdot m^{-2}}{(m/s^2)(m)} = \frac{kg \cdot m \cdot s^{-2} \cdot m^{-2}}{(m/s^2)(m)} = \frac{kg}{m^3}$$

as expected for a density.

25. **INTERPRET** This problem involves calculating the depth at which water pressure is 1 MPa.

DEVELOP Equation 15.3 $p = p_0 + \rho g h$ gives the pressure as a function of depth. For this problem, we can assume that $p_0 = 1$ atm $= 101.3$ kPa and that we are dealing with fresh water with a density of $\rho = 10^3$ kg/m^3. Thus, we can solve for the depth h at which the pressure $p = 1 \times 10^6$ Pa.

EVALUATE Solving for the depth h, we find

$$h = \frac{p - p_0}{\rho g} = \frac{(1 - 0.1013) \times 10^6 \ Pa}{(1.0 \times 10^3 \ kg/m^3)(9.8 \ m/s^2)} = 92 \ m$$

ASSESS The depth is a little less in salt water because its density is slightly greater.

27. **INTERPRET** This problem involves calculating the pressure difference needed between that pushing down on the water in the cup and that pushing down on the water in the straw to generate the force needed to lift the water up the given amount.

DEVELOP Make a sketch of the situation (see figure below) and apply Equation 15.3, which for this situation takes the form

$$p_{atm} = p_{mouth} + \rho g h$$

where $h = 0.75$ m and $p_{atm} = 101.3$ kPa.

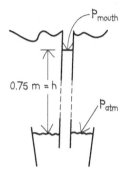

EVALUATE Solving the expression above for $p_{atm} - p_{mouth}$ gives

$$p_{atm} - p_{mouth} = \rho g h = \left(1.0 \times 10^3 \text{ kg/m}^3\right)\left(9.8 \times 10^3 \text{ N/m}^3\right)(0.75 \text{ m}) = 7.4 \text{ kPa}$$

ASSESS This corresponds to a reduction of

$$\frac{7.4 \text{ kPa}}{101.3 \text{ kPa}} = 7.3\%$$

Section 15.3 Archimedes' Principle and Buoyancy

29. **INTERPRET** This problem involves the buoyancy force, which will help us to carry a concrete block if it is submerged in water. We can use Archimedes's principal and Newton's second law to calculate the most massive concrete block we could lift underwater. We are given the mass of the largest block we can carry on land and the density of the concrete.

    DEVELOP Make a free-body diagram of the situation (see Figure 15.8). Applying Newton's second law to the concrete block gives

$$\vec{F}_{net} = m_w \vec{a}$$
$$F_b + F_{app} - m_w g = 0$$
$$F_{app} = m_w g - F_b$$

where the subscript w indicates the mass we can carry under water. The maximum force we can apply is $F_{app} = m_L g$, where $m_L = 25$ kg is the maximum mass we can carry on land. The buoyancy force on a block is $F_b = \rho_w g V_c$, where $\rho_w = 1.0 \times 10^3$ kg/m^3 is the density of water and V_c is the volume of the concrete block, which is given by $V_c = m_w/\rho_c$ where $\rho_c = 2200$ kg/m^3 is the density of the concrete. We can solve this expression for the maximum mass m_w that we can carry in water.

    EVALUATE Inserting the given quantities in the expression for F_{app} and solving for m_w gives

$$m_L g = m_w g - \rho_w g V_c = m_w g - \rho_w g \left(\frac{m_w}{\rho_c}\right) = m_w g \left(1 - \frac{\rho_w}{\rho_c}\right)$$

$$m_w = m_L \left(\frac{\rho_c}{\rho_c - \rho_w}\right) = 46 \text{ kg}$$

    ASSESS We can check this solution by looking at what happens if $\rho_c = \rho_w$. In this case, the "block" would have neutral buoyancy and we would be able to lift any size.

31. **INTERPRET** This problem involves the buoyancy force (Archimedes's principal), which applies to objects in air just as it does to objects in water. We can use this to find the fractional error that occurs when weighing Styrofoam in air as opposed to in a vacuum.

DEVELOP In a vacuum, the weight of a given volume V of Styrofoam is

$$w = mg = \rho_s V g$$

In air, the buoyancy force F_b acts against gravity, so that the apparent weight w' of the same volume of Styrofoam is

$$w' = w - F_b = \rho_s V g - \rho_w V g$$

The fractional error Δ is

$$\Delta = \frac{w - w'}{w}$$

EVALUATE Inserting the expression for the weights gives

$$\Delta = \frac{w - w'}{w} = \frac{\rho_s V g - (\rho_s V g - \rho_a V g)}{\rho_s V g} = \frac{\rho_a}{\rho_s} = \frac{1.2 \text{ kg/m}^3}{160 \text{ kg/m}^3} = 0.75\%$$

ASSESS This error of almost 1% may not be negligible in all situations

Sections 15.4 and 15.5 Fluid Dynamics and Applications

33. **INTERPRET** This problem involves calculating the flow rate of an incompressible fluid in a pipe with a varying cross section. We use the principal of conservation of mass to find the speed of fluid flow in the narrow section, given the speed of flow in the wide section and the diameters of each section.

    DEVELOP The continuity Equation 15.5 for a liquid is vA = constant. Applied to the pipe in question gives $v_1 A_1 = v_2 A_2$, where $A_1 = \pi(d_1/2)^2$ and $A_2 = \pi(d_2/2)^2$, with $d_1 = 2.5$ cm and $d_2 = 2.0$ cm. Given that the speed in the wide section is $v_1 = 1.8$ m/s, we can solve for v_2.

    EVALUATE Inserting the given quantities into the expression derived using the continuity equation gives

    $$v_1 A_1 = v_2 A_2$$
    $$v_2 = v_1 \left(\frac{A_1}{A_2} \right) = v_1 \left(\frac{\pi d_1^2}{\pi d_2^2} \right) = v_1 \left(\frac{d_1}{d_2} \right)^2 = (1.8 \text{ m/s}) \left(\frac{2.5 \text{ cm}}{2.0 \text{ cm}} \right)^2 = 2.8 \text{ m/s}$$

    ASSESS As expected, the speed increases in the narrow section.

35. **INTERPRET** This problem deals with fluid flow rate. The key concepts involved are conservation of mass and the continuity equation.

    DEVELOP The mass flow rate is given by Equation 15.4, $R_m = \rho v A$ and the volume flow rate is given by Equation 15.5: $R_V = vA$. The speed v of the flow can be determined once the flow rate R_m and R_V and the corresonding cross-sectional areas A are known.

    EVALUATE (a) Inserting the given quantities, the volume flow rate for the Mississippi River is

    $$R_V = \rho v = \frac{R_m}{\rho} = \frac{1.8 \times 10^7 \text{ kg/s}}{1.0 \times 10^3 \text{ kg/m}^3} = 1.8 \times 10^4 \text{ m}^3/\text{s}$$

    (b) At a point in the river where the cross-sectional area is given, the average speed of flow is

    $$v = \frac{R_V}{A} = \frac{1.8 \times 10^4 \text{ m}^3/\text{s}}{(2.0 \times 10^3 \text{ m})(6.1 \text{ m})} = 1.5 \text{ m/s}$$

    ASSESS The flow speed we find is reasonable. Note that the actual flow rate of any river varies with the season, local weather, vegetation conditions, human water consumption, etc.

37. **INTERPRET** This problem deals with flow speed of a fluid, which in this case is the blood in the artery. The key concepts involved are mass conservation and the continuity equation.

    DEVELOP Apply the continuity Equation 15.5 vA = constant. Without the clot, we have $v_1 A_1$ = constant. With the clot, we have $v_2 A_2$ = constant, where $A_2 = 0.20 A_1$. Because the constant is the same, we can equate these two expressions for the volume flow rate and solve for v_2.

EVALUATE   Solving for v_2 and inserting the given quantities gives

$$v_2 = v_1\left(\frac{A_1}{A_2}\right) = v_1\left(\frac{A_1}{0.20A_1}\right) = \frac{(35 \text{ cm/s})}{0.20} = 1.8 \text{ cm/s}$$

ASSESS   The flow speed of blood increases in the region where the cross-sectional area of the artery has been reduced due to clotting. Notice that the initial diameter of the artery is not needed to solve this problem; it suffices to know the ratio of the arterial cross sections.

PROBLEMS

39.   **INTERPRET**   This problem involves calculating the area needed for a given pressure to produce a given force. We are given the mass and the gauge pressure of the tires, and we want to find the total tire area that's in contact with the road.

DEVELOP   As shown in Equation 15.1, pressure measures the normal force per unit area exerted by a fluid. For this problem, the fluid is air. The force exerted on the road by the tires is the weight of the car, $F = mg$.

EVALUATE   With a gauge pressure of $p = 230$ kPa, the contact area is

$$A = \frac{F_g}{p} = \frac{mg}{p} = \frac{(1950 \text{ kg})(9.8 \text{ m/s}^2)}{23 \times 10^4 \text{ Pa}} = 0.0830 \text{ m}^2 = 830 \text{ cm}^2$$

ASSESS   Our result implies that the contact area of each wheel is about 200 cm², or the area of a 25×8 cm² rectangular surface, which seems reasonable.

41.   **INTERPRET**   We have an open tube filled with water on the bottom and oil on the top of water. The two fluids do not mix. We want to find the gauge pressures at the oil-water interface as well as at the bottom of the tube.

DEVELOP   The pressure pushing down on the oil at the top of the tube is atmospheric pressure, p_{atm}. The gauge pressure at the interface of the oil and water is the difference between the absolute pressure and the atmospheric pressure, or $\Delta p = p_i - p_{atm} = \rho_{oil}gh_{oil}$. To find h_{oil}, note that

$$m_{oil} = \rho_{oil}V_{oil} = \rho_{oil}A_{tube}h_{oil} = 5.0\text{g}$$

where V_{oil} is the volume of oil and A_{tube} is the cross-sectional area of the tube. The gauge pressure at the bottom is the total weight of fluid divided by the cross-sectional area of the tube, which is

$$\Delta p_{bot} = \frac{m_w g}{A_{tube}} + \frac{m_{oil}g}{A_{tube}}$$

EVALUATE   (a) The gauge pressure at the interface is

$$\Delta p = p_i - p_{atm} = \rho_{oil}gh_{oil} = \frac{m_{oil}g}{A_{tube}} = \frac{(0.0050 \text{ kg})(9.8 \text{ m/s}^2)}{\pi(0.010 \text{ m})^2/4} = 620 \text{ Pa}$$

(b) The gauge pressure at the bottom is the total weight of fluid divided by the cross-sectional area of the tube,

$$\Delta p_{bot} = (m_w + m_{oil})\frac{g}{A_{tube}} = (0.0050 \text{ kg} + 0.0050 \text{ kg})\left(\frac{9.8 \text{ m/s}^2}{\pi(0.010 \text{ m})^2/4}\right) = 1.2 \text{ kPa}$$

ASSESS   Oil floats on top of water because its density is lower than that of water. The gauge pressure at the bottom of the tube is due to the weight of both the oil and the water. The absolute pressure there would be equal to the sum of the gauge pressure and the atmospheric pressure. Note that because $m_w = m_{oil}$ in this problem, the gauge pressure at the bottom is just twice that at the interface.

43.   **INTERPRET**   The U tube contains two liquids, oil and water, in hydrostatic equilibrium. We want to find their height difference.

DEVELOP   The pressure at point 2 in the figure below, which is the oil-water interface, is

$$p_2 = p_{atm} + \rho_{oil}gl$$

where $l = 2.0$ cm. The pressure at point 1, which is at the same height as point 2, is

$$p_1 = p_{atm} + \rho_w g(l - h)$$

From Equation 15.3, $p = p_0 + \rho gh$, we see that the pressure at points at the same height are the same, so $p_1 = p_2$. Using the information that $\rho_{oil} = 0.82\rho_w$ allows us to solve for h.

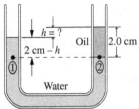

EVALUATE   Equating the two pressures leads to $\rho_{oil}gl = \rho_w g(l-h)$ or

$$h = \left(1 - \frac{\rho_{oil}}{\rho_w}\right)l = \left(1 - \frac{0.82\rho_w}{\rho_w}\right)l = (1 - 0.82)(2.0 \text{ cm}) = 3.6 \text{ mm}$$

ASSESS   Note that the final answer does not depend on the atmospheric pressure, p_{atm}, because this pressure pushes down equally on both the oil and the water. The U tube can be used to measure the density of a fluid, if we know the height difference h and the density of the other fluid.

45.   **INTERPRET**   This problem involves Pascal's law, which we can use to calculate the maximum mass the hydraulic lift can support.

DEVELOP   If we neglect the variation of pressure with height in the hydraulic system (which is usually small compared to the applied pressure), the fluid pressure is the same throughout the system:

$$p = \frac{F_1}{A_1} = \frac{F_2}{A_2}$$

where F_1 is the applied force and F_2 is the resulting force. The mass the system can support is

$$m = \frac{F_2}{g}$$

EVALUATE   Solving the equations above for m gives

$$m = \frac{F_2}{g} = \frac{pA_2}{g} = \frac{\left(50 \times 10^4 \text{ Pa}\right)\pi(0.45 \text{ m})^2}{4\left(9.8 \text{ m/s}^2\right)} = 8100 \text{ kg}$$

ASSESS   Because the fluid pressure is essentially constant throughout the system, the force is scaled by the ratio of the surface areas in question. Energy, however, is conserved in this process.

47.   **INTERPRET**   You're asked to determine how much the accused person drank, given the change in the buoyant force of a keg of beer.

DEVELOP   By Archimedes' principle, the buoyant force on the keg is equal to the weight of the fluid displaced. The keg is probably made of aluminum and is filled with a mixture of beer and air. Let's assume that the level of the beer in the keg was initially L_i, and the keg was submerged in the water to a depth of H_i, as shown in the figure below.

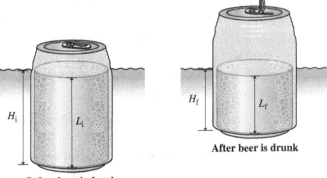

Before beer is drunk

After beer is drunk

As such, the initial volume of beer would be $\pi r^2 L_i$, and the initial volume of water displaced would be $\pi r^2 H_i$ (where we have assumed the aluminum shell is thin enough that the interior and exterior radii are essentially the same). The total weight of the keg and beer (neglecting the weight of the air) is balanced by the buoyant force, which equals the weight of the water displaced.

$$F_g = \left(m_{keg} + \rho_{beer}\left(\pi r^2 L_i\right)\right)g = F_b = \rho_{water}\left(\pi r^2 H_i\right)g$$

A similar equation can be written for the keg at the end of the day, using L_f for the final level of beer and H_f for the final depth. Subtracting these two equations gives

$$L_i - L_f = H_i - H_f$$

where we have used the fact that $\rho_{beer} \approx \rho_{water}$.

EVALUATE   You are told that the keg rose by 1.2 cm, so the level of beer fell by that same amount. This corresponds to a volume of beer:

$$\Delta V = \pi r^2\left(L_i - L_f\right) = \pi\left(\tfrac{1}{2} \cdot 40 \text{ cm}\right)^2(1.2 \text{ cm}) = 1.5 \text{ L}$$

where we've used the conversion $1 \text{ cm}^3 = 1 \text{ mL}$. In terms of English units $(1 \text{ L} = 33.8 \text{ oz})$, the accused drank 51 oz, which would imply that he/she was legally impaired.

ASSESS   The defendant may want a more careful analysis, taking into account the thickness of the keg's outer shell, but this would actually increase the estimated beer volume that he/she drank.

49.   **INTERPRET**   This problem involves applying Archimedes' principle to find the minimum water depth for the load-carrying ship to navigate.

DEVELOP   Archimedes' principle states that the buoyancy force is equal to the weight of the water displaced by the floating supertanker:

$$F_b = m_w g = \rho_w g V = \rho_w g A y$$

where we have used the fact that the volume V of water displaced is proportional to its draft (depth y in the water); $V = Ay$, where A is the cross-sectional area (see figure below).

EVALUATE   Because the total mass of the full supertanker is three times that when empty, three times the buoyancy force is needed to support the ship. From the expression above, we see that the buoyancy force is proportional to the draft y, so the y must increase three fold. Thus, $y_{full} = 3y_{empty} = 3(9.0 \text{ m}) = 27 \text{ m}$.

ASSESS   Our result is independent of the mass of the supertanker. The heavier the supertanker plus load, the deeper it will submerge.

51.   **INTERPRET**   This problem is about the buoyancy force provided by the helium balloon and the hot-air balloon, which we can use to calculate how much He is needed to lift a given mass.

DEVELOP   For the balloon to lift off, the buoyancy force must exceed the weight of the load (mass M, including the balloon) plus the gas (mass m):

$$F_b \geq (M + m)g$$

where the buoyancy force is simply equal to $F_b = \rho_{air} g V$. If we neglect the volume of the balloon's skin etc. compared to that of the gas it contains, then $V = m/\rho_{gas}$. Therefore,

$$m = \rho_{gas} V = \rho_{gas}\frac{F_b}{\rho_{air} g} \geq \frac{\rho_{gas}}{\rho_{air}}(M + m)$$

$$m \geq \left(\frac{\rho_{gas}}{\rho_{air} - \rho_{gas}}\right)M$$

EVALUATE   **(a)** When the gas is helium, the density ratio is $\rho_{air}/\rho_{He} = (1.2 \text{ kg/m}^3)/(0.18 \text{ kg/m}^3) = 6.67$, and

$$m \geq \left(\frac{\rho_{gas}}{\rho_{air} - \rho_{gas}}\right)M = \left(\frac{1}{6.67 - 1}\right)(280 \text{ kg}) = 49 \text{ kg}$$

(b) For hot air, $\rho_{gas} = 0.9\rho_{air}$, and

$$m \geq \left(\frac{\rho_{gas}}{\rho_{air} - \rho_{gas}}\right) M = \left(\frac{0.9}{1 - 0.9}\right)(280 \text{ kg}) = 2500 \text{ kg}$$

ASSESS   These masses correspond to gas volumes of 275 m³ for helium and 2330 m³ for hot air, which are reasonable for a helium-filled balloon and a hot-air balloon.

53.   **INTERPRET**   This problem deals with flow speed of a fluid, which in this case is the blood in the artery. The key point involved here is Bernoulli's equation.

DEVELOP   The continuity equation, $vA = \text{constant}$, as given in Equation 15.5, is a reasonable approximation for blood circulation in an artery. Neglecting any pressure differences due to height, we find, from Bernoulli's equation, that

$$p + \tfrac{1}{2}\rho v^2 = p' + \tfrac{1}{2}\rho v'^2$$

EVALUATE   We're told that the clot reduces the cross-sectional area by 80%, so $A' = 0.20A$, and

$$v' = v\left(\frac{A}{A'}\right) = (0.35 \text{ m/s})\frac{A}{0.20A} = 1.75 \text{ m/s}$$

From Bernoulli's equation, the gauge pressure at the clot is

$$p' = p + \tfrac{1}{2}\rho\left(v^2 - v'^2\right) = 16 \text{ kPa} + \tfrac{1}{2}\left(1.06\frac{\text{g}}{\text{cm}^3}\right)\left[(0.35 \text{m/s})^2 - (1.75 \text{m/s})^2\right] = 14 \text{ kPa}$$

ASSESS   The flow speed of blood increases in the region where the cross-sectional area of the artery has been reduced due to clotting. Since $p + \tfrac{1}{2}\rho v^2 = \text{constant}$, the gauge pressure must decrease.

55.   **INTERPRET**   This problem involves the flow of water, which we can consider to be an incompressible fluid. Bernoulli's equation allows us to find the maximum height reached by the water coming out from the hose.

DEVELOP   Make a sketch of the situation (see figure below). The flow of water in the hose can be described by Bernoulli's equation (Equation 15.6):

$$p + \frac{1}{2}\rho v^2 + \rho gy = \text{constant}$$

The pressure, velocity, and height of the water in the hose (point 1) are $p_1 = p_{atm} + \Delta p_1 = p_{atm} + 140 \text{ kPa}$, $v_1 \approx 0$ and $y_1 = 0$. At the highest point attained by a jet of water emerging from a hole (point 2), $p_2 = p_{atm}$, $v_2 \approx 0$, and $y_2 = h$. We can equate the result of Bernoulli's equation at points 1 and 2 to find h.

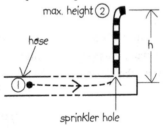

EVALUATE   Using Bernoulli's equation we have

$$p_1 = p_2 + \rho gh$$
$$p_{atm} + \Delta p_1 = p_{atm} + \rho gh$$
$$h = \frac{\Delta p_1}{\rho g} = \frac{140 \text{ kPa}}{\left(1.0 \times 10^3 \text{ kg/m}^3\right)\left(9.8 \text{ m/s}^2\right)} = 14 \text{ m}$$

ASSESS   At the maximum height, all the work done by pressure has been converted to potential energy of the fluid. Energy is conserved in the process (ignoring dissipative forces such as air resistance).

57.   INTERPRET   A narrower section is placed in a pipe carrying an incompressible fluid. We are to find the flow speed in the pipe and the volume flow rate, given the pressure difference between the fluid in the pipe and the fluid in the narrow section. We will assume that the flow is non-turbulent, and use Bernoulli's equation. The velocity is related to the cross-sectional area by the continuity equation.

DEVELOP   The pressure difference between the venturi and the unrestricted pipe is $\Delta P = 17$ kPa. The radius of the unconstricted pipe is $r_1 = 0.010$ m, and the radius of the constricted region is $r_2 = 0.0050$ m. We will assume that any height changes are negligible, and take the density of water to be $\rho = 1.0 \times 10^3$ kg/m^3. Bernoulli's equation is $p + \rho g h + \frac{1}{2}\rho v^2 = $ constant, and the continuity equation for incompressible fluids such as water is $v_1 A_1 = v_2 A_2$. We equate the result of Bernoulli's equation for the unrestricted pipe equal to that for the restricted pipe, with $h_1 = h_2$.

$$p_1 + \rho g h_1 + \frac{1}{2}\rho v_1^2 = p_2 + \rho g h_2 + \frac{1}{2}\rho v_2^2$$

Use the continuity Equaiton 15.5 (for liquid) $v_2 = v_1 A_1 / A_2$ and solve for v_1:

EVALUATE   (a) The flow speed v_1 of the water is

$$p_1 + \frac{1}{2}\rho v_1^2 = p_2 + \frac{1}{2}\left(\frac{v_1 A_1}{A_2}\right)^2$$

$$p_1 - p_2 = \frac{1}{2}\rho v_1^2 \left(\frac{A_1^2}{A_2^2} - 1\right)$$

$$v_1 = \pm\sqrt{\frac{2(p_1 - p_2)}{\rho\left[(r_1/r_2)^4 - 1\right]}} = \pm\sqrt{\frac{2(17\,\text{kPa})}{(1.0 \times 10^3\,\text{kg/m}^3)(2.0^4 - 1)}} = \pm 1.5\,\text{m/s}$$

where the positive and negative signs indicate the flow may be in either direction. Without loss of generality, we can use the positive value.

(b) To find the volume flow rate, in m^3/s, we multiply the flow speed by the area of the pipe, which gives

$$R_V = v_1 A_1 = v_1 \pi \frac{d_1^2}{4} = \frac{\pi(1.5\,\text{m/s})(0.020\,\text{m})^2}{4} = 4.7 \times 10^{-4}\,\text{m}^3/\text{s} = 0.47\,\text{L/s}$$

ASSESS   Both the flow speed and volume flow rate seem reasonable for a small pipe such as this.

59.   INTERPRET   This problem deals with blood flow through an artery that is obstructed by a plaque.

DEVELOP   We'll assume there's no appreciable change in the gravitational potential, so Bernoulli's equation can be written as:

$$p + \tfrac{1}{2}\rho v^2 = p' + \tfrac{1}{2}\rho v'^2$$

We're told that the pressure drops by 5% $(p' = 0.95p)$. The blood flow can be approximated by the continuity equation, $vA = v'A'$. We are given p, ρ, and v, and we want to find the fraction of the area that is obstructed: $(A - A')/A$.

EVALUATE   Putting together the information that we have:

$$p - p' = 0.05p = \tfrac{1}{2}\rho\left[\left(\frac{A}{A'}v\right)^2 - v^2\right]$$

Rearranging the terms, the fraction of the area that is obstructed is

$$\frac{\Delta A}{A} = 1 - \left(1 + \frac{0.1p}{\rho v^2}\right)^{-1/2} = 1 - \left(1 + \frac{0.1(10\,\text{kPa})}{(1.06\,\text{g/cm}^3)(30\,\text{cm/s})^2}\right)^{-1/2} = 0.70$$

ASSESS   This is a rather large blockage, but surprisingly the pressure only drops by 5%.

61.   INTERPRET   This problem involves flow of juice which we take to be an incompressible fluid. We apply both the continuity equation and Bernoulli's equation to solve the problem.

DEVELOP   For steady incompressible fluid flow, the continuity equation (Equation 15.5) is

$$v_1 A_1 = v_2 A_2 \quad \rightarrow \quad v_1 \left(\frac{\pi D_1^2}{4} \right) = v_2 \left(\frac{\pi D_2^2}{4} \right)$$

and Bernoulli's equation reads

$$p_1 + \frac{1}{2}\rho v_1^2 + \rho g y_1 = p_2 + \frac{1}{2}\rho v_2^2 + \rho g y_2$$

Eliminating v_2 using the first equation, we obtain

$$\Delta p = p_1 - p_2 = \frac{1}{2}\rho v_1^2 \left[\left(\frac{A_1}{A_2} \right)^2 - 1 \right] + \rho g (y_2 - y_1) = \frac{1}{2}\rho v_1^2 \left[\left(\frac{D_1}{D_2} \right)^4 - 1 \right] + \rho g (y_2 - y_1)$$

EVALUATE   (a) We assume the juice density to be that of water. When $y_2 - y_1 = 6.5$ cm, the pressure difference is

$$\Delta p = \rho \left(\frac{1}{2} v_1^2 \left[\left(\frac{D_1}{D_2} \right)^4 - 1 \right] + g(y_2 - y_1) \right)$$

$$= (1000 \text{ kg/m}^3) \left(\tfrac{1}{2}(0.002 \text{ m/s})^2 \left[\left(\frac{8}{0.3} \right)^4 - 1 \right] + (9.8 \text{ m/s}^2)(0.065 \text{ m}) \right)$$

$$= 1.65 \times 10^3 \text{ Pa} = (1.63\%) \, p_{atm}$$

Therefore, the pressure in the mouth is 98% less than atmospheric pressure.

(b) For a constant pressure difference, $y_2 - y_1$ attains its maximum value when $v_1 = 0$. Thus,

$$(y_2 - y_1)_{max} = \frac{\Delta p}{\rho g} = \frac{1.65 \times 10^3 \text{ Pa}}{(1000 \text{ kg/m}^3)(9.8 \text{ m/s}^2)} = 17 \text{ cm}$$

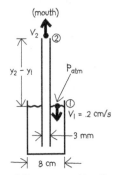

ASSESS   As the juice level drops, the pressure difference and/or flow speed may change.

63.   INTERPRET   This problem involves finding the force on a suction cup due to atmospheric pressure. Given this force, we can find the force due to friction that allows the suction cup to support objects.

DEVELOP   The normal force on the suction cup is the result of atmospheric pressure. We assume a perfect vacuum inside the cup. The force f_s due to static friction that supports the object of mass m must satisfy

$$mg = f_s \le \mu_s n = \mu_s p_{atm} A = \mu_s p_{atm} \left(\pi d^2 / 4 \right)$$

EVALUATE   The maximum mass that can be supported by the suction cup is therefore

$$m_{max} = \frac{\mu_s p_{atm} \pi d^2}{4g} = \frac{(0.72)(101.3 \text{ kPa}) \pi (0.050 \text{ m})^2}{4(9.8 \text{ m/s}^2)} = 15 \text{ kg}$$

ASSESS   The maximum value of 15 kg is about the mass of a toddler. The force on the cup due to the atmospheric pressure is quite large.

65. **INTERPRET** You want to verify the power output of a proposed wind farm.

DEVELOP From the text, you know that the theoretical maximum power per unit area that can be extracted from the with wind is $\frac{8}{27}\rho v^3$, where the air density is given by $\rho = 1.2 \text{ kg/m}^3$. The plan is to build a 800 turbines, each with blade diameter of 95 m, in a area where the average wind speed is 12 m/s.

EVALUATE If you assume that the turbines on average generate 30% of the theoretical maximum power, the total power that the wind farm could produce is

$$P_{\text{tot}} = N(0.3)\left(\tfrac{8}{27}\rho v^3\right)\left(\pi r^2\right) = (800)(0.3)\left[\tfrac{8}{27}\left(1.2\,\tfrac{\text{kg}}{\text{m}^3}\right)\left(12\,\tfrac{\text{m}}{\text{s}}\right)^3\right]\left(\pi\left(\tfrac{1}{2}95\text{m}\right)^2\right) = 1.0 \text{ GW}$$

Yes, the wind farm could conceivably replace a 1-GW nuclear power plant.

ASSESS Even if the wind farm can on average generate the desired power, there will be fluctuations in the output due to changing weather conditions. Utility companies are still reluctant to entirely abandon coal and nuclear, which supply a more stable baseline of power.

67. **INTERPRET** In this problem we want to find the time it takes for the can to drain out all its water through its hole. An integral is needed since the water level is continuous, from 0 to height h.

DEVELOP Let y be the height of the water above the bottom of the can, then $-dy/dt$ is the magnitude of the flow speed of the top surface of the water draining out (y decreases as a function of time). The continuity equation gives

$$-\frac{dy}{dt}A_0 = v_1 A_1 \quad \rightarrow \quad dt = -\left(\frac{A_0}{A_1}\right)\frac{dy}{v_1}$$

where subscript 1 refers to the small hole in the bottom. For most of the time, $v_1 \approx \sqrt{2gy}$ (see Example 15.6) and we assume the top of the can is open).

EVALUATE Carrying out the integration, we find the total time required to be

$$t = \int dt \approx -\int_h^0 \left(\frac{A_0}{A_1}\right)\frac{dy}{\sqrt{2gy}} = \frac{A_0}{A_1\sqrt{2g}}2\sqrt{y}\,\Big|_0^h = \frac{A_0}{A_1}\sqrt{\frac{2h}{g}}$$

ASSESS This result is approximate since dy/dt cannot be neglected compared to v_1 when y is small. If we use Bernoulli's equation without this approximation, then

$$\frac{1}{2}\rho\left(\frac{dy}{dt}\right)^2 + \rho gy = \frac{1}{2}\rho v_1^2$$

since the pressure is atmospheric pressure at both the top of the can and the hole. Combining with the continuity equation gives

$$v_1 = \sqrt{2gy + (dy/dt)^2} = -\left(\frac{A_0}{A_1}\right)\frac{dy}{dt} \quad \rightarrow \quad \frac{dy}{dt} = -\sqrt{\frac{2gy}{(A_0/A_1)^2 - 1}}$$

Integration of this yields a more exact outflow time of $t = \sqrt{\frac{2h}{g}[(A_0/A_1)^2 - 1]}$.

69. **INTERPRET** Using the fact that the density and pressure in Earth's atmosphere are proportional, we are to use the result of the previous problem to express the atmospheric density as a function of height and to find the height below which half the Earth's atmospheric mass lies.

DEVELOP From Problem 15.68, we have the atmospheric pressure is $p = p_0 e^{-h/h_0}$. Combining this with the given information that $\rho = p/(h_0 g)$, we can express the density as a function of height h.

EVALUATE **(a)** The atmospheric density as a function of height is

$$\rho(h) = \frac{p}{h_0 g} = \frac{p_0}{h_0 g}e^{-h/h_0} = \rho_0 e^{-h/h_0}$$

(b) The mass of atmosphere contained in a thin spherical shell of thickness dh, at height h, is

$$dm = \rho dV = \left(\rho_0 e^{-h/h_0}\right)4\pi\left(R_E + h\right)^2 dh = 4\pi\rho_0\left(R_E + h\right)^2 e^{-h/h_0} dh$$

where R_E is the radius of the Earth and $R_E + h$ is the radius of the shell. The mass of atmosphere below height h_1 is

$$M(h_1) = \int_0^{h_1} dm = 4\pi\rho_0 R_E^2 \int_0^{h_1} \left(1 + 2\frac{h}{R_E} + \frac{h^2}{R_E^2}\right) e^{-h/h_0}\, dh$$

The integrals can be evaluated easily enough with the use of the table of integrals in Appendix B. However, if $h_1/R_E \ll 1$, only the first term is important. (Even if h_1 is large, the exponential term is negligibly small for $h_1 \ll h_0$ and none of the terms contribute significantly for large h.) To a good approximation, therefore

$$M(h_1) \approx 4\pi\rho_0 R_E^2 \int_0^{h_1} e^{-h/h_0}\, dh = 4\pi\rho_0 R_E^2 h_0 \left(1 - e^{-h_1/h_0}\right)$$

The total mass of the atmosphere is approximately $M(\infty) = 4\pi\rho_0 R_E^2 h_0$, so the height bounding half the total mass is given by the equation $M/2 = M\left(1 - e^{-h_1/h_0}\right)$ or

$$h_1 = h_0 \ln(2) = (8.2 \text{ km}) \ln(2) = 5.8 \text{ km}$$

ASSESS   This is the same result as we obtained for Problem 15.68.

71.   **INTERPRET**   We're asked to find the density at the center of a non-uniform sphere.

DEVELOP   We can integrate the formula for the density over the volume and equate it to the total mass of the sphere: $M = \int \rho dV$. Since the density varies with radius, we can choose to integrate over spherical shells of radius r and volume $dV = 4\pi r^2 dr$.

EVALUATE   The integral we have to do has the form:

$$M = \int \rho dV = \int_0^R \left(\rho_0 e^{r/R - 1}\right)\left(4\pi r^2 dr\right) = 4\pi\rho_0 \int_0^R r^2 e^{r/R - 1}\, dr$$

This will require integration by parts, as described in Appendix A.

$$\int u\, dv = uv - \int v\, du$$

Let $u = r^2$, so that $du = 2r \cdot dr$, and let $dv = e^{r/R - 1}\, dr$, so that $v = Re^{r/R - 1}$. Plugging these into the above formula, we get:

$$\int_0^R r^2 e^{r/R - 1}\, dr = r^2 Re^{r/R - 1}\Big|_0^R - \int_0^R Re^{r/R - 1}\left(2r \cdot dr\right) = R^3 - 2R \int_0^R re^{r/R - 1}\, dr$$

To evaluate the remaining integral, we do integration by parts again with $u = r$, and $dv = e^{r/R - 1}\, dr$:

$$-2R \int_0^R re^{r/R - 1}\, dr = -2R\left[rRe^{r/R - 1}\Big|_0^R - \int_0^R Re^{r/R - 1}\, dr\right]$$

$$= -2R^3 + 2R^3\left(1 - e^{-1}\right) = -2e^{-1}R^3$$

Solving for the central density:

$$\rho_0 = \frac{M}{4\pi R^3 \left(1 - 2e^{-1}\right)} \approx \frac{M}{1.06 \cdot \pi R^3}$$

ASSESS   The units are correct for a density term.

73.   **INTERPRET**   In Problem 15.42, we determined the force on a dam due to the water behind the dam. In this problem, we find the torque around the bottom edge of the same dam. We will use $F = pA$ and $\tau = yF$, as shown in the figure below.

DEVELOP   The pressure varies with depth, according to $p(y) = \rho g(H - y)$. We will find the force $dF = pdA$, and thus the torque $d\tau = ydF$ from each horizontal strip across the dam. Integrating $d\tau$ gives us the total torque. The dam has width $w = 1500$ m and the water is $H = 95$-m deep. The density of water is $\rho = 1000 \text{ kg/m}^3$.

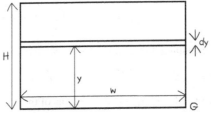

EVALUATE   The pressure $d\tau = ydF = y\big[p(y)\big]dA = y\big[\rho g(H-y)\big](wdy)$ Integrate this from $y = 0$ to $y = H$.

$$\tau = \rho g w \int_0^H y(H-y)\,dy = \rho g w \left(\frac{H}{2}y^2 - \frac{y^3}{3}\right)\Bigg|_0^H = \rho g w \left(\frac{H^3}{2} - \frac{H^3}{3}\right) = \frac{1}{6}\rho g w H^3$$

$$= \frac{1}{6}\big(1.0 \times 10^3 \text{ kg/m}^3\big)\big(9.8 \text{ m/s}^2\big)(1500 \text{ m})(95 \text{ m})^3 = 2.1 \times 10^{12} \text{ N}\cdot\text{m}$$

ASSESS   The units in our final equation are

$$\tau = \left(\frac{\text{kg}}{\text{m}^3}\right)\left(\frac{\text{m}}{\text{s}^2}\right)\cdot\text{m}\cdot\text{m}^3 = \overbrace{\text{kg m/s}^2}^{\text{N}}\cdot\text{m} = \text{N}\cdot\text{m}$$

which work out correctly.

75.   **INTERPRET**   We are to find the expected density of a mix of immiscible liquids, and compare it with a measured density to see if the mix is what it should be. The density of a mix of liquids should be the total mass divided by the total volume.

DEVELOP   The "official" dressing is 1 part vinegar to 3 parts oil, measured by volume. So the dressing should have volume $4V$ and mass $m_{\text{vinegar}} + m_{\text{oil}} = \rho_{\text{vinegar}}V + 3\rho_{\text{oil}}V$. Calculate the density of this mix, and compare it with the measured density $\rho' = 0.97$ g/cm^3. If the density of the sample is higher than it should be, then it has probably been diluted with water. The density of oil is $\rho_{\text{oil}} = 0.92$ g/cm^3, and the density of vinegar is $\rho_{\text{vinegar}} = 1.0$ g/cm^3.

EVALUATE

$$\rho = \frac{\rho_{\text{vinegar}}V + 3\rho_{\text{oil}}V}{4V} = \frac{\rho_{\text{vinegar}} + 3\rho_{\text{oil}}}{4} = 0.94 \text{ g/cm}^3$$

ASSESS   The dressing has been altered.

77.   **INTERPRET**   You are asked to find the maximum load that a ship can hold, given the size and shape of the hull, and the initial waterline of the ship.

DEVELOP   By Archimedes' principle, the buoyant force is equal to the weight of the water displaced by the ship. To find the volume of water displaced, we'll need the formula for the area of an isosceles triangle with apex angle θ and height h: $A = h^2 \tan\theta/2$.

EVALUATE   When the ship is empty, the buoyant force only needs to support the weight of the ship:

$$m_{\text{ship}}g = \rho_{\text{H}_2\text{O}}Vg = \rho_{\text{H}_2\text{O}}\left(Lh_1^2 \tan\frac{\theta}{2}\right)g$$

When the maximum cargo load, m_{max}, is placed on the ship, the entire hull is submerged:

$$\big(m_{\text{ship}} + m_{\text{max}}\big)g = \rho_{\text{H}_2\text{O}}\left(Lh_0^2 \tan\frac{\theta}{2}\right)g$$

We subtract these two equations to find the maximum load:

$$m_{\text{max}} = \rho_{\text{H}_2\text{O}}L\tan\frac{\theta}{2}\big(h_0^2 - h_1^2\big)$$

ASSESS   A bigger concern for your design might be whether this shape of hull is stable. It is not immediately obvious that the center of gravity is below the center of buoyancy, see Fig. 15.10.

79.   **INTERPRET**   We're asked to consider some of the physics of arterial stenosis.

DEVELOP   The flow speed has to change in order to keep the volume flow rate constant.

EVALUATE   From Equation 15.5: $v' = \big(A/A'\big)v$, so if the artery wall's thicken and the area decreases: $A' < A$, then the flow speed must increase: $v' > v$.

The answer is (c).

ASSESS   If the speed didn't increase, the blood would begin piling up in front of the stenosis.

81. **INTERPRET**   We're asked to consider some of the physics of arterial stenosis.

DEVELOP   As pointed out in Problem 15.79, the flow speed in the stenosis is $v' = (A/A')v$. Since the area is related to the diameter by $A = \frac{\pi}{4}D^2$, the flow speed goes as $v' = (D/D')^2 v$.

EVALUATE   If the diameter decreases by half, the flow speed increases by a factor of four. The answer is (e).

ASSESS   If we use the values from Problem 15.59: $\rho = 1.06$ g/cm^3, $p = 10$ kPa, and $v = 30$ cm/s, then the pressure in the stenosis will drop by 7%.

TEMPERATURE AND HEAT

16

EXERCISES

Section 16.1 Heat, Temperature, and Thermodynamic Equilibrium

15. **INTERPRET** This problem involves converting temperature from the Fahrenheit scale to the Celsius scale.

DEVELOP The two temperature scales are related by Equation 16.2:

$$T_F = \frac{9}{5}T_C + 32$$

EVALUATE Inserting $T_F = 68°F$ and solving the above equation for the Celsius temperature, we obtain

$$T_C = \frac{5}{9}(T_F - 32) = \frac{5}{9}(68°F - 32°F) = 20°C$$

ASSESS This is a useful result to remember since 20°C or 68°F is a typical room temperature.

17. **INTERPRET** Given both Fahrenheit and Celsius scales, we want to know when T_F and T_C are numerically equivalent.

DEVELOP The two temperature scales are related by Equation 16.2:

$$T_F = \frac{9}{5}T_C + 32$$

The condition that the readings are numerically equivalent is

$$T_F = \frac{9}{5}T_C + 32 = T_C$$

EVALUATE The above equation can be solved to give

$$T_C = -\frac{5}{4}(32) = -40 = T_F$$

ASSESS This is the only temperature in which both scales yield the same reading: $-40°F = -40°C$.

19. **INTERPRET** This problem is about converting temperature from the Celsius scale to the Fahrenheit scale.

DEVELOP The two temperature scales are related by Equation 16.2: $T_F = (9/5)T_C + 32$

EVALUATE Solving the above equation for the Fahrenheit temperature, we obtain

$$T_F = \frac{9}{5}(39.1) + 32 = 102°F$$

ASSESS The temperature is way above the normal body temperature of 98.6°F (or 37°C). Call the doctor immediately!

Section 16.2 Heat Capacity and Specific Heat

21. **INTERPRET** We are to find the energy necessary to change the temperature of an object by a given amount. This involves the heat capacity of the object and the temperature change.

DEVELOP Apply Equation 16.3 $Q = mc\Delta T$. The mass of the aluminum block is $m = 2.0$ kg, the specific heat (from Table 16.1) is $c = 900$ J/(kg·K), and the temperature change is $\Delta T = 18$ C° = 18 K (see Equation 16.1).

EVALUATE Inserting the given quantities gives

$$Q = (2.0 \text{ kg}) \left[90 \times 10^1 \text{ J}/(\text{kg} \cdot \text{K}) \right] (18 \text{ K}) = 32 \text{ kJ}$$

ASSESS The same value would be the heat released by the aluminum if it cooled 18 C°.

23. **INTERPRET** The problem involves calculating the average power output of the human body, given the information about the energy acquired in a day from an average diet. Recall that power is energy per unit time.

    DEVELOP In a single day, the energy gained from the diet is

    $$\Delta E = (2 \times 10^6 \text{ cal})(4.184 \text{ J/cal}) = 8.37 \times 10^6 \text{ J}$$

    where we have used the conversion factor 1 cal = 4.184 J (see Appendix C). If the body expends all this energy (and does not store any of it), then the energy expended must be this same value (by conservation of energy). Therefore, the average power output of the body is $\overline{P} = \Delta E / \Delta t$, where $\Delta t = (1 \text{ day})(86{,}400 \text{ s/day}) = 86{,}400 \text{ s}$.

    EVALUATE The average power output is

    $$\overline{P} = \frac{\Delta E}{\Delta t} = \frac{8.37 \times 10^6 \text{ J}}{86{,}400 \text{ s}} = 96.9 \text{ W} = 100 \text{ W}$$

    to a single significant figure.

    ASSESS The average power output by the human body at rest is about 80 W, the same as a bright light bulb, so this result seems reasonable.

25. **INTERPRET** Given the energy it takes to warm the wrench by the given temperature difference, we want to find its heat capacity, as well as the specific heat of the metal from which it is made.

    DEVELOP The heat capacity of an object is given by $C = \Delta Q / \Delta T$, where ΔQ is the amount of heat transfer that results in a temperature change $\Delta T = 15 \text{ C}° = 15 \text{ K}$. Comparing this expression with Equation 16.3, we see that the specific heat of a material is $c = C/m$ (i.e., the heat capacity per unit mass).

    EVALUATE **(a)** Inserting the given quantities gives the average heat capacity as

    $$C = \frac{\Delta Q}{\Delta T} = \frac{2.52 \times 10^3 \text{ J}}{15 \text{ K}} = 170 \text{ J/K}$$

    to two significant figures.

    (b) The average specific heat of the material is

    $$c = \frac{C}{m} = \frac{168 \text{ J/K}}{0.350 \text{ kg}} = 480 \text{ J}/(\text{kg} \cdot \text{K})$$

    ASSESS The wrench is probably made of iron which has a specific heat of $448 \text{ J/kg} \cdot \text{K}$.

Section 16.3 Heat Transfer

27. **INTERPRET** This problem is about converting heat loss expressed in Btu/h to SI units.

    DEVELOP One Btu (British thermal unit) is equal to 1054 J (see Appendix C), which is the amount of heat that is needed to raise the temperature of 1 lb of water from 63°F to 64°F.

    EVALUATE The conversion to SI units is

    $$1.00 \left(\frac{\text{Btu}}{\text{h}} \right) = \left(\frac{1.00 \text{ Btu}}{\text{h}} \right) \overbrace{\left(\frac{1054 \text{ J}}{\text{Btu}} \right)}^{=1} \overbrace{\left(\frac{1.00 \text{ h}}{3.60 \times 10^3 \text{ s}} \right)}^{=1} = 0.293 \text{ W}$$

    ASSESS Our result shows that 1 W is about 3.4 Btu/h. The power output of air conditioners is commonly given in terms of Btu/h.

29. **INTERPRET** This involves calculating the rate of heat conduction through the stove top, given the stove's dimensions and the inside and outside temperatures.

    DEVELOP Take the positive-x direction to be upward. We assume a steady flow of heat through the an area $A = 90 \text{ cm} \times 40 \text{ cm} = 0.36 \text{ m}^2$, with no flow through the edges. The rate of heat flow is given by Equation 16.5:

    $$H = -kA \frac{\Delta T}{\Delta x}$$

The temperature difference is $\Delta T = T_{outside} - T_{inside} = 295°C - 310°C = 15°C = -15$ K (see Equation 16.1) and $\Delta x = x_{outside} - x_{inside} = 0.0045$ m.

EVALUATE   From Table 16.2, we find the thermal conductivity of steel to be $k = 46$ W/(m·K). Thus, the rate of heat conduction is

$$H = -kA\frac{\Delta T}{\Delta x} = -\left[46 \text{ W/(m·K)}\right]\left(0.36 \text{ m}^2\right)\frac{(-15 \text{ K})}{0.0045 \text{ m}} = 55 \text{ kW}$$

ASSESS   The heat flow is positive, for x going from the inside of the stove to the outside, because the temperature gradient, $\Delta T/\Delta x$, is negative. This means that the thermal energy is flowing from the inside of the stove to the outside of the stove.

31.   **INTERPRET**   This problem involves calculating the rate of heat conduction through the concrete slab, given the temperature difference between the two sides of the slab and the dimensions of the slab.

DEVELOP   Take the downward direction as the positive-x direction. We assume a steady flow of heat through the area $A = 8.0 \text{ m} \times 12 \text{ m} = 96 \text{ m}^2$, with no flow through the edges. The rate of heat flow is given by Equation 16.5:

$$H = -kA\frac{\Delta T}{\Delta x}$$

The temperature difference is $\Delta T = T_{outside} - T_{inside} = 10°C - 20°C = -10°C = -10$ K (see Equation 16.1) and $\Delta x = x_{outside} - x_{inside} = 0.23$ m.

EVALUATE   From Table 16.2, we find the thermal conductivity of concrete to be $k = 1$ W/(m·K). Thus, the rate of heat conduction is

$$H_{floor} = -kA\frac{\Delta T}{\Delta x} = -\left[1 \text{ W/(m·K)}\right]\left(96 \text{ m}^2\right)\frac{(-10\text{K})}{0.23 \text{ m}} = 4 \text{ kW}$$

which is reported to a single significant figure because the thermal conductivity of concrete is given to one significant figure.

ASSESS   The energy loss through the floor by conduction is substantial. That's why carpeting can prevent heat loss and keeps the house warm during winter season.

33.   **INTERPRET**   This problem is an exercise in calculating the $\mathcal{R}$ factors for various materials of 1-inch thickness.

DEVELOP   The $\mathcal{R}$-factor of a material is given by Equation 16.8:

$$\mathcal{R} = RA = \frac{\Delta x}{k}$$

where R is the thermal resistance and k is the thermal conductivity of a material having a thickness Δx. We will calculate the $\mathcal{R}$ factors in SI units, using $\Delta x = 1$ in $= 25.4$ mm $= 0.0254$ m.

EVALUATE   Using Table 16.2, with $k_{air} = 0.026$ W/(m·K) for air, we have

$$\mathcal{R}_{air} = \frac{0.0254 \text{ m}}{0.026 \text{ W/(m·K)}} = 0.98 \text{ m}^2 \cdot \text{K/W}$$

Similarly, with $k_{concrete} = 1$, $k_{fiberglass} = 0.042$ $\overline{k}_{glass} = 0.8$, $k_{Styrofoam} = 0.029$ and $k_{pine} = 0.11$ [all in units of W/(m·K)], the $\mathcal{R}$-factors are

$$\mathcal{R}_{concrete} = \frac{0.0254 \text{ m}}{1 \text{ W/(m·K)}} = 0.03 \text{ m}^2 \cdot \text{K/W}$$

$$\mathcal{R}_{fiberglass} = \frac{0.0254 \text{ m}}{0.042 \text{ W/(m·K)}} = 0.60 \text{ m}^2 \cdot \text{K/W}$$

$$\mathcal{R}_{glass} = \frac{0.0254 \text{ m}}{0.8 \text{ W/(m·K)}} = 0.03 \text{ m}^2 \cdot \text{K/W}$$

$$\mathcal{R}_{Styrofoam} = \frac{0.0254 \text{ m}}{0.029 \text{ W/(m·K)}} = 0.88 \text{ m}^2 \cdot \text{K/W}$$

$$\mathcal{R}_{\text{pine}} = \frac{0.0254 \text{ m}}{0.11 \text{ W}/(\text{m} \cdot \text{K})} = 0.23 \text{ m}^2 \cdot \text{K/W}$$

ASSESS   The $\mathcal{R}$-factor of a material is inversely proportional to the thermal conductivity. Good thermal insulators such as Styrofoam or wood have large $\mathcal{R}$-factors.

Section 16.4 Thermal-Energy Balance

35.   **INTERPRET**   This is an energy-balance problem involving a stove. We are given the energy loss per unit time per degree temperature difference, and the temperature difference. Note that we are not given the heat-loss mechanism(s), although we can assume it is primarily convection and radiation. We wish to find the rate of energy loss, which by energy balance must be the power required to maintain the temperature.

DEVELOP   The thermal energy leaving the oven is $H_T \Delta T$, which must be balanced by the power P supplied to the oven in order to maintain thermal-energy balance. We multiply the energy loss rate per degree by the temperature difference in degrees. We can therefore write

$$P = H_T \Delta T$$

EVALUATE   Inserting $H_T = 14 \text{ W/C}°$ and $\Delta T = (180°\text{C} - 20°\text{C}) = 160 \text{ C}°$ gives

$$P = (14 \text{ W/°C})(160 \text{ C}°) = 2.2 \text{ kW}$$

ASSESS   2 kW is a reasonable power requirement for an oven.

37.   **INTERPRET**   This problem involves radiative heat loss and the Stefan-Boltzmann law.

DEVELOP   Apply the Stefan-Boltzmann law, Equation 16.9, which is $P = e\sigma A T^4$. The power is P = 100 W, the temperature is T = 3000 K, and s = 5.67×10^{-8} W/(m$^2 \cdot$K^4), so we can solve for the area A. We will assume that the emissivity is $e \approx 1$.

EVALUATE   Inserting the given quantities into the Stefan-Boltzmann law gives

$$P = e\sigma A T^4$$

$$A = \frac{P}{e\sigma T^4} = \frac{100 \text{ W}}{(1)\left[5.67 \times 10^{-8} \text{ W}/\left(\text{m}^2 \cdot \text{K}^4\right)\right](3000 \text{ K})^4} = 2 \times 10^{-5} \text{ m}^2$$

ASSESS   This is about 20 square millimeters, which seems reasonable for the total area of a light bulb filament.

PROBLEMS

39.   **INTERPRET**   This problem is about finding the pressure at different temperatures, given its pressure at a reference temperature and that the volume is held constant.

DEVELOP   For a constant-volume system, the pressure will be linear in temperature (see Figure 16.3). Therefore, we can write

$$\frac{p}{T} = \frac{p_{\text{ref}}}{T_{\text{ref}}} \quad \Rightarrow \quad p = \left(\frac{T}{T_{\text{ref}}}\right) p_{\text{ref}}$$

If we use the given values at the normal melting point of ice, then the pressure-temperature relationship is

$$p = \left(\frac{T}{T_{\text{ref}}}\right) p_{\text{ref}} = T\left(\frac{101 \text{ kPa}}{273.15 \text{ K}}\right)$$

EVALUATE   (a) When the temperature is the normal boiling point of water $T = 100 \text{ °C} = 373.15$ K, the pressure is

$$p = (373.15 \text{ K})\left(\frac{101 \text{ kPa}}{273.15 \text{ K}}\right) = 138 \text{ kPa}$$

(b) If the temperature is the normal boiling point of oxygen (90.2 K), then

$$p = (90.2 \text{ K})\left(\frac{101 \text{ kPa}}{273.15 \text{ K}}\right) = 33.4 \text{ kPa}$$

(c) If the temperature is the normal boiling point of mercury (630 K), then

$$p = (630 \text{ K})\left(\frac{101 \text{ kPa}}{273.15 \text{ K}}\right) = 233 \text{ kPa}$$

ASSESS   These results hold only if the volume is held constant while the temperature varies.

41. **INTERPRET** In this problem, we are asked to calculate the boiling point of SO_2, given the height difference between the liquid levels in a constant-volume gas thermometer.

DEVELOP The thermometric equation for an ideal constant-volume gas thermometer is (see Problem 16.39)

$$p = \left(\frac{T}{T_{ref}}\right) p_{ref}$$

where T is measured in the Kelvin scale. Since the pressure in the constant-volume gas thermometer shown is proportional to h, the temperature of the boiling point of SO_2 is

$$T = T_3 \frac{p}{p_3} = T_3 \frac{h}{h_3}$$

EVALUATE From the equation above, we find the boiling point of SO_2 to be

$$T = (273.16 \text{ K})\left(\frac{57.8 \text{ mm}}{60.0 \text{ mm}}\right) = 263 \text{ K} = -10.0°C$$

ASSESS For a constant-volume gas thermometer, p/T is constant. Since pressure can be measured in mm of mercury $(p = \rho g h)$, it is also true that h/T is constant.

43. **INTERPRET** This problem involves calculating the amount of energy a body uses to run a marathon and, assuming that fat is converted to energy with 100% efficiency, converting this energy to an equivalent mass of fat.

DEVELOP The energy expended in running a marathon for a person with the given mass is

$$\Delta Q = (125 \text{ kcal/mi})(26.2 \text{ mi}) = 3.28 \times 10^3 \text{ kcal}$$

Knowing the amount of energy per gram of fat allows us to answer the question.

EVALUATE Since typical fats contain about 9 kcal per gram, ΔQ is equivalent to the energy content of

$$\frac{3.28 \times 10^3 \text{ kcal}}{9 \text{ kcal/g}} = 364 \text{ g}$$

or about 13 oz of fat.

ASSESS Running a marathon is a good way to burn the fat stored in the body.

45. **INTERPRET** We are interested in the energy needed to raise the temperature of a system. We can solve this problem using the specific heat of the given substances.

DEVELOP The energy Q required to increase the temperature by ΔT is given by Equation 16.3: $Q = mc\Delta T$, where c is the specific heat and m is the mass of the material. The specific heats of some common materials can be found in Table 16.1.

EVALUATE **(a)** When just the pan is heated, with $c_{Cu} = 386$ J/(kg·K), the energy required is

$$\Delta Q = m_{Cu} c_{Cu} \Delta T = (0.8 \text{ kg})\left[386 \text{ J/(kg·K)}\right](90 \text{ K} - 15 \text{ K}) = 23.2 \text{ kJ}$$

(b) If the pan contains water and both are heated between the same temperatures, we then have

$$\Delta Q = \left(m_{Cu} c_{Cu} + m_{H_2O} c_{H_2O}\right) \Delta T = 23.2 \text{ kJ} + (1.0 \text{ kg})\left[4184 \text{ J/(kg·K)}\right](75 \text{ K}) = 337 \text{ kJ}$$

(c) With $m_{Hg} = 4$ kg of mercury replacing the water,

$$\Delta Q = \left(m_{Cu} c_{Cu} + m_{Hg} c_{Hg}\right) \Delta T = 23.2 \text{ kJ} + (4.0 \text{ kg})\left[140 \text{ J/(kg·K)}\right](75 \text{ K}) = 65.2 \text{ kJ}$$

ASSESS The energy required is proportional to the specific heat c. In this problem,

$$c_{Hg}\left[140 \text{ J/(kg·K)}\right] < c_{Cu}\left[386 \text{ J/(kg·K)}\right] < c_{H_2O}\left[4184 \text{ J/(kg·K)}\right]$$

47. **INTERPRET** You wish to know how long it will take a microwave to heat a cup of water to the boiling temperature.

DEVELOP The heat needed to bring the water to the point of boiling can be found with Equation 16.3: $Q = mc\Delta T$, where $c = 4184$ J/kg·K from Table 16.1. The mass of 330 mL of water can be found from the density: $\rho = 1$ g/cm$^3 = 1$ g/mL. The temperature change is $\Delta T = 100°C - 10°C = 90°C$. Note: We don't have to convert to Kelvin, since the change in degrees Celsius is the same as the change in Kelvin. The time it takes the water to absorb this much heat comes from the energy divided by the power.

EVALUATE The time to heat the water to the boiling temperature is

$$t = \frac{Q}{P} = \frac{\rho V c \Delta T}{P} = \frac{(1 \text{ g/mL})(330 \text{ mL})(4184 \text{ J/kg} \cdot \text{K})(90 \text{ K})}{(900 \text{ W})} = 138 \text{ s}$$

ASSESS A little over two minutes to bring the water to boil sounds about right.

49. **INTERPRET** You want to compare the rate at which water is heated by a microwave in a paper cup to on a stovetop in a pan. The hitch is that the stovetop has to heat the pan too.

DEVELOP The temperature rise per second is equal to the heat absorbed per second divided by the heat capacity:

$$\frac{\Delta T}{\Delta t} = \frac{Q/C_{tot}}{\Delta t} = \frac{\bar{P}}{C_{tot}}$$

where $\bar{P}$ is the average power supplied, and $C_{tot} = C_{H_2O} + C_{cnt}$ is the total heat capacity from both the water and the container. This assumes that the water and container both have the same instantaneous temperature. The water's heat capacity is $C_{H_2O} = mc$, where $c = 4184 \text{ J/kg} \cdot \text{K}$ from Table 16.1. For the paper cup used in the microwave oven, $C_{cnt} \approx 0$, whereas for the pan used on the stove burner, $C_{cnt} = 1.4 \text{ kJ/K}$.

EVALUATE If you equate the rates at which the temperatures rise,

$$\frac{\bar{P}_{micro}}{mc} = \frac{\bar{P}_{stove}}{mc + C_{cnt}}$$

You can then solve for the mass:

$$m = \frac{C_{cnt}/c}{\left(\bar{P}_{stove}/\bar{P}_{micro} - 1\right)} = \frac{(1.4 \text{ kJ/K})/(4.184 \text{ kJ/kg} \cdot \text{K})}{\left((1000 \text{ W})/(625 \text{ W}) - 1\right)} = 0.56 \text{ kg}$$

ASSESS This is a little over half a liter. Your own experience may confirm this. For heating a cup of tea, the microwave oven seems to work faster. But for heating a big bowl of soup, the stove will take less time.

51. **INTERPRET** Given the power output of the stove and the amount of time it takes to heat up the water, we want to know how much water is in the kettle. This problem involves specific heat.

DEVELOP The energy supplied by the stove burner heats the kettle and the water in it from 20°C to 100°C, so $\Delta T = 80 \text{ K}$. If we neglect any heat losses and the heat capacity of the burner, this energy is just the burner's power output times the time:

$$\Delta Q = \bar{P} \Delta t = \left(m_w c_w + m_K c_K\right) \Delta T$$

This equation can be used to solve for m_w.

EVALUATE Since all of these quantities are given except for the mass of the water, we can solve for m_w:

$$m_w = \frac{1}{c_w}\left(\frac{\bar{P}\Delta t}{\Delta T} - m_K c_K\right) = \frac{1}{4184 \text{ J/(kg} \cdot \text{K)}}\left(\frac{(2.0 \text{ kW})(5.4 \times 60 \text{ s})}{80 \text{ K}} - (1.2 \text{ kg})\left[447 \text{ J/(kg} \cdot \text{K)}\right]\right)$$
$$= 1.8 \text{ kg}$$

ASSESS We find that m_w is proportional to Δt. This makes sense because the more water in the kettle, the more time we would expect it takes to heat up the water.

53. **INTERPRET** The objects of interest are the steel disks of the disk brakes. The problem deals with transformation of energy from the kinetic energy of the car to the thermal energy of the brake disks, which we can calculate knowing the specific heat of the disk-brake material.

DEVELOP By energy conservation, the loss of kinetic energy of the car is equal to the thermal energy gained by the four brakes:

$$Q = \Delta K \quad \Rightarrow \quad 4m_{brake} c \Delta T = \frac{1}{2}m_{car} v^2$$

EVALUATE From the equation above, with $v = 40 \text{ km/h} = 11.1 \text{ m/s}$, the change of temperature is

$$\Delta T = \frac{m_{car}v^2/2}{4m_{brake}c} = \frac{(1500 \text{ kg})(11.1 \text{ m/s})^2}{8(5.0 \text{ kg})[502 \text{ J/(kg·K)}]} = 9.2 \text{ K}$$

ASSESS This is a big increase in temperature. The brakes can get very hot depending on how fast the car was moving initially.

55. **INTERPRET** Our system consists of two materials, water and copper, which are initially at different temperatures. They are brought together and reach a thermal equilibrium. We want to find the mass of the copper, for which we can use the specific heat of copper.

DEVELOP Let us assume that all the heat lost by the copper is gained by the water, with no heat transfer to the container or its surroundings. Then $-Q_{Cu} = Q_w$ (as in Example 16.2). Expressing each side of this equation using Equation 16.3, we find

$$-m_{Cu}c_{Cu}(T - T_{Cu}) = m_w c_w (T - T_w)$$

The specific heats of copper and water can be found in Table 16.1.

EVALUATE Expressing all the temperatures in the Kelvin scale and solving for m_{Cu}, one finds

$$m_{Cu} = \frac{m_w c_w (T - T_w)}{c_{Cu}(T_{Cu} - T)} = \frac{(1.0 \text{ kg})[4184 \text{ J/(kg·K)}](298 \text{ K} - 293 \text{ K})}{[386 \text{ J/(kg·K)}](573 \text{ K} - 298 \text{ K})} = 0.20 \text{ kg}$$

ASSESS Since the water has much greater mass and higher specific heat, its temperature change is less compared to copper.

57. **INTERPRET** This problem involves the thermal resistance of a material, which we can use to calculate the rate of heat lost through the material given the temperature difference between the different sides of the material.

DEVELOP The total surface area (sides, top, and bottom) of the cooler is
$A = 2(3.0 \times 2.0 + 3.0 \times 2.3 + 2.0 \times 2.3) \text{ m}^2 = 35 \text{ m}^2$. A thickness of 8 cm of Styrofoam of this area has a thermal resistance of $R = Dx/(kA)$ (Equation 16.6), and the heat-flow Equation 16.7 gives

$$P = H = \frac{|\Delta T|}{R} = kA\frac{|\Delta T|}{\Delta x}$$

EVALUATE Using $k = 0.029$ W/(m·K) from Table 16.2 gives

$$P = kA\frac{|\Delta T|}{\Delta x} = \frac{(0.029)(35 \text{ m}^2)(20°\text{C} - 4.0°\text{C})}{0.080 \text{ m}} = 2.0 \times 10^2 \text{ W}$$

to two significant figures.

ASSESS The power requires is equivalent to about 3 60-W light bulbs.

59. **INTERPRET** You want to see if the power output from the party guests can compensate for the heat-loss from the house.

DEVELOP Combined, the 36 people will generate 3600 W of heat. The house will be in energy balance when the inside temperature results in a heat loss that matches what the people produce:

$$P_{loss} = (320 \text{ W/°C})(T_{inside} - 8°\text{C}) = P_{people} = 3600 \text{ W}$$

EVALUATE Solving for the inside temperature

$$T_{inside} = 8°\text{C} + \frac{3600 \text{ W}}{320 \text{ W/°C}} = 19°\text{C}$$

This is equal to about 66°F, which means the house will remain at a comfortable temperature.

ASSESS If you wanted the house even a little warmer, you could ask some of the people to do a little light exercise to generate more than a 100 W of heat.

61. **INTERPRET** This problem involves thermal energy balance. The source of power is the electric current that heats the wire, and the loss mechanism is by radiation for part (a), and by thermal energy conduction for part (b).

DEVELOP The strip is in energy balance between the input power and the net power radiated [the only transfer mechanism available for part (a)]. Thus, according to Equation 16.9,

$$P_{in} = P_{rad} = e\sigma A\left(T_1^4 - T_2^4\right)$$

where $P_{in} = 50$ W and $A = 2\left[(0.50)(5.0) + (0.010)(5.5) + (0.50)(0.010)\right](cm^3) = 5.12 \times 10^{-4}$ m^2. This equation allows us to determine the temperature of the strip, T_1. For part (b), the input power is the same, but the output heat loss is only through thermal conduction, so

$$P_{in} = H = -\frac{T_2 - T_1}{R}$$

EVALUATE (a) Inserting the given quantities into the energy-balance equation above gives

$$T_1 = \left(\frac{P_{in}}{e\sigma A} + T_2^4\right)^{1/4} = \left(\frac{50 \text{ W}}{(1.0)\left[5.67 \times 10^{-8} \text{ W/}(m^2 \cdot K^4)\right](5.12 \times 10^{-4} \text{ m}^2)} + (300 \text{ K})^4\right)^{1/4} = 1.2 \times 10^3 \text{ K}$$

(b) Solving the equation above for part (b) for the temperature T_1 gives

$$T_1 = T_2 + RP_{in} = 300 \text{ K} + (8.0 \text{ K/W})(50 \text{ W}) = 700 \text{ K}$$

ASSESS We get a higher temperature of the strip when heat transfer is caused by radiation than by conduction. At these temperatures, conduction transfers more thermal energy than radiation. However, radiation dominates at higher temperatures because of its T^4 dependence.

63. **INTERPRET** Our system consists of two materials, water and an iron horseshoe, which are initially at different temperatures. They are brought together and reach a thermal equilibrium. We want to find the equilibrium temperature.

    DEVELOP Let us assume that all the heat lost by the horseshoe is gained by the water, with no heat transfer to the container or its surroundings. In this case, $-Q_{Fe} = Q_w$ (as in Example 16.2). Using Equation 16.4 gives

    $$-m_{Fe}c_{Fe}\left(T - T_{Fe}\right) = m_w c_w\left(T - T_w\right)$$

    The specific heats of copper and water can be found in Table 16.1.

    EVALUATE Solving for T, one finds

    $$T = \frac{m_{Fe}c_{Fe}T_{Fe} + m_w c_w T_w}{m_{Fe}c_{Fe} + m_w c_w}$$

    $$= \frac{(1.1 \text{ kg})\left[0.107 \text{ kcal/}(kg \cdot {}^\circ C)\right](550^\circ C) + (15 \text{ kg})\left[1.0 \text{ kcal/}(kg \cdot {}^\circ C)\right](20^\circ C)}{(1.1 \text{ kg})\left[0.107 \text{ kcal/}(kg \cdot {}^\circ C)\right] + (15 \text{ kg})\left[(1.0 \text{ kcal/kg} \cdot {}^\circ C)\right]} = 24^\circ C$$

    ASSESS The change of water temperature is $\Delta T_w = T - T_w = 24.1^\circ C - 20^\circ C = 4.1^\circ C$, while the change of temperature of the iron horseshoe is $|\Delta T_{Fe}| = 525.9^\circ C$. Because there is more water (by mass) and it has a much higher specific heat, its temperature changes less compared to the horseshoe.

65. **INTERPRET** This problem is about the radiation emitted by a burning log. Given its emissivity and its radiating power, we are to calculate its temperature.

    DEVELOP If we neglect the radiation absorbed by the log from its environment (which should be negligible because the temperature of the log is much, much greater than room temperature), then the net power radiated by the log is just that given by the Stefan-Boltzmann law (Equation 16.9):

    $$P = e\sigma AT^4$$

    Knowing the surface area of the log allows us to determine T.

EVALUATE The surface area of the log is

$$A = \pi dL + \pi d^2/2 = \pi d\left(L + d/2\right) = \pi\left(0.15 \text{ m}\right)\left(0.65 \text{ m} + 0.075 \text{ m}\right) = 0.342 \text{ m}^2$$

Solving for T, we find

$$T = \left(\frac{P}{e\sigma A}\right)^{1/4} = \left(\frac{34 \times 10^3 \text{ W}}{\left[5.67 \times 10^{-8} \text{ W/}\left(\text{m}^2 \cdot \text{K}^4\right)\right]\left(0.342 \text{ m}^2\right)}\right)^{1/4} = 1.2 \times 10^3 \text{ K}$$

ASSESS When a burning log is glowing red hot, its temperature is above 1000°C. If the temperature continues to rise, its color will turn orange, then yellow, then white when it emits over a broad range of visible wavelengths.

67. **INTERPRET** This problem is about the heat loss through various structural parts of the house via conduction.
DEVELOP Follow the approach outlined in Example 16.4. By Equation 16.8, 16.6, and 16.5, the heat-flow rate is related to the $\mathcal{R}$-factor as

$$H = -kA\frac{\Delta T}{\Delta x} = -A\frac{\Delta T}{\Delta x/k} = -A\frac{\Delta T}{\mathcal{R}}$$

The window area here is $A_{\text{window}} = 10\left(2.5 \text{ ft} \times 5.0 \text{ ft}\right) = 125 \text{ ft}^2$, and the wall area is 125-ft^2 less than in Example 16.4, or $A_{\text{walls}} = 1506 \text{ ft}^2 - 125 \text{ ft}^2 = 1381 \text{ ft}^2$. Thus, the heat lost through these structural parts are:

$$H_{\text{walls}} = \left(\frac{1}{12.37}\frac{\text{Btu}}{\text{h} \cdot \text{ft}^2 \cdot {}^\circ\text{F}}\right)\left(1381 \text{ ft}^2\right)\left(50{}^\circ\text{F}\right) = 5583 \text{ Btu/h}$$

$$H_{\text{roof}} = \left(\frac{1}{31.37}\frac{\text{Btu}}{\text{h} \cdot \text{ft}^2 \cdot {}^\circ\text{F}}\right)\left(1164 \text{ ft}^2\right)\left(50{}^\circ\text{F}\right) = 1855 \text{ Btu/h}$$

$$H_{\text{windows}} = \left(\frac{1}{0.90}\frac{\text{Btu}}{\text{h} \cdot \text{ft}^2 \cdot {}^\circ\text{F}}\right)\left(125 \text{ ft}^3\right)\left(50{}^\circ\text{F}\right) - 4\left(30\frac{\text{Btu}}{\text{h} \cdot \text{ft}^2}\right)\left(12.5 \text{ ft}^2\right) = 5444 \text{ Btu/h}$$

where we have included the heat gain by solar energy (= 1500 Btu/h) in H_{windows}. Thus, the rate of thermal energy loss from the entire house is

$$H_{\text{total}} = \left(5583 + 1855 + 5444\right) \text{ Btu/h} = 12.88 \times 10^3 \text{ Btu/h}$$

EVALUATE (a) The monthly fuel bill is

$$\left(12.88 \times 10^3 \text{ Btu/h}\right)\left(24 \times 30 \text{ h/mo}\right)\left(1 \text{ gal/}10^5 \text{ Btu}\right)\left(\$2.20\text{/gal}\right) = \$200\text{/mo}$$

to two significant figures.
(b) The solar gain from the south windows is worth

$$\left(1500 \text{ Btu/h}\right)\left(24 \times 30 \text{ h/mo}\right)\left(1 \text{ gal/}10^5 \text{ Btu}\right)\left(\$2.20\text{/gal}\right) = \$24\text{/mo}$$

ASSESS This is an expensive fuel bill. You probably would want to improve the insulation.

69. **INTERPRET** This problem is about radiation received by Pluto from the Sun. Treating Pluto as a blackbody, we want to find its average surface temperature.

DEVELOP From Example 16.5, the power radiated by the Sun is $P_S = 3.9 \times 10^{26}$ W. This radiation spreads evenly out to the orbital radius of Pluto, $r_p = 5.91 \times 10^9$ km (from Appenix E). If we assume that Pluto absorbs the fraction of radiation falling on its cross-sectional area $\left(A_{cs} = \pi R_P^2\right)$, then Pluto's heat input from the Sun is

$$P_{\text{in}} = P_S\frac{\pi R_P^2}{4\pi r_P^2}$$

Pluto will be radiating away this heat, according to Stefan Boltzmann's law: $P_{\text{out}} = e\sigma A T^4$, where the area in this case is the total surface area, $A = 4\pi R_P^2$. The surface temperature, T, will settle to a value where the outgoing radiation matches the incoming radiation.

EVALUATE   Equating the two powers gives the following for the surface temperature:

$$T = \left(\frac{P_S}{16\pi\sigma r_P^2}\right)^{1/4} = \left(\frac{3.9\times10^{26}\,\text{W}}{16\pi\left(5.67\times10^{-8}\,\text{W/m}^2\cdot\text{K}^4\right)\left(5.91\times10^{12}\,\text{m}\right)^2}\right)^{1/4} = 44\,\text{K}$$

ASSESS   Astronomers have recently estimated the temperature on Pluto to be about 43 K, so this answer is in agreement with that. One effect that we didn't account for is Pluto's albedo, i.e., how much of the incoming sunlight gets reflected away instead of absorbed.

71.   INTERPRET   You want to check whether the Sun's recent increase in power output can explain the rise in the global average temperature. This is your friend's argument against human-induced global warming.

DEVELOP   From the Application "The Greenhouse Effect and Global Warming," you were told that the Earth currently absorbs energy from the Sun at a rate of $S = 960\,\text{W/m}^2$, averaged over the cross-sectional area of the planet, πR_E^2. Using energy balance arguments and a assuming the Earth's emissivity is 1, a formula was derived for the Earth's average temperature:

$$T = \left(\frac{S}{4e\sigma}\right)^{1/4} = \left(\frac{960\,\text{W/m}^2}{4(1)\left(5.67\times10^{-8}\,\text{W/m}^2\cdot\text{K}^4\right)}\right)^{1/4} = 255\,\text{K} = -18^\circ\text{C}$$

This is too cold. The global average temperature is actually about 15°C, due to the greenhouse effect. Effectively, the greenhouse effect reduces the Earth's emissivity to about 0.61. Let's assume that the emissivity has been constant since the start of the industrial era. Then $T \propto S^{1/4}$, and we can verify if the change in the solar flux can account for the measured temperature change since the start of the industrial era.

EVALUATE   The solar flux has increased by 0.04% since pre-industrial time, which can be expressed as $S = (1.0004)\,S_{\text{pre}}$. The temperature should correspondingly be higher due to this change:

$$T = T_{\text{pre}}\left(\frac{S}{S_{\text{pre}}}\right)^{1/4} \quad\rightarrow\quad \Delta T = T - T_{\text{pre}} = \left(1 - (1.0004)^{-1/4}\right)T = 1.00\times10^{-4}\,T$$

In Kelvin, the current global average temperature is $T = 288$ K, so the temperature change from the solar flux increase is $\Delta T = 0.029^\circ\text{C}$. This only accounts for about 4% of the measured temperature increase, so your friend is wrong.

ASSESS   The argument for human-induced global warming is that the temperature increase is due to a decrease in the effective emissivity. Rising levels of greenhouse gases since the beginning of the industrial era allow less of the infrared radiation from the Earth's surface to be emitted into space.

73.   INTERPRET   This is an energy balance problem. The rabbit hutch loses energy at a given rate. In equilibrium, the heat lamp provides energy at the same rate that the hutch loses energy. You will find the equilibrium temperature difference to see if the interior temperature can stay above freezing.

DEVELOP   The thermal resistance is given as $R = 0.25$ K/W, which means the hutch loses heat at a rate of $H = -\Delta T/R$. The power of the heater is $P = 50$ W, so in equilibrium the temperature difference is $\Delta T = PR$.

EVALUATE   Since the outside temperature is −15°C, the interior temperature of the hutch is

$$T_{\text{in}} = T_{\text{out}} + PR = -15^\circ\text{C} + (50\,\text{W})(0.25\,\text{K/W}) = -2.5^\circ\text{C}$$

The rabbit's water will freeze.

ASSESS   You need to get a bigger heater, or insulate the cage better, otherwise your niece's rabbit will not fare very well.

75.   INTERPRET   We are to show that the equation for conductive heat transfer through a conical solid is as given. To do this, we will integrate the conductive heat transfer through thin circular disks normal to the cone axis.

DEVELOP   We will integrate the heat flow through the cone, treating the cone as a stack of circular disks. The radius of any disk depends on x as

$$r(x) = R_1 + \frac{R_2 - R_1}{L} x$$

so the area of a disk is

$$A(x) = \pi r(x)^2 = \pi \left(R_1 + \frac{R_1 + R_2}{L} x \right)^2$$

The heat transfer rate is given by the differential form of Equation 16.3, which is $H = -kA(dT/dx)$. Inserting the expression for area gives

$$H dx = -kA(x) dT = -k\pi \left(R_1 + \frac{R_1 + R_2}{L} x \right)^2 dT$$

$$\frac{H dx}{\left[R_1 + x(R_1 + R_2)/L \right]^2} = -k\pi dT$$

which we can integrate to find the heat transfer rate H.

EVALUATE   Integrating both sides gives

$$H \int_0^L \frac{dx}{\left(R_1 + \frac{R_2 - R_1}{L} x \right)^2} = -k\pi \int_{T_1}^{T_2} dT$$

$$H \left[\frac{L}{R_2 (R_1 - R_2)} - \frac{L}{R_1 (R_1 - R_2)} \right] = -k\pi (T_2 - T_1)$$

$$H \left[\frac{L (R_1 - R_2)}{R_1 R_2 (R_1 - R_2)} \right] = -k\pi (T_2 - T_1)$$

$$H = -k\pi (T_2 - T_1) \left(\frac{R_1 R_2}{L} \right)$$

ASSESS   We have shown what was required.

77.   **INTERPRET**   We're asked to compute the temperature inside a greenhouse given a time-varying solar input.

DEVELOP   We'll assume the Sun's energy only enters through the windows $\left(A_w = 250\,\text{ft}^2 \right)$, in which case the rate of heat gain from the Sun is

$$P_{in} = SA_w = \left(40\ \text{Btu/h/ft}^2 \cdot \sin^2 \left(\tfrac{\pi}{24} t \right) \right) \left(250\ \text{ft}^2 \right) = 1.0 \times 10^4\ \text{Btu/h} \cdot \sin^2 \left(\tfrac{\pi}{24} t \right)$$

The rate of heat loss was computed in Example 16.7:

$$P_{out} = H_{tot} = \left(149\ \text{Btu/h/}^\circ\text{F} \right) \left(T - T_{out} \right)$$

where T is the indoor temperature, and we assume that the outdoor temperature remains constant throughout the day: $T_{out} = 15^\circ\,\text{F}$. The net heat exchange will cause the indoor temperature to change according to

$$P_{in} - P_{out} = \frac{dQ}{dt} = C \frac{dT}{dt} = \left(1500\ \text{Btu/}^\circ\text{F} \right) \frac{dT}{dt}$$

This is a linear first-order differential equation. We set $y = T - T_{out}$, such that:

$$\frac{dy}{dt} + Ay = B \sin^2 (\omega t)$$

where $A = 0.0993\ \text{h}^{-1}$, $B = 6.67\ ^\circ\text{F/h}$, and $\omega = \pi / 24\text{h}$.

EVALUATE   One can solve the differential equation with a computer program or a calculator. We will solve it analytically. If we multiply both sides of the equation by e^{At}, then the solution for $y(t)$ has the form

$$y(t) = e^{-At} \left[\int e^{At} \cdot B \sin^2 (\omega t) \, dt + D \right]$$

where D is an integration constant. One can find the integral in a table:

$$y(t) = \frac{BA}{A^2 + 4\omega^2} \left[\sin^2 (\omega t) - \frac{2\omega}{A} \sin (\omega t) \cos (\omega t) + \frac{2\omega^2}{A^2} \right] + De^{-At}$$

We will neglect the exponential term because it will decay away, so we are left with

$$y(t) = \frac{B}{2A} \left[1 - \frac{A^2 \cos \omega t + 2A\omega \sin \omega t}{A^2 + 4\omega^2} \right]$$

To find the maximum and minimum of $y(t)$, we take the derivative and set it to zero. The extrema occur when $\tan \omega t = 2\omega / A$, which corresponds to $\omega t = 1.21$ and $\omega t = 4.35$. Substituting these values back into the original equation, we find the minimum and maximum values of $y(t)$ are $22°F$ and $45°F$, respectively. Adding these values to the outdoor temperature, the minimum and maximum indoor temperatures are $37°F$ and $60°F$.

ASSESS   The average temperature in the greenhouse is $48.5°F$, which is $33.5°F$ above the outdoor temperature. Notice that this is exactly half the temperature difference found in Example 16.7 $(\Delta T = 67°F)$. This makes sense, since the average solar input in this problem is half of what it was in Example 16.7:

$$\langle S \rangle = \left\langle 40 \text{ Btu/h/ft}^2 \cdot \sin^2 \left(\tfrac{\pi}{24} t \right) \right\rangle = 20 \text{ Btu/h/ft}^2$$

Here, we've used the fact that the average of $\sin^2$ is ½.

79.   **INTERPRET**   We consider the physical properties of fiberglass insulation.

DEVELOP   Aluminum foil has a very high thermal conductivity, $k = 237 \text{ W/m} \cdot \text{K}$, so its definitely not being used to reduce heat loss by conduction. It will help prevent air from flowing through the fiberglass, but that's usually not a problem in an attic or a wall, where the air is pretty still.

EVALUATE   Aluminum is a good reflector of radiation, so it will reflect back radiation emitted from the fiberglass. This will help to reduce heat loss from radiation.

The answer is (c).

ASSESS   The reflectivity is a measure of how good a material is at reflecting radiation. It is equal to $1 - e$, where e is the emissivity. Since e is a measure of absorption as well as emission, we can understand that a good reflector is a bad absorber. Aluminum foil has an emissivity of 0.03, which is why it is a good reflector.

81.   **INTERPRET**   We consider the physical properties of fiberglass insulation.

DEVELOP   Squeezing a fiberglass sheet will reduce the amount of air trapped between the glass fibers. By cramming two sheets into the space of one, we would essentially be replacing trapped air with glass fibers.

EVALUATE   As we argued in Problem 16.78, the trapped air is providing a large part of the insulation thanks to its low thermal conductivity. Therefore, squeezing the air out will reduce the overall $\mathcal{R}$-factor.

The answer is (c).

ASSESS   One might imagine that the best insulation would be a layer of air, with only a thin shell to keep it in place. In fact, that's the logic behind double-pane windows. However, if the air layer is too thick, you start to have convection, which vastly reduces the insulation quality.

THE THERMAL BEHAVIOR OF MATTER

17

EXERCISES

Section 17.1 Gases

17. **INTERPRET**   This problem involves the ideal-gas law, which we can use to find the volume of 1 mol of Martian atmosphere given its temperature and pressure.

DEVELOP   Apply Equation 17.2 $PV = nRT$ with $P = 0.01P_E$, $T = 215$ K, $n = 1.0$ mol, and $R = 8.314$ J/K.

EVALUATE   Solving the ideal-gas law for the volume and inserting the given quantities gives

$$V = \frac{nRT}{P} = \frac{(1.0 \text{ mol})(8.314 \text{ J/K})(215 \text{ K})}{(0.01)(1.01 \times 10^5 \text{ Pa})} = 1.8 \text{ m}^3$$

ASSESS   The dimensions of this expression are

$$\frac{\text{mol}(\text{J/K})\text{K}}{\text{N/m}^2} = \frac{\text{mol}(\text{N} \cdot \text{m})}{\text{N/m}^2} = \text{mol} \cdot \text{m}^3$$

but a mole is a dimensionless number, so the final dimensions are m³, as expected for a volume.

19. **INTERPRET**   This problem involves an ideal gas, so we can apply the ideal-gas law to find the pressure of the gas at the given temperature and volume.

DEVELOP   In terms of moles, the ideal-gas law is given by Equation 17.2, $PV = nRT$. The volume is $V = 2.0$ L $= 2.0 \times 10^{-3}$ m³, and $T = -150°C = 123$ K.

EVALUATE   Solving for the pressure and inserting the given quantities gives

$$P = \frac{nRT}{V} = \frac{(3.5 \text{ mol})(8.314 \text{ J/K})(123 \text{ K})}{2.0 \times 10^{-3} \text{ m}^3} = 1.8 \times 10^6 \text{ Pa}$$

ASSESS   This is about 20 times standard atmospheric pressure.

21. **INTERPRET**   This problem involves an ideal gas, so we can apply the ideal-gas law. We are to find the volume of an ideal gas given its temperature and pressure, then find its new temperature if the pressure is increased and the volume is cut in half.

DEVELOP   In terms of moles, the ideal-gas law takes the form of Equation 17.2, $pV = nRT$. For part (a), $p = 1.5$ atm $= 1.5 \times 1.01 \times 10^5$ Pa $= 1.515 \times 10^5$ Pa, $T = 250$ K, and $n = 2.0$. For part (b), we can take the ratio of the ideal-gas law applied to part (a) an that applied to part (b) to get

$$\frac{p_a V_a}{p_b V_b} = \frac{nRT_a}{nRT_b} = \frac{T_a}{T_b}$$

which we can solve for T_b given that $V_b = V_a/2$ and $P_b = 4.0$ atm. Note that we have used the fact that the number of moles does not change.

EVALUATE   (a) The volume V_a of the gas is

$$V = \frac{nRT}{P} = \frac{(2.0 \text{ mol})\left[8.314 \text{ J/(mol·K)}\right](250 \text{ K})}{1.515 \times 10^5 \text{ Pa}}$$
$$= 2.7 \times 10^{-2} \text{ m}^3 = 27 \text{ L}$$

(b) Upon compressing the gas and increasing its pressure, the new temperature is

$$T_b = \left(\frac{4.0 \text{ atm}}{1.5 \text{ atm}}\right)\left(\frac{0.5V_1}{V_1}\right)(250 \text{ K}) = 330 \text{ K}$$

to two significant figures.

ASSESS   As expected, the temperature increases if we compress the gas.

23.   **INTERPRET**   This problem in an exercise in calculating the thermal speed of ideal-gas molecules at a given temperature.

DEVELOP   The thermal speed (also called the rms, or root-mean-square speed) is, from Equation 17.4,

$v_{th} = \sqrt{3kT/m}$, where m is the mass of a molecule. From Appendix C we estimate the mass of a H_2 molecule to be $2 \times 1.66 \times 10^{-27}$ kg.

EVALUATE   Inserting the given quantities into Equation 17.4 gives

$$v_{th} = \sqrt{\frac{3(1.38 \times 10^{-23} \text{ J/K})(800 \text{ K})}{2(1.66 \times 10^{-27} \text{ kg})}} = 3.16 \text{ km/s}$$

ASSESS   This is about 100 times faster than the standard speed of sound (343 m/s).

Section 17.2 Phase Changes

25.   **INTERPRET**   This problem involves the latent heat of fusion, which is the energy it takes to liberate the molecules that compose the ice to form water. We are asked to find the energy required to melt a 65-g ice cube.

DEVELOP   The energy required for a solid-liquid phase transition at the normal melting point of water (0°C) is (Equation 17.5) $Q = mL_f$, where m = 0.065 kg and L_f = 334 kJ/kg (from Table 17.1).

EVALUATE   The heat required to melt the ice cube is

$$Q = mL_f = (0.065 \text{ kg})(334 \text{ kJ/kg}) = 22 \text{ kJ}$$

ASSESS   This is equivalent to 5.2 kcal.
(See Table 17.1 for the heats of transformation.)

27.   **INTERPRET**   This problem involves the latent heat of vaporization, which is the energy required to pass from the liquid to the gas phase. Given the latent heat of vaporization (Table 17.1) and the energy required to vaporize the substance, we are to calculate the mass of the substance.

DEVELOP   Assuming the vaporization takes place at the normal boiling point for oxygen at atmospheric pressure, we may use Equation 17.5, $Q = L_v m$, where L_v = 213 kJ/kg (from Table 17.1) and Q = 840 kJ.

EVALUATE   The mass of oxygen in the sample is

$$m = \frac{Q}{L_v} = \frac{840 \text{ kJ}}{213 \text{ kJ/kg}} = 3.9 \text{ kg}$$

ASSESS   Given that O_2 is approximatley 32 g/mol, this corresponds (at standard temperature and pressure) to a volume of

$$V = \frac{nRT}{p} = \frac{(3900 \text{ g})(8.134 \text{ J/K})(273 \text{ K})}{(32 \text{ g/mol})(10^5 \text{ Pa})} = 2.7 \text{ m}^3$$

29. **INTERPRET**   This problems involves the phase transformation of a liquid to a gas, so the latent heat of vaporization comes into play. Because the liquid is at its boiling point, any heat added to the liquid will cause vaporization instead of a temperature rise.

   DEVELOP   From Table 17.1, the latent heat of vaporization for O_2 is 213 kJ/kg. Use this in Equation 17.5, $Q = L_v m$, to find the heat needed to vaporize 28 kg of O_2.

   EVALUATE   Inserting the given quantities into Equation 17.5 gives $Q = mL_v = (28 \text{ kg})(213 \text{ kJ/kg}) = 6.0 \text{ MJ}$ to two significant figures.

   ASSESS   At standard temperature and pressure (STP, $T = 273.15$ K, $p = 10^5$ Pa), 28 kg of O_2 would occupy a volume of

$$V = \frac{nRT}{p} = \frac{(28 \times 10^3 \text{ g})(8.134 \text{ J/K})(273 \text{ K})}{(32 \text{ g/mol})(10^5 \text{ Pa})} = 19 \text{ m}^3$$

Section 17.3 Thermal Expansion

31. **INTERPRET**   This is a problem in thermal expansion. We know the initial volume, the material, and the change in temperature, and we need to find the new volume.

   DEVELOP   The change in volume is given by Equaiton 17.6,

$$\beta = \frac{\Delta V}{V \Delta T}$$

   The initial volume is $V = 1.00$ L. The material is ethyl alcohol, which has a volume coefficient of expansion of $\beta = 75 \times 10^{-5} \text{ K}^{-1}$, and the change in temperature is $\Delta T = -18$ K. The new volume V' of the liquid is

$$V' = V + \Delta V = V(1 + \beta \Delta T)$$

   EVALUATE   Inserting the given quantities gives

$$V' = V(1 + \beta \Delta T) = (1.00 \text{ L})\left[1 + (75 \times 10^{-5})(-18 \text{ K})\right] = 0.987 \text{ L}$$

   ASSESS   This is enough to observe in your own refrigerator: Seal a volume of liquid at room temperature, put it in the refrigerator, and observe the deformation of the container when it cools.

33. **INTERPRET**   This problem deals with thermal expansion of a steel washer. The quantity of interest is the diameter of the washer, so the relevant quantity is the coefficient of linear expansion, α.

   DEVELOP   The coefficient of linear expansion is defined as (see Equation 17.7):

$$\alpha = \frac{\Delta L / L}{\Delta T}$$

   For steel, its value is (see Table 17.2) $\alpha = 12 \times 10^{-6} \text{ K}^{-1}$. Solve this equation for ΔT.

   EVALUATE   From the equation above, we get

$$\Delta T = \frac{\Delta L}{\alpha L} = \frac{9.55 \text{ mm} - 9.52 \text{ mm}}{(12 \times 10^{-6} \text{ K}^{-1})(9.52 \text{ mm})} = 263 \text{ K}$$

   Since the initial temperature is 0°C we must heat the washer to 263°C.

   ASSESS   Since α is very small, a large increase in temperature results in a small increase in the washer's diameter.

PROBLEMS

35. **INTERPRET**   The system of interest is the solar corona, which we treat as an ideal gas. The quantity of interest is the number density of air molecules.

   DEVELOP   The number density implied by the ideal-gas law (Equation 17.1) is

$$pV = NkT \quad \rightarrow \quad \frac{N}{V} = \frac{p}{kT}$$

EVALUATE   Applying the above equation to the solar corona, we obtain

$$\left(\frac{N}{V}\right)_{corona} = \frac{p}{kT} = \frac{3 \times 10^{-2}\,\text{Pa}}{(1.38 \times 10^{-23}\,\text{J/K})(2 \times 10^{6}\,\text{K})} = 1 \times 10^{15}\,\text{m}^{-3}$$

If we assume the Earth's atmosphere has standard temperature and pressure, the particle density is

$$\left(\frac{N}{V}\right)_{STP} = \frac{p}{kT} = \frac{1.013 \times 10^{5}\,\text{Pa}}{(1.38 \times 10^{-23}\,\text{J/K})(273\,\text{K})} = 2.7 \times 10^{25}\,\text{m}^{-3}$$

So the corona is over 10 billion times less dense than on Earth.

ASSESS   Scientists are still not entirely certain how the corona ends up being so hot.

37.   **INTERPRET**   The object of interest is the cylinder compressed with air. We are given the pressure, temperature, and volume, and want to find the number of moles (i.e., the number of air molecules) in the cylinder.

DEVELOP   We shall treat the air as an ideal gas (although this is somewhat risky at 180 atm) and use the ideal-gas law $PV = nRT$ given in Equation 17.2, to find the number n of moles. The volume of the cylinder is

$$V = \pi\left(\frac{d^2}{4}\right)h = \pi(0.10\,\text{m})^2(1.0\,\text{m}) = 0.01\,\pi\,\text{m}^3$$

EVALUATE   (a) Applying the ideal-gas law gives

$$n = \frac{pV}{RT} = \frac{(180\,\text{atm})(1.013 \times 10^{5}\,\text{Pa/atm})(0.01\pi\,\text{m}^3)}{(8.314\,\text{J/K}\cdot\text{mol})(293\,\text{K})} = 235\,\text{mol}$$

where we have used $T = 20°\text{C} = 293$ K as the room temperature.

(b) If the pressure is $p' = 1$ atm, then the volume would be

$$V' = \frac{nRT}{p'} = \left(\frac{p}{p'}\right)V = \left(\frac{180\,\text{atm}}{1\,\text{atm}}\right)(0.01\pi\,\text{m}^3) = 5.65\,\text{m}^3$$

ASSESS   When temperature is held constant, $PV = $ constant for an ideal gas. Therefore, decreasing the pressure increases the volume in a proportional amount.

39.   **INTERPRET**   The object of interest is the flask filled with air, which we treat as an ideal gas. We explore the effect of changing temperature and pressure. The maximum pressure in the flask will occur when the gas inside the flask, which is initially at STP, is heated to the boiling point of water (100°C). To find the number of moles that escape when the flask is opened, we consider that the gas escapes so fast that the temperature of the gas can be considered to be constant on this timescale.

DEVELOP   When the flask is immersed in boiling water, its volume remains fixed. Therefore, the ideal-gas law $pV = nRT$ (Equation 17.2) applied at each temperature gives

$$\left.\begin{matrix} p_1 V = nRT_1 \\ p_2 V = nRT_2 \end{matrix}\right\} \quad \frac{p_2}{p_1} = \frac{T_2}{T_1}$$

The initial conditions of the gas are $p_1 = 1$ atm, $V = 3.00$ L $= 3.00 \times 10^{-3}$ m^3, and $T_1 = 293$ K. The maximum pressure in the flask occurs when $T_2 = 100°\text{C} = 373$ K. For part (b), we first calculate the number of moles of gas initially in the flask. Again applying the ideal-gas law, we find the number of molecules to be

$$n_1 = \frac{p_1 V}{RT_1} = \frac{(1\,\text{atm})(1.013 \times 10^{5}\,\text{Pa/atm})(3 \times 10^{-3}\,\text{m}^3)}{[8.314\,\text{J/(K}\cdot\text{mol})](293\,\text{K})} = 0.125\,\text{mol}$$

When the flask is opened at $T_2 = 373$ K the pressure rapidly decreases to $p_2 = 1$ atm, so the quantity of gas remaining in the flask is

$$n_2 = \frac{p_1 V_1}{RT_2} = \left(\frac{T_1}{T_2}\right)n_1$$

so the number of moles that escaped from the flask is $\Delta n = n_1 - n_2$. After the flask is closed and cooled back down to $T_3 = 20°C = 293$ K, we again apply the ideal-gas law using n_2 to find the new pressure. This gives

$$p_3 = \frac{n_2 R T_1}{V} = \left(\frac{n_2}{n_1}\right) p_1$$

EVALUATE   **(a)** From the equation above, we find the maximum pressure reached in the flask to be

$$p_2 = \left(\frac{T_2}{T_1}\right) p_1 = \left(\frac{373\text{ K}}{293\text{ K}}\right)(1\text{ atm}) = 1.27\text{ atm}$$

(b) After opening the flask the quantity of gas left in the flask is

$$n_2 = \left(\frac{T_1}{T_2}\right) n_1 = \left(\frac{293\text{ K}}{373\text{ K}}\right)(0.125\text{ mol}) = 0.0980\text{ mol}$$

Therefore, the amount that escaped is $\Delta n = n_1 - n_2 = 0.0268$ mol.

(c) After sealing the flask and cooling it to 293 K, the pressure of the gas in the flask is

$$p_3 = \frac{n_2 R T_1}{V} = \left(\frac{n_2}{n_1}\right) p_1 = \left(\frac{0.098\text{ mol}}{0.125\text{ mol}}\right)(1\text{ atm}) = 0.786\text{ atm}$$

ASSESS   As expected, the pressure in the flask is greatest for part (a), when the temperature is highest and there are the most moles of gas in the flask, and the pressure is the lowest for part (c), when the reverse is true. Pressure is proportional to the number of molecules in the volume, so after some gas molecules escape from the flask, the pressure decreases.

41.   **INTERPRET**   This problem involves finding the time required to transform the given amount of 100-°C water to gas with the given rate of heating.

     DEVELOP   From Equation 17.5, the energy absorbed during the vaporization of water at its boiling point is $Q = L_v m$. If this energy is supplied at in a time t, then the power must be $P = Q/t$, so $t = Q/P$. Given that $P = 1500$ W, we can solve for the time t.

     EVALUATE   The time it takes to boil away the water is

$$t = \frac{Q}{P} = \frac{L_v m}{P} = \frac{(2257\text{ kJ/kg})(1.1\text{ kg})}{1.5\text{ kW}} = 1.66 \times 10^3\text{ s} = 27.6\text{ min}$$

     ASSESS   This seems like a reasonable time, and it explains why it is a good idea to add water occasionally to steamers when cooking with steam.

43.   **INTERPRET**   You want to know how long it will take your camping stove to melt snow. Note that the stove here is the same as the one in Problem 16.56 that was used to boil water.

     DEVELOP   You can integrate the given power, P, to find the total heat that the snow has absorbed. You can then equate that to the amount of energy needed to melt the snow. Since the snow starts off around 0°C, it doesn't need to be warmed up to the freezing point. From Equation 17.5: $Q = L_f m$, where $L_f = 334$ kJ/kg from Table 17.1.

     EVALUATE   The heat absorbed by the snow over a given time is:

$$Q = \int_0^t P(t')\, dt' = at + \tfrac{1}{2}bt^2 = (1.1\text{kW})t + (1.15\text{ W/s})t^2$$

You want to know how long until this absorbed heat melts the snow

$$Q = L_f m = (334\text{ kJ/kg})(5.0\text{ kg}) = 1670\text{ kJ}$$

This requires solving a quadratic equation with the quadratic formula from Appendix A:

$$t = \frac{-(1100\ \text{W}) + \sqrt{(1100\ \text{W})^2 + 4(1.15\ \text{W/s})(1670\ \text{kJ})}}{2(1.15\ \text{W/s})} = 818\ \text{s} \approx 14\ \text{min}$$

ASSESS   In Problem 16.56, it took the same stove 9 minutes to boil water 2.5 kg of water. If we double this result, we see that it takes about 20% less time to melt snow than to boil the same amount of water.

45.   **INTERPRET**   This problem involves the latent heat of fusion, with which we can calculate the time it takes to freeze water that is at 0°C if we remove energy a the rate given.

DEVELOP   From Equation 17.5, the refrigerator must extract the energy $Q = L_f m$, where $m = 0.75$ kg and $L_f = 334$ kJ/kg (from Table 17.1). At the rate $P = Q/t$, where P is the power applied by the refrigerator, it will take a time $t = Q/P$ to freeze the water.

EVALUATE   Inserting the given quantities into the expression above gives

$$t = \frac{Q}{P} = \frac{L_f m}{P} = \frac{(334\ \text{kJ/kg})(0.75\ \text{kg})}{0.095\ \text{kW}} = 2.64 \times 10^3\ \text{s} = 43.9\ \text{min}$$

ASSESS   This is a good figure to keep in mind if you need to make ice cubes for a party.

47.   **INTERPRET**   This problem involves a change in temperature and a phase change (solid to liquid). We will apply the concepts of specific heat and heat of fusion to find temperature of the system in equilibrium.

DEVELOP   From Example 17.4, we expect that the 50 g of ice at −10°C will completely melt in the 1.0 kg of water at 15°C. To find the equilibrium temperature, consider that the water must first warm the ice from −10°C to 0°C, which will require an energy $Q_1 = m_{ice} c_{ice} \Delta T$, with $\Delta T = 10$ K, $m_{ice} = 0.050$ kg, and $c_{ice} = 2050$ J/(kg·K) (see Table 16.1). Next, the water will melt the ice, which costs an energy $Q_2 = L_f m_{ice}$, with $L_f = 334$ kJ/kg (see Table 17.1), and finally, the water will warm up the newly melted water from 0°C to the equilibrium temperature T, which will cost an energy $Q_3 = m_{ice} c_{water} T$. The sum of these energies must equate to the energy *lost* by the water, or

$$m_{water} c_{water} \left(T_{0,water} - T \right) = Q_1 + Q_2 + Q_3 = m_{ice} \left(c_{ice} \Delta T + L_f + c_{water} T \right)$$

where $T_{0,water} = 15$°C and $c_{water} = 4184$ J/(kg·K) (see Table 16.1).

EVALUATE   Solving the expression above for the equilibrium temperature T gives

$$T = \frac{m_{water} c_{water} T_{0,water} - m_{ice} \left(c_{ice} \Delta T + L_f \right)}{\left(m_{ice} + m_{water} \right) c_{water}}$$

$$= \frac{(1.0\ \text{kg})\left[4184\ \text{J/(kg·K)} \right](15°\text{C}) - (0.050\ \text{kg})\left\{ \left[2050\ \text{J/(kg·K)} \right](10°\text{C}) + 334\ \text{kJ/kg} \right\}}{(0.050\ \text{kg} + 1.0\ \text{kg})\left[4184\ \text{J/(kg·K)} \right]}$$

$$= 10°\text{C}$$

ASSESS   As expected, the water has not reached 0°C, so all the ice melts and our initial assumption is confirmed.

49.   **INTERPRET**   This problem involves raising the temperature of water, which involves the specific heat of water, then changing its phase from liquid to gas, which involves the latent heat of vaporization. Using these two concepts, we are to find the initial temperature of the water given that the 90% of the energy needed to boil the water away is used to change the phase, with only 10% being used to raise its temperature.

DEVELOP   Equation 16.3 $Q_1 = mc\Delta T$ gives the energy needed to raise the water's temperature, where $\Delta T = 100$°C − T where T is the initial temperature of the water. The energy needed to boil the water (i.e., change it from the liquid phase to the gas phase) is $Q_2 = mL_v$. The quantities L_v and c can be found in Tables 17.1 and 16.1, respectively. The problem statement says that $Q_1 = Q_2/10$, so

$$mc\Delta T = mc\left(100°C - T\right) = \frac{mL_v}{10}$$

$$T = \left(100°C\right) - \frac{L_v}{10c}$$

we can solve for the initial water temperature T.

EVALUATE   Inserting the specific heat and latent heat of vaporization into the expression above gives

$$T = \left(100°C\right) - \frac{2257 \text{ kJ/kg}}{10.0\left[4.184 \text{ kJ/}\left(\text{kg}\cdot\text{K}\right)\right]} = 46.1°C$$

ASSESS   Much more heat is required to boil the water away (i.e., change its phase from liquid to gas) than to raise its temperature from 46°C to 100°C.

51.   **INTERPRET**   This problem involves mixing ice with water and letting the mixture come to equilibrium. We are to calculate the minimum amount of ice needed so that the final equilibrium mixture is at 0°C. This calculation will involve the specific heat of water to calculate the energy required to raise the temperature of the ice to 0°C and the latent heat of fusion to calculate the energy required to melt the ice.

DEVELOP   For the final equilibrium temperature in Example 17.4 to be 0°C, the original 1.0 kg of water must lose at least $Q_2' = 62.8$ kJ of heat energy (see Example 17.4). It could lose more energy, if some or all of it froze, but this would clearly require a greater amount of ice. The amount of ice needed to absorb this thermal energy and just melt, without exceeding 0°C, is given by summing the energy needed to raise its temperature to zero ($Q_1 = mc\Delta T$, where $\Delta T = 10°C$) and the energy needed to melt the ice ($Q_2 = L_v m$). Equating Q_2' with $Q_1 + Q_2$ gives

$$Q_2' = Q_1 + Q_2 = m\left(c\Delta T + L_v\right)$$

which we can solve for m.

EVALUATE   Solving for m gives

$$m = \frac{c\Delta T + L_v}{Q_2'} = \frac{\left[2.050 \text{ kJ/}\left(\text{kg}\cdot\text{K}\right)\right]\left(10°C\right) + 2257 \text{ kJ/kg}}{62.8 \text{ kJ}} = 177 \text{ g}$$

ASSESS   The amount of original ice that could produce a final temperature of 0°C and freeze all the original water is

$$m_{ice} = \frac{m_w\left(c_w\Delta T_w + L_f\right)}{c_{ice}\Delta T_{ice}} = \frac{\left(1.0 \text{ kg}\right)\left(62.8 + 334\right) \text{ kJ/kg}}{20.5 \text{ kJ/kg}} = 19 \text{ kg}$$

The amount of ice that would produce a final temperature of 0°C with none of the ice melted and none of the water frozen is

$$m_{ice} = \frac{m_w c_w \Delta T_w}{c_{ice}\Delta T_{ice}} = \frac{\left(1.0 \text{ kg}\right)\left[4.184 \text{ kJ/}\left(\text{kg}\cdot\text{K}\right)\right]\left(15°C\right)}{\left[2.050 \text{ kJ/}\left(\text{kg}\cdot\text{K}\right)\right]\left(10°C\right)} = 3.1 \text{ kg}$$

For 177 g $< m_{ice} <$ 3.1 kg, some of the original ice melts, and for 3.1 kg $< m_{ice} <$ 19 kg, some of the original water freezes.

53.   **INTERPRET**   This problem involves mixing an equal mass of ice and water and letting the mixture reach equilibrium. We are to calculate at what temperature the water must be if the final mixture is to contain equal amounts of ice and water.

DEVELOP   Assume that all the heat gained by the ice was lost by the water, with no heat transfer to the container or the surroundings. An equilibrium mixture of ice and water (at atmospheric pressure) must be at 0°C, and if the masses of ice and water start out and remain equal, there is no net melting or freezing. Thus any energy spent raising the temperature of the ice and melting it must be balanced by energy lost by the water as its temperature lowers and it freezes. These energies are given by Equations 16.3 (for the temperature change), $Q_1 = mc$ and (for the melting or freezing) $Q_2 = L_f m$. We can therefore sum these energies for both ice and water and equate the result, which gives

$$Q_{ice} = \overbrace{m_{ice}c_{ice}(0°C - T_{ice})}^{Q_1^{ice}} + \overbrace{L_f m_{ice}}^{Q_2^{ice}} = -Q_w = \overbrace{m_w c_w (T_w - 0°C)}^{-Q_1^w} + \overbrace{L_f m_w}^{-Q_2^w}$$

where $T_{ice} = -10°C$. Given that $m_{ice} = m_w$, we can solve for the water temperature T_w.

EVALUATE   Solving for T_w and inserting the given quantities gives

$$T_w = \frac{c_{ice}(-T_{ice})}{c_w} = \frac{2.050 \text{ kJ}/(\text{kg} \cdot \text{K})}{4.184 \text{ kJ}/(\text{kg} \cdot \text{K})}(10°C) = 4.9°C$$

ASSESS   Because any thermal energy spent melting the ice is balanced by freezing the water, only the specific heats of water and ice come into play.

55.   **INTERPRET**   This problem involves the latent heat of fusion of water, which is the energy required per unit mass to change ice to water (or vice-versa, but with the opposite sign). We can use this concept to find the energy required to melt 20 kg of ice in 6 min, and from there find the power.

DEVELOP   If the melting occurs at atmospheric pressure and if the ice is at 0°C, the energy required to melt the ice is given by Equation 17.3, $Q = L_f m$, where L_f is the latent heat of fusion, which is given in Table 17.1. To melt the ice in a time $t = 6$ min would require a power $P = Q/t$.

EVALUATE   The power required is

$$P = \frac{Q}{t} = \frac{L_f m}{t} = \frac{(20 \text{ kg})(334 \text{ kJ/kg})}{6 \times 60 \text{ s}} = 19 \text{ kW}.$$

ASSESS   This is equivalent to the power needed to light 190 100-W bulbs.

57.   **INTERPRET**   This problem involves a change in temperature and phase (liquid to gas) for water, so both the specific heat and the latent heat of vaporization come into play. We use these to calculate how long it takes, given a constant input power of 200 MW, to boil away half of the water.

DEVELOP   The reactor must first raise the temperature of the water to the boiling point, which requires an energy

$$Q_1 = mc\Delta T$$

(Equation 16.3), where $\Delta T = 100°C - 10°C = 90$ K. Next, half the water must be boiled off, which requires an energy

$$Q_2 = \frac{L_v m}{2}$$

(Equation 17.5), where L_v is the latent heat of vaporization (see Table 17.1). The total energy required is the sum of these two, so at the given power P, it will take a time $t = (Q_1 + Q_2)/P$ to boil off half the water.

EVALUATE   Inserting the given quantities, we find

$$t = \frac{m(c\Delta T + L_v/2)}{P} = \frac{(4.5 \times 10^5 \text{ kg})\{[4.184 \text{ kJ}/(\text{kg} \cdot \text{K})](90 \text{ K}) + 0.5[2257 \text{ kJ/kg}]\}}{20 \times 10^4 \text{ kW}} = 3.4 \times 10^3 \text{ s} = 56 \text{ min}$$

ASSESS   For this solution, we have assumed that the water is heated uniformly, which may or may not be true depending on the geometry of the situation. If the water does not circulate well, it is possible that half the water boils off before the entire mass of water is heated to 100°C. In this case, the time would be less than we found above, which is therefore the maximum time it can take for half the water to boil off.

59.   **INTERPRET**   This problem involves the linear thermal expansion of steel. We are to find the temperature at which the hole in the steel plate will become large enough to allow the marble to pass through it.

DEVELOP   The diameter of the hole expands with the coefficient of linear expansion of steel. Using Equation 17.7

$$\alpha\Delta T = \frac{\Delta L}{L}$$

with $L = 1.000$, $L_0 = 0.997$, we can solve for ΔT.

EVALUATE   Solving for T and inserting the given quantities gives

$$\Delta T = \frac{\Delta L}{L\alpha} = \frac{0.003 \text{ cm}}{(0.997 \text{ cm})(12\times10^{-6} \text{ K}^{-1})} = 251 \text{ K}$$

so the plate must be heated to 251 K above room temperature.

ASSESS   Note that we have treated the problem as if we were considering a steel disk of initial diameter 0.997 cm, instead of a hole in a steel plate. The treatment is valid because such a hole must expand at the same rate as the disk because if we put the disk in the hole, it must fit at all temperatures.

61.   **INTERPRET**   This problem involves the linear expansion of Pyrex and steel. We are to find the temperature at which the diameter of the Pyrex tube is 2 μm greater than the diameter of the steel ball.

DEVELOP   Since the coefficient of linear expansion of steel is greater than that of Pyrex glass, the unit must be cooled to provide clearance. The difference in the contraction of steel and Pyrex must be twice the given clearance on one side, so

$$|\Delta L_{\text{steel}}| - |\Delta L_{\text{pyrex}}| = 2 \ \mu\text{m} = \left(\alpha_{\text{steel}} - \alpha_{\text{pyrex}}\right)L \ |\Delta T|$$

which we can solve for ΔT (find the values for α_{steel} and α_{pyrex} in Table 17.2).

EVALUATE   Solving for ΔT and inserting the known quantities gives

$$|\Delta T| = \frac{2.0\times10^{-4} \text{ cm}}{\left(\alpha_{\text{steel}} - \alpha_{\text{pyrex}}\right)L} = \frac{2.0\times10^{-4} \text{ cm}}{\left[(12-3.2)\times10^{-6} \text{ K}\right](1.0 \text{ cm})} = 22.7 \text{ K}$$

Since we must cool the system by this amount, the final temperature of the system will be T = 330 K − 23 K = 307 K.

ASSESS   Reversing this process is a good technique to create tightly fitted parts.

63.   **INTERPRET**   This problem involves the linear expansion of a rod as it is heated, so the coefficient of linear expansion will come into play. We are asked to calculate the height d of the apex of the triangle formed by the rod that cracks upon expanding because it is fixed between two immovable walls.

DEVELOP   If the two straight pieces in Fig. 17.11 are of equal length, the Pythagorean Theorem gives

$$d = \sqrt{(L/2)^2 - (L_0/2)^2}$$

where $L = L_0(1 + \alpha\Delta T)$ is the total expanded length of the rod.

EVALUATE   Inserting the expression for L gives

$$d = \frac{L_0}{2}\sqrt{2\alpha\Delta T + \alpha^2\Delta T^2}$$

ASSESS   Checking the limits, we see that for $\Delta T \to 0$ K , $d \to 0$, as expected.

65.   **INTERPRET**   This problem involves the temperature change and phase change (solid-liquid) of Glauber salt. Given its thermal parameters, we are to calculate how long the salt takes to cool to 60°F, and how long the salt takes to solidify at 90°F.

DEVELOP   In cooling from 95°F to 90°F, the liquid expels a heat $Q_1 = mc_{\text{liq}}\Delta T_{\text{liq}}$ (Equation 16.3). Changing phase at 90°F from liquid to solid expels a further amount $Q_2 = L_f m$ (Equation 17.5). Finally, cooling the solid salt from 90°F to 60°F expels the heat $Q_3 = mc_{\text{sol}}\Delta T_{\text{sol}}$. Thus, the total heat expelled is

$$Q = Q_1 + Q_2 + Q_3 = mc_{\text{liq}}\Delta T_{\text{liq}} + L_f m + mc_{\text{sol}}\Delta T_{\text{sol}}$$

Given that the house loses heat at a rate of $P = 20,000$ Btu/h, the time t it takes to cool to 60°F is $t = Q/P$.

EVALUATE   (a) Inserting the given quantities gives

$$t = \frac{mc_{\text{liq}}\Delta T_{\text{liq}} + L_f m + mc_{\text{sol}}\Delta T_{\text{sol}}}{P}$$

$$= \left(\frac{5.0 \times 2000 \text{ lbs}}{2.0 \times 10^4 \text{ Btu/h}}\right)\left\{(5°F)\left[0.68 \text{ Btu/}(\text{lb}\cdot°F)\right] + 104 \text{ Btu/lb} + \left[0.46 \text{ Btu/}(\text{lb}\cdot°F)\right](30°F)\right\}$$

$$= 61 \text{ h}$$

(b) The time spent during just the solidification at 90°F is

$$\frac{mL_f}{P} = \frac{(5.0 \times 2000 \text{ lb})(104 \text{ Btu/lb})}{2.0 \times 10^4 \text{ Btu/h}} = 52 \text{ h}$$

ASSESS   Most of the time the salt is at 90° because the latent heat of fusion is much greater than the heat liberated by temperature change.

67.   **INTERPRET**   For this problem, we are to show at what temperature between 0°C and 20°C water has its greatest density. We are given the expression for the coefficient of volume expansion as a function of temperature.

DEVELOP   We do not actually need to differentiate the density or the volume [$\rho(t)$ = constant mass/$V(T)$] because Equation 17.6 shows that $dV/dT = \beta V = 0$ when $\beta(T) = 0$. Thus, the maximum density (or minimum volume) occurs for a temperature satisfying $a + bT + cT^2 = 0$, which allows us to solve for T.

EVALUATE   The quadratic formula gives

$$T = \frac{-b \pm \sqrt{b^2 - 4ac}}{2c}$$

or, since both a and c are negative,

$$T = \frac{b \mp \sqrt{b^2 - 4|a||c|}}{2c}$$

Canceling a factor of 10^{-5} from the given coefficients, we find

$$T = \frac{1.70 \mp \sqrt{(1.70)^2 - 4(6.43)(0.0202)}°C}{0.0404} = 3.97°C$$

ASSESS   The second root, 80.2°C, can be discarded because it is outside the range of validity, $0 \leq T \leq 20°C$, of the original function $\beta(T)$. Thus, the maximum density of water occurs at a temperature close to 4°C. That this represents a minimum volume can be verified by plotting $V(T)$, or from the second derivative,

$$\frac{d^2V}{dT^2} = V\frac{d\beta}{dT} + \beta\frac{dV}{dT} = V\left(\beta^2 + \frac{d\beta}{dT}\right)$$

$$= V\left(\beta^2 + b + 2cT\right) > 0$$

for $T = 3.97°C$.

69.   **INTERPRET**   This problem involves the conversion of gravitational potential energy into thermal energy. We are to find the gravitational potential energy equivalent to the thermal energy needed to melt a falling ice cube on impact.

DEVELOP   The assumptions stated in the problem (no air resistance or heat exchange with the environment) imply that the change in the gravitational potential energy of the ice cube, per unit mass, must equal the heat of transformation of the ice cube. Expressed mathematically, this gives

$$mgh = mL_f$$

which we can solve for the height h.

EVALUATE   Solving for h, we find

$$h = \frac{L_f}{g} = (334 \text{ kJ/kg})(9.81 \text{ m/s}^2) = 34.1 \text{ km}$$

ASSESS   Of course, the expression for potential energy difference, mgh, is not valid over such a large range, but $mgy\, R_E/(R_E + y)$ only changes this result to 34.3 km. The thermal energies of ordinary macroscopic objects are very large compared to their mechanical energies.

71.   **INTERPRET**   We are asked to derive the given equation for the volume expansion coefficient β.

DEVELOP   Equation 17.6 gives the volume expansion coefficient as

$$\beta = \frac{1}{V}\frac{\Delta V}{\Delta T}$$

For infinitesimally small changes, this becomes

$$\beta = \frac{1}{V}\frac{dV}{dT}$$

Using the chain rule in the expression for β gives

$$\beta = \frac{1}{V}\frac{dV}{dT} = \frac{1}{L^3}\frac{dV}{dL}\frac{dL}{dT}$$

Given that

$$\frac{dV}{dL} = \frac{d}{dL}\left(L^3\right) = 3L^2$$

and, from Equation 17.7,

$$\frac{dL}{dT} = \alpha L$$

we can evaluate the chain-rule expression for β in terms of α.

EVALUATE   Inserting the expressions above for the derivatives gives volume expansion coefficient as

$$\beta = \frac{1}{L^3}\frac{dV}{dL}\frac{dL}{dT} = \frac{1}{L^3}\left(3L^2\right)\left(\alpha L\right) = 3\alpha$$

ASSESS   Alternatively, use the binomial approximation for $\Delta V = (L + \Delta L)^3 - L^3 = 3L^2\,\Delta L$, keeping only the lowest order term in ΔL. Since $\Delta V = \beta V\,\Delta T$ and $\Delta L = \alpha L \Delta T$, one finds $3L^2\left(\alpha L\,\Delta T\right) = \beta L^3\,\Delta T$, or $\beta = 3\alpha$.

73.   **INTERPRET**   We consider what goes on inside a pressure cooker.

DEVELOP   The line connecting the triple point to the critical point in Figure 17.9 (the liquid-gas boundary) designates all of the situations where the combination of pressure and temperature is right for water to boil.

EVALUATE   If the pressure is higher than normal in a pressure cooker, then the temperature at which water boils will be higher as well. By definition, this combination of elevated pressure and temperature must lie on the line connecting the triple point to the critical point.

The answer is (b).

ASSESS   If we started with boiling water at normal atmospheric pressure, and suddenly increased the pressure, the water would stop boiling. This corresponds to moving vertically upwards from the liquid-gas boundary in the phase diagram of Figure 17.9. For the water to start boiling again, the temperature will have to increase until the (p, T) combination is again located on the liquid-gas boundary.

75.   **INTERPRET**   We consider what goes on inside a pressure cooker.

DEVELOP   From the information given we can estimate the slope of the line marking the liquid-gas boundary.

EVALUATE   Between 1 and 2 atm, the average slope of the line is $\Delta p / \Delta T = 1/20$ atm/K, whereas between 2 and 3 atm, the average slope is $\Delta p / \Delta T = 1/14$ atm/K. Since the slope is getting larger as we move to the right on the graph, the line should be concave upward.

The answer is (a).

ASSESS   If the boundary line were a straight line, i.e. $p = \alpha T$ for some constant α, this would imply that the gas density remains constant as the temperature and pressure increase (see the previous problem). In fact, the gas density increases, reflecting a higher number of water molecules in the gas phase at higher pressures.

18

HEAT, WORK, AND THE FIRST LAW OF THERMODYNAMICS

EXERCISES

Section 18.1 The First Law of Thermodynamics

15. **INTERPRET** We identify the system as the water in the insulated container. The problem involves calculating the work done to raise the temperature of a system, so the first law of thermodynamics comes into play.

   DEVELOP Because the container is a perfect thermal insulator, no heat enters or leaves the water inside of it. Thus, $Q = 0$ and the first law of thermodynamics in Equation 18.1 gives $\Delta U = Q + W = W$, where W is the work done by shaking the container. The change in the internal energy of the water is determined from its temperature rise, $\Delta U = mc\,\Delta T$ (see comments in Section 16.1 on internal energy). Combining these two expressions for the internal energy change allows us to find the work done.

   EVALUATE The work done on the water is

$$W = \Delta U = mc\,\Delta T$$
$$= (1.0\ \text{kg})\left[4.184\ \text{kJ/(kg}\cdot\text{K)}\right](7.0\ \text{K}) = 29\ \text{kJ}$$

   ASSESS According to the convention adopted for the first law of thermodynamics, positive work signifies that work is done on the water.

17. **INTERPRET** The system of interest is the gas that undergoes expansion. The problem involves calculating the change in internal energy of a system, so the first law of thermodynamics comes into play.

   DEVELOP The heat added to the gas is $Q = Pt = (40\ \text{W})(25\ \text{s}) = 1000\ \text{J}$. In addition, the system does 750 J of work on its surroundings, so the work done by the surroundings on the system is $W = -750\ \text{J}$. The change in internal energy can be found by using the first law of thermodynamics given in Equation 18.1.

   EVALUATE Using Equation 18.1 we find

$$\Delta U = Q + W = 1000\ \text{J} - 750\ \text{J} = 250\ \text{J}$$

   ASSESS Since $\Delta U > 0$, we conclude that the internal energy has increased.

19. **INTERPRET** This problem is about heat and mechanical energy, which are related by the first law of thermodynamics. The system of interest is the automobile engine.

   DEVELOP Since we are dealing with rates, we make use of the dynamic form of the first law of thermodynamics, Equation 18.2:

$$\frac{dU}{dt} = \frac{dQ}{dt} + \frac{dW}{dt}$$

   If we assume that the engine system operates in a cycle, then $dU/dt = 0$. The engine's mechanical power output dW/dt can then be calculated once the heat output is known.

   EVALUATE The above conditions yield $\left(dQ/dt\right)_{\text{out}} = 68\ \text{kW}$ and $\left(dW/dt\right) = 0.17\left(dQ/dt\right)_{\text{in}}$. Equation 18.2 then gives

$$\frac{dW}{dt} = -\frac{dQ}{dt} = \left(\frac{dQ}{dt}\right)_{\text{out}} - \left(\frac{dQ}{dt}\right)_{\text{in}} = \left(\frac{dQ}{dt}\right)_{\text{out}} - \frac{1}{0.17}\frac{dW}{dt}$$

   or

$$\frac{dW}{dt} = \frac{(dQ/dt)_{out}}{1-(0.17)^{-1}} = \frac{68 \text{ kW}}{1-(0.17)^{-1}} = -14 \text{ kW}$$

ASSESS   We find the mechanical power output dW/dt to be proportional to the heat output, $(dQ/dt)_{out}$. In addition, dW/dt also increases with the percentage of the total energy released in burning gasoline that ends up as mechanical work. The mechanical output is negative because the system is doing work on its surroundings.

Section 18.2 Thermodynamic Processes

21.   **INTERPRET**   The expansion of the ideal gas involves two stages: an isochoric (constant-volume) process and an isobaric (constant-pressure) process. We are asked to find the total work done by the gas.

DEVELOP   For an isochoric process, $\Delta V = 0$ so $W_1 = 0$ (see Equation 18.3 and Table 18.1). On the other hand, for an isobaric process, the work done is $W_2 = p\Delta V$ which is the negative of Equation 18.7 because we are interested in the work done by the gas (Equation 18.7 gives the work done on the gas).

EVALUATE   Summing the two contributions to find the total work gives

$$W_{tot} = W_1 + W_2 = p_2(V_2 - V_1) = 2p_1(2V_1 - V_1) = 2p_1V_1$$

ASSESS   In the pV diagram of Fig. 18.19, the area under AC is zero, and that under CB, a rectangle, is $2p_1V_1$. The work done by the gas is the area under the pV curve.

23.   **INTERPRET**   The constant temperature of 300 K indicates that the process is isothermal. We are to find the increase in volume of the balloon and the work done by the gas, given the pressure difference in experiences.

DEVELOP   We assume the gas to be ideal and apply the ideal-gas law given in Equation 17.2: $pV = nRT$. For an isothermal process, $T = $ constant, so we obtain $p_1V_1 = p_2V_2$, from which we can find the fractional volume increase. The total work done by the gas can be calculated using the negative of Equation 18.4:

$$W = nRT \ln\left(\frac{V_2}{V_1}\right)$$

because we are interested in the work done by the gas, not on the gas.

EVALUATE   (a) For the isothermal expansion process, the volume increases by a factor of

$$\frac{V_2}{V_1} = \frac{p_1}{p_2} = \frac{100 \text{ kPa}}{75 \text{ kPa}} = \frac{4}{3}$$

(b) Using Equation 18.4, the work done by the gas is

$$W = nRT \ln\left(\frac{V_2}{V_1}\right) = (0.30 \text{ mol})\left[8.314 \text{ J}/(\text{mol} \cdot \text{K})\right](300 \text{ K})\ln\left(\frac{4}{3}\right) = 220 \text{ J}$$

to two significant figures.

ASSESS   Because $V_2 > V_1$, we find the work to be positive, $W > 0$. This makes sense because the gas inside the balloon must do positive work to expand outward.

25.   **INTERPRET**   The thermodynamic process here is adiabatic, so no heat flows between the system (the gas) and its environment. We are to find the volume change needed to double the temperature.

DEVELOP   In an adiabatic process, $Q = 0$, so the first law of thermodynamics (Equation 18.1) becomes $\Delta U = W$. The temperature and volume are related by Equation 18.11b:

$$T_1 V_1^{\gamma-1} = \text{constant} = T_2 V_2^{\gamma-1}$$

The temperature doubles, so $T_2 = 2T_1$ and $\gamma = 1.4$, so we can solve for the fractional change in volume.

EVALUATE   From the equation above, we have

$$T_1 V_1^{\gamma-1} = T_2 V_2^{\gamma-1} \quad \Rightarrow \quad \frac{V_2}{V_1} = \left(\frac{T_1}{T_2}\right)^{1/(\gamma-1)}$$

Thus, for the temperature to double, the volume change must be

$$\frac{V_2}{V_1} = \left(\frac{T_1}{T_2}\right)^{1/(\gamma-1)} = \left(\frac{1}{2}\right)^{1/(1.4-1)} = 0.177$$

ASSESS   We see that increasing the temperature along the adiabat is accompanied by a volume decrease. In addition, since $pV^\gamma =$ constant, the final pressure is also increased:

$$p_2 = p_1\left(\frac{V_1}{V_2}\right)^\gamma = p_1\left(\frac{1}{0.177}\right)^{1.4} = 11.3\,p_1$$

Section 18.3 Specific Heats of an Ideal Gas

27.   INTERPRET   The problem is about the specific heat of a mixture of gases. We want to know what fraction of the molecules is monatomic, given the its specific-heat ratio.

DEVELOP   The internal energy of a mixture of two ideal gases is

$$U = f_1 N \overline{E}_1 + f_2 N \overline{E}_2$$

where f_i is the fraction of the total number of molecules, N, of type i, and $\overline{E}_i$ is the average energy of a molecule of type i. Classically, $\overline{E} = g\left(\frac{1}{2}kT\right)$, where g is the number of degrees of freedom (the equipartition theorem). From Equation 18.6, the molar specific heat at constant volume is

$$C_V = \frac{1}{n}\frac{dU}{dT} = nR\frac{d}{dT}\left(f_1 g_1 \frac{1}{2}T + f_2 g_2 \frac{1}{2}T\right) = \frac{1}{2}R\left(f_1 g_1 + f_2 g_2\right)$$

Suppose that the temperature range is such that $g_1 = 3$ for the monatomic gas, and $g_2 = 5$ for the diatomic gas, as discussed in Section 18.3. Then

$$C_V = \frac{1}{2}R\left(3f_1 + 5f_2\right) = R\left(2.5 - f_1\right)$$

where $f_2 = 1 - f_1$ since the sum of the fractions of the mixture is unity. Now, C_V can also be specified by the ratio

$$\gamma = \frac{C_p}{C_V} = \frac{C_V + R}{C_V} = 1 + \frac{R}{C_V}$$

$$C_V = \frac{R}{\gamma - 1}$$

Equating the two expressions allows us to solve for f_1.

EVALUATE   Solving, we find

$$2.5 - f_1 = \frac{1}{\gamma - 1} = \frac{1}{0.52} \text{ or}$$
$$f_1 = 57.7\%$$

ASSESS   From the equation above, we see that the specific-heat ratio can be written as

$$\gamma = 1 + \frac{1}{2.5 - f_1}$$

In the limit where all the gas molecules are monatomic, $f_1 = 1$, and $\gamma = 1.67$. On the other hand, if all the molecules are diatomic, then $f_1 = 0$ and the specific-heat ratio is $\gamma = 1.4$. The equation yields the expected results in both limits.

29.   INTERPRET   The thermodynamic process is adiabatic, and we want to know the temperature change when work is done on a monatomic gas and a diatomic gas.

DEVELOP   In an adiabatic process, $Q = 0$, so from the first law of thermodynamics (Equation 18.1) $\Delta U = W$. where W is the work done on the gas. From Equation 18.6, $\Delta U = nC_V \Delta T$, the change in temperature is

$$\Delta T = \frac{\Delta U}{nC_V} = \frac{W}{nC_V}$$

If the work done per mole on the gas is $W/n = 2.5$ kJ/mol, then $\Delta T = (2.5 \text{ kJ/mol})/C_V$

EVALUATE (a) For an ideal monatomic gas, $C_V = \frac{3}{2}R = \frac{3}{2}\left[8.314 \text{ J}/(\text{mol} \cdot \text{K})\right]$, so $\Delta T = 200$ K.

(b) For an ideal diatomic gas (with five degrees of freedom), $C_V = \frac{5}{2}R$ so $\Delta T = 120$ K.

ASSESS Since the diatomic gas has a greater specific heat C_V, its temperature change is less than that of the monatomic gas.

PROBLEMS

31. **INTERPRET** We're asked to find the rate that heat is produced in the body when cycling. This involves the first law of thermodynamics.

    DEVELOP We're dealing with the rates of work and heat production, so we'll use Equation 18.2: $\frac{dU}{dt} = \frac{dQ}{dt} + \frac{dW}{dt}$. If the body is releasing stored food energy, that corresponds to a *decrease* in the body's internal energy: $dU/dt = -500$ W. Likewise, the mechanical power quoted is for work done *by* the body, so $dW/dt = -120$ W. We are looking for the rate at which the body produces heat, which is technically heat that it is losing ($-dQ$).

    EVALUATE The rate of heat production is the negative heat absorbed, so

    $$-\frac{dQ}{dt} = -\frac{dU}{dt} + \frac{dW}{dt} = -(-500 \text{ W}) + (-120 \text{ W}) = 380 \text{ W}$$

    ASSESS All the signs can be confusing, but essentially the body burns stored energy and some of it is used to do work and the rest is released as heat.

33. **INTERPRET** Assume the air inside the spherical bubble behaves like an ideal gas at constant temperature, so the process is isothermal. We need to find the diameter of the bubble at maximum pressure and the work done on the gas in compressing it.

    DEVELOP Apply the ideal-gas law given in Equation 17.2: $pV = nRT$. For an isothermal process, T = constant, which leads to $p_1 V_1 = p_2 V_2$. Since the volume of a spherical bubble of diameter d is

    $$V = \frac{4\pi}{3}\left(\frac{d}{2}\right)^3 = \frac{\pi d^3}{6}$$

    the relationship between the diameter and the pressure is

    $$p_1\left(\frac{\pi d_1^3}{6}\right) = p_2\left(\frac{\pi d_2^3}{6}\right)$$

    $$\frac{d_2}{d_1} = \left(\frac{p_1}{p_2}\right)^{1/3}$$

    EVALUATE (a) Using the equation above, we find the diameter at the maximum pressure to be

    $$d_2 = \left(\frac{p_1}{p_2}\right)^{1/3} d_1 = \left[\frac{(80+760) \text{ mm of Hg}}{(125+760) \text{ mm of Hg}}\right]^{1/3} (1.52 \text{ mm}) = 1.49 \text{ mm}$$

    (b) The work done *on* the air is given by Equation 18.4, or

    $$W_{\text{on air}} = -nRT \ln\left(\frac{V_2}{V_1}\right) = -p_1 V_1 \ln\left(\frac{p_1}{p_2}\right) = p_1 V_1 \ln\left(\frac{p_2}{p_1}\right)$$

    $$= (840 \text{ mm of Hg})\left(\frac{101.3 \text{ kPa}}{760 \text{ mm of Hg}}\right)\frac{\pi}{6}(1.52 \text{ mm})^3 \ln\left(\frac{885 \text{ mm of Hg}}{885 \text{ mm of Hg}}\right)$$

    $$= 10.7 \ \mu\text{J}$$

    ASSESS Positive work is done by the blood in compressing the air bubble.

35. **INTERPRET** The thermodynamic process here is adiabatic, with no heat flowing between the system (the gas) and its environment. We are to find the specific-heat ratio γ given the fraction increase in pressure of the gas.

    DEVELOP In an adiabatic process, $Q = 0$ and the first law of thermodynamics becomes $\Delta U = W$. The pressure and volume are related by Equation 18.11a: PV^γ = constant. This implies

$$p_1V_1^\gamma = p_2V_2^\gamma \quad \Rightarrow \quad \frac{p_2}{p_1} = \left(\frac{V_1}{V_2}\right)^\gamma$$

so we can solve for γ.

EVALUATE   Taking the natural logarithm on both sides of the above to solve for γ, we obtain

$$\ln\left(\frac{p_2}{p_1}\right) = \gamma \ln\left(\frac{V_1}{V_2}\right) \quad \Rightarrow \quad \gamma = \frac{\ln(p_2/p_1)}{\ln(V_1/V_2)} = \frac{\ln(2.55)}{\ln(2)} = 1.35$$

ASSESS   The value of γ indicates that gas consists of polyatomic molecules (see Problems 27 and 28).

37.   **INTERPRET**   This problem involves a cyclic process. The three processes that make up the cycle are: isothermal (AB), isochoric (BC), and isobaric (CA). We are given the pressure at point A and are to find the pressure at point B and the net work done on the gas.

DEVELOP   Along the isotherm AB $T =$ constant, so the ideal-gas law (Equation 17.2, $pV = nRT$) gives $p_AV_A = p_BV_B$. For an isothermal process, the work W done on the gas is (Equation 18.4):

$$W = -nRT \ln\left(\frac{V_2}{V_1}\right)$$

EVALUATE   **(a)** If AB is an isotherm, with $V_A = 5$ L and $V_B = 1$ L, then the ideal-gas law gives

$$P_B = \left(\frac{V_A}{V_B}\right)P_A = \left(\frac{5}{1}\right)(60 \text{ kPa}) = 300 \text{ kPa}$$

(b) The work W done *on* the gas in the isothermal process AB is

$$W_{AB} = -nRT_A \ln\left(\frac{V_B}{V_A}\right) = -p_AV_A \ln\left(\frac{V_B}{V_A}\right) = -(300 \text{ J}) \ln\left(\frac{1.0 \text{ L}}{5.0 \text{ L}}\right) = 483 \text{ J}$$

The process BC is isochoric (constant volume) so $W_{BC} = 0$. The process CA is isobaric (constant pressure) so the work done by the gas is (see Table 18.1)

$$W'_{CA} = p_A(V_A - V_C) = (60 \text{ kPa})(5.0 \text{ L} - 1.0 \text{ L}) = 240 \text{ J}$$

or the work done *on* the gas is $W_{CA} = -240$ J. The total work done by the gas is the sum of these three contributions, or

$$W_{ABCA} = W_{AB} + W_{BC} + W_{CA} = 483 \text{ J} + 0 - 240 \text{ J} = 243 \text{ J} \approx 240 \text{ J}$$

to two significant figures.

ASSESS   Since the process is cyclic, the system returns to its original state, there's no net change in internal energy, so $\Delta U = 0$. This implies that $Q = -W_{ABCA} = -240$ J. That is, 240 J of heat must come *out* of the system.

39.   **INTERPRET**   We identify the thermodynamic process here as adiabatic compression.

DEVELOP   In an adiabatic process, $Q = 0$, and the first law of thermodynamics becomes $\Delta U = W$. The temperature and volume are related by Equation 18.11b:

$$TV^{\gamma-1} = \text{constant}$$

From the equation above, we obtain

$$T_1V_1^{\gamma-1} = T_2V_2^{\gamma-1} \quad \Rightarrow \quad T_2 = T_1\left(\frac{V_1}{V_2}\right)^{\gamma-1}$$

where V_1/V_2 is the compression ratio (for T and V at maximum compression).

EVALUATE   Inserting the values given gives

$$T_2 = T_1\left(\frac{V_1}{V_2}\right)^{\gamma-1} = (303 \text{ K})(8.5)^{0.4} = 713 \text{ K} = 440°\text{C}$$

ASSESS   Note that the temperature T appearing in the gas laws is the absolute temperature. The higher the compression ratio V_1/V_2, the greater the temperature at the maximum compression, and hence a higher thermal efficiency.

41.   **INTERPRET**   We identify the thermodynamic process here as adiabatic compression. We are to find the temperature and pressure in the cylinder when it is at maximum compression.

DEVELOP   In an adiabatic process, $Q = 0$, and the first law of thermodynamics becomes $\Delta U = W$. Because we are dealing with an adiabatic process, the temperature and volume are related by Equation 18.11b:

$$TV^{\gamma-1} = \text{constant}$$

Applying this expression before (subscript 1) and after (subscript 2) compression gives

$$T_1 V_1^{\gamma-1} = T_2 V_2^{\gamma-1} \quad \Rightarrow \quad T_2 = T_1 \left(\frac{V_1}{V_2} \right)^{\gamma-1}$$

where $V_1/V_2 = 10.2$ is the compression ratio (for T and V at maximum compression) and $\gamma = 1.4$. In addition, since $pV^{\gamma} = \text{constant}$ (Equation 18.11a), the final pressure is $p_2 = p_1 \left(V_1/V_2 \right)^{\gamma}$.

EVALUATE   (a) Substituting the values given in the problem statement, we find the air temperature at the maximum compression to be

$$T_2 = T_1 \left(\frac{V_1}{V_2} \right)^{\gamma-1} = (320 \text{ K})(10.2)^{0.4} = 810 \text{ K}$$

(b) The corresponding pressure is

$$p_2 = p_1 \left(V_1/V_2 \right)^{\gamma} = (101.3 \text{ kPa})(10.2)^{1.4} = 2.62 \text{ MPa} = 25.8 \text{ atm}$$

ASSESS   The higher the compression ratio V_1/V_2, the greater the temperature and pressure at the maximum compression, and hence, a higher thermal (fuel) efficiency.

43.   **INTERPRET**   This problem explores how different ways of adding heat (isothermal, isochoric, or isobaric) affects the final temperature of the system. Starting with the given quantity of gas at the given temperature, we are to find the work done by the gas upon adding heat to the gas via these different processes.

DEVELOP   In an isothermal process, the temperature T is kept constant, so $\Delta U = 0$. The first law of thermodynamics (Equation 18.1) therefore gives $Q = -W$. In an isochoric process, $\Delta V = 0$ and $W = 0$ (see Equation 18.7), so the first law of thermodynamics gives $Q = \Delta U = nC_V \Delta T$. Finally, in an isobaric process, $\Delta p = 0$ and

$$Q = nC_p \Delta T = n(C_V + R)\Delta T$$

Use these expressions to solve for ΔT and W for each case.

EVALUATE   (a) With $\Delta U = 0$, $W = -Q = -1.5$ kJ. For an isothermal process, the temperature is constant so $T_2 = 300$ K.

(b) Solving the expression above for ΔT, we find for an isochoric process

$$\Delta T = \frac{Q}{nC_V} = \frac{1.5 \text{ kJ}}{(2 \text{ mol})(5R/2)} = 36 \text{ K}$$

Therefore, $T_2 = 300 \text{ K} + \Delta T = 336 \text{ K}$. Because the volume does not change, $W = 0$.

(c) In an isobaric process,

$$\Delta T = \frac{Q}{nC_p} = \frac{Q}{n(C_V + R)} = \frac{1.5 \text{ kJ}}{(2 \text{ mol})(5R/2 + R)} = 26 \text{ K}$$

and $T_2 = 326$ K. The work done is

$$W = p\Delta V = nR\Delta T = \frac{R}{C_p} Q = \frac{R}{(7R/2)} Q = \frac{2(1.5 \text{ kJ})}{7} = 430 \text{ J}$$

ASSESS   Comparing all three cases, we find

$$\Delta T : \text{ isothermal} < \text{isobaric} < \text{isochoric}$$
$$W: \text{ isochoric } < \text{isobaric} < \text{isothermal}$$

The results agree with that illustrated in Table 18.1.

45.   INTERPRET   The problem involves a cyclic process with three separate stages: adiabatic, isochoric, and isothermal. We are given the initial volume and pressure of the gas, and its adiabatic exponent γ, and are to find the pressure at points B and C and the net work done on the gas.

DEVELOP   For the adiabatic process AB, $Q = 0$ and the first law of thermodynamics becomes $\Delta U = -W$. The pressure and volume are related by Equation 18.11a: $pV^\gamma = $ constant , which gives

$$p_A V_A^\gamma = p_B V_B^\gamma \quad \Rightarrow \quad p_B = \left(\frac{V_A}{V_B}\right)^\gamma p_A$$

Point C lies on an isotherm (constant temperature) with A, so the ideal-gas law (Equation 17.2) yields

$$p_C = \frac{p_A V_A}{V_C}$$

To find the net work done on the gas, sum the contributions from each stage of the cycle. The contribution W_{AB} may be found from Equation 18.12 for an adiabatic process:

$$W_{AB} = \frac{p_B V_B - p_A V_A}{\gamma - 1}$$

For an isochoric (constant volume) process $W_{BC} = 0$ (see Equation 18.7), and for the isothermal process

$$W_{CA} = -nRT_A \ln\left(\frac{V_A}{V_C}\right)$$

EVALUATE   (a) From the equation above, the pressure at point B is

$$p_B = \left(\frac{V_A}{V_B}\right)^\gamma p_A = \left(250 \text{ kPa}\right)\left(\frac{1}{3}\right)^{1.67} = 40 \text{ kPa}$$

(b) The pressure at point C is

$$p_C = p_A\left(\frac{V_A}{V_C}\right) = \frac{250 \text{ kPa}}{3} = 83 \text{ kPa}.$$

(c) The net work done *by* the gas is $W_{ABCA} = W_{AB} + W_{BC} + W_{CA}$. The work W_{AB} for the adiabatic segment is

$$W_{AB} = \frac{p_B V_B - p_A V_A}{\gamma - 1} = \frac{\left(39.9 \text{ kPa}\right)\left(3.0 \text{ m}^3\right) - \left(250 \text{ kPa}\right)\left(1.00 \text{ m}^3\right)}{0.67} = -194 \text{ kJ}$$

W_{BC} is for an isochoric process and equals zero. Finally, W_{CA} is for an isothermal process and is

$$W_{CA} = -nRT_A \ln\left(\frac{V_A}{V_C}\right) = \left(250 \text{ kJ}\right)\ln\left(\frac{1}{3}\right) = 275 \text{ kJ}$$

Summing these contributions gives

$$W_{ABCA} = W_{AB} + W_{BC} + W_{CA} = -194 \text{ kJ} + 0 + 275 \text{ kJ} = 80 \text{ kJ}$$

so the work done *on* the gas 80 kJ.

ASSESS   Since the process is cyclic, the system returns to its original state, there's no net change in internal energy, so $\Delta U = 0$. This implies that $Q = -W_{ABCA} = -80$ kJ. That is, 80 kJ of heat must come *out* of the system.

47.   INTERPRET   We are to find the work done in the given heat cycle, which consists of an isochoric doubling in pressure, and adiabatic compression, an isochoric cooling to 300 K, and an isothermal expansion.

DEVELOP   The net (or total) work done on the gas is the sum of the work done for each process in the cycle (see figure below), which we give here:

(AB) It is heated at constant volume until the pressure is doubled. Because $\Delta V = 0$, $W_{AB} = 0$ (see Equation 18.7).

(BC) It is compressed adiabatically until its volume is $\frac{1}{4}$ the initial value. This is analogous to the process AB of Problem 18.46, so the result is

$$W_{BC} = p_B V_B \frac{\left(V_B/V_C\right)^{\gamma-1} - 1}{\gamma - 1}$$

where $V_B/V_C = 1/4$ and $p_B V_B = 800$ J (see figure below).

(CD) It is cooled at constant volume to a temperature of 300 K. No work is done because $\Delta V = 0$, so $W_{CD} = 0$.

(DA) It is expanded isothermally until it returns to the original state. This is analogous to the process CA of Problem 18.46, so we use that result:

$$W_{DA} = -p_A V_A \ln\left(\frac{V_A}{V_D}\right)$$

where $V_A/V_D = 4$ and $p_A V_A = 400$ J (see figure).

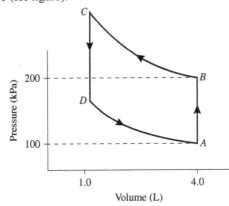

EVALUATE Summing the nonzero contributions to the work gives

$$W_{tot} = W_{BC} + W_{DA} = p_B V_B \frac{\left(V_B/V_C\right)^{\gamma-1} - 1}{\gamma - 1} - p_A V_A \ln\left(\frac{V_A}{V_D}\right)$$

$$= \left(800 \text{ J}\right)\frac{4^{0.4} - 1}{0.4} - \left(400 \text{ J}\right)\ln\left(4\right) = 930 \text{ J}$$

ASSESS Work is done on the gas. From the figure above, we see that the area under the entire curve is negative, because the gas goes around the cycle counterclockwise. As per Figure 18.6, the work done by the gas is the negative of the area under the pV curve.

49. **INTERPRET** We're asked to derive the relation between pressure and temperature in an adiabatic process.

    DEVELOP We just need to combine Equations 18.11a and 18.11b: $pV^\gamma = $ constant, and $TV^{\gamma-1} = $ constant.

    EVALUATE Isolating the volume in both equations and matching the exponents gives

$$\left. \begin{aligned} V^{\gamma(\gamma-1)} &= \frac{\text{constant}}{p^{(\gamma-1)}} \\ V^{\gamma(\gamma-1)} &= \frac{\text{constant}}{T^\gamma} \end{aligned} \right\} \qquad p^{-(\gamma-1)}T^\gamma = p^{1-\gamma}T^\gamma = \text{constant}$$

The "constant" term is just a place-holder. You can raise the constant to a power and it's still just a constant. Or you can multiply the constant by another constant, and the result is still just a constant.

ASSESS For $\gamma > 1$, the result says the pressure and temperature will increase or decrease together during an adiabatic process.

51. **INTERPRET** The pump handle is pressed rapidly, so you can assume that there's no time for heat to flow into or out of the gas in the pump. This means that process is adiabatic.

    DEVELOP You want to check that the temperature rise in the pump is less than 50°C. Equation 18.11b tells you how to relate the final temperate and volume to the initial temperature and volume: $TV^{\gamma-1} = T_0 V_0^{\gamma-1}$. Therefore, the temperature rise is

$$\Delta T = T - T_0 = T_0 \left[\left(\frac{V_0}{V} \right)^{\gamma-1} - 1 \right]$$

The gas in the cylinder is air $(\gamma = 1.4)$ initially at $T_0 = 20°\text{C} = 293 \text{ K}$.

EVALUATE   Since the cross-sectional area of the cylinder remains unchanged during the compression, the volume ratio is just equal to the ratio in the cylinder's height:

$$\Delta T = T_0 \left[\left(\frac{V_0}{V} \right)^{\gamma-1} - 1 \right] = (293 \text{ K}) \left[\left(\frac{30 \text{ cm}}{17 \text{ cm}} \right)^{1.4-1} - 1 \right] = 75 \text{ K}$$

This does not meet your 50°C criteria.

ASSESS   There's no obvious way to avoid this temperature rise. The cylinder can be made of a material with a high thermal conductivity, such as aluminum, so that heat can flow out as fast as possible.

53. **INTERPRET**   For this problem, we are given the heat transferred from a monatomic gas to the surroundings and the change in internal energy of the gas, and we are to find the work done in the process.

DEVELOP   Apply the first law of thermodynamics, Equation 18.1, $\Delta U = Q + W$, where $Q = -15 \text{ kJ}$. The internal energy change can be calculated using the constant-volume specific heat for a monatomic gas, which is $C_V = 3R/2$ (see Equation 18.13). Thus,

$$\Delta U = nC_V \Delta T$$

where $n = 21$ mol and $\Delta T = 160$ K.

EVALUATE   Combining the above expressions and inserting the given quantities gives

$$W = \Delta U - Q = nC_V \Delta T - Q = (21 \text{ mol}) \left(\frac{3}{2} \right) \left[8.314 \text{ J/(mol·K)} \right] (160 \text{ K}) - 15 \text{ kJ} = 57 \text{ kJ}$$

ASSESS   The work is positive so the work is done on the gas.

55. **INTERPRET**   This problem involves an adiabatic compression of a gas, for which we know the initial temperature and pressure and the final pressure. We are to find the final temperature of the gas.

DEVELOP   For an adiabatic process, the temperature and pressure are related by Equation 18.11a:

$$pV^\gamma = \text{consant}$$

Rearranging this equation and using the ideal-gas law (Equation 17.2, $pV = nRT$) gives

$$pV^\gamma = p^{1-\gamma} \left(pV \right)^\gamma = p^{1-\gamma} \left(nRT \right)^\gamma = \text{constant}$$
$$pT^{\gamma/(1-\gamma)} = \text{another constant}$$

Applying this to the gas both before and after the compression gives $T = T_0 (P_0/P)^{(1/\gamma)-1}$

EVALUATE   Inserting the given quantities into the expression above for temperature gives

$$T = T_0 (P_0/P)^{(1/\gamma)-1} = (273 \text{ K}) \left(\frac{100 \text{ kPa}}{240 \text{ kPa}} \right)^{(1/1.3)-1} = 330 \text{ K}$$

to two signficant figures.

ASSESS   Because the gas is compressed adiabatically, its temperature rises.

57. **INTERPRET**   This problem is similar to the preceding one, except that the order of segments in the cycle are changed. The cycle proceeds along the 350-K isotherm from 50 kPa to 250 kPa (AC), then decreases its pressure to 50 kPa via an isochoric process (CD), then returns to the starting point via a 50-kPa isobar (DA). We are to find the total work done on the gas for the complete cycle and the heat transferred in segment CD.

DEVELOP   The formulas are the same as for the previous problem, but the initial and final point for each segment are different. The work done on the gas over segment AC will be the opposite of the result we found for the segment CA in Problem 18.56, so $W_{AC} = 402$ J. The work done over segment CD is zero because the volume is constant (see Equation 18.7). The work done over segment DA is

$$W_{BC} = -p_D(V_A - V_D)$$

The heat transfer can be calculated as per Problem 18.56, which gives

$$Q_{CD} = \frac{p_D V_D - p_C V_C}{\gamma - 1}$$

EVALUATE   (a) The total work is

$$W_{ACD} = W_{AC} + \overbrace{W_{CD}}^{=0} + W_{DA} = W_{AC} - p_D(V_A - V_D) = 402\ \text{J} - (50.0\ \text{kPa})(5.00\ \text{L} - 1.00\ \text{L}) = 202\ \text{J}$$

(b) The heat transfer over segment CD is

$$Q_{CD} = \frac{p_D V_D - p_C V_C}{\gamma - 1} = \frac{(50\ \text{kPa} - 250\ \text{kPa})(1\ \text{L})}{0.4} = -500\ \text{J}$$

Because $Q_{CD} < 0$, we interpret the result as 500 J of heat transferred out of the gas.

ASSESS   At constant volume, the gas must be cooled to lower its pressure.

59. **INTERPRET**   For this problem, we are to find the ratio of triatomic molecules to monatomic molecules that will result in a gas that has the specific heat at constant volume of a diatomic gas.

DEVELOP   The specific heat of a mixture of two gases is $C_V = f_1 C_{V_1} + f_2 C_{V_2}$, where the f_i are the number fractions of the gases. Gas 1 is monatomic ($C_V = \frac{3}{2}R$), gas 2 is triatomic (with $C_V = 3R$, as given in the problem statement), and we wish to have mixture with $C_V = \frac{5}{2}R$, as for a diatomic gas. In this case, then

$$\tfrac{5}{2}R = \tfrac{3}{2}R f_1 + 3R f_2, \text{ or } 5 = 3f_1 + 6f_2.$$

Given that $f_1 + f_2 = 1$ and $f_1 = 10$ mol, we can solve for f_2.

EVALUATE   From the equations above, we find $f_1 = 1/3$ and $f_2 = 2/3$. With 10 mol of monatomic gas, one needs 20 mol of triatomic gas.

ASSESS   There is twice the number of triatomic molecules as there are monatomic molecules.

61. **INTERPRET**   This problem deals with an isothermal expansion of a gas from an unknown pressure to the ambient pressure of 1 atm. The expansion extracts heat from the surrounding ice-water bath so that 10 g of ice are created in the process. We are to find the original pressure of the gas.

DEVELOP   For an isothermal expansion of an ideal gas, Equation 18.4 and the ideal-gas law (PV = nRT = constant for an isothermal expansion) give

$$Q = nRT \ln\left(\frac{V_2}{V_1}\right) = nRT \ln\left(\frac{p_2}{p_1}\right)$$

From this,

$$p_1 = p_2 e^{Q/nRT}$$

where p_1 is the original gas pressure, $p_2 = 1$ atm, and n = 0.30 mol. The heat Q is provided by the freezing of 10 g of ice that is already at 0°C, which is (see Equation 17.5)

$$Q = m_{ice} L_f = (0.01\ \text{kg})(334\ \text{kJ/kg}) = 3.34\ \text{kJ}$$

EVALUATE   Inserting the known quantities into the expression for p1 gives

$$p_1 = (1.0\ \text{atm})\exp\left\{\frac{3.34\ \text{kJ}}{(0.30\ \text{mol})\left[8.314\ \text{J}/(\text{mol}\cdot\text{K})\right](273\ \text{K})}\right\} = (1.0\ \text{atm})e^{4.91} = 140\ \text{atm}$$

to two significant figures.

ASSESS   Notice that the original quantity of ice is not involved, provided we know the ice-water bath is at 0°C and the quantity of the ice that is produced by the energy requirements of the isothermal expansion.

63.   **INTERPRET**   For this problem, we are to derive the work done by an adiabatic process by applying Equation 18.3, which describes the work done by changes in a volume of gas.

DEVELOP   For an adiabatic process, pressure and volume are related by Equation 18.11a, $pV^\gamma =$ constant. Applying this to a gas before and after an arbitrary adiabatic process

$$p_1 V_1^\gamma = pV^\gamma$$

$$p = \frac{p_1 V_1^\gamma}{V^\gamma}$$

Insert this expression into Equation 18.3 to find the work done by an adiabatic process.

EVALUATE   The work in going from a pressure p_1 to a pressure p_2 via an adiabatic process is

$$W_{12} = -\int_{V_1}^{V_2} pdV = -\int_{V_1}^{V_2} \left(p_1 V_1^\gamma\right)\frac{dV}{V^\gamma} = -p_1 V_1^\gamma\left(\frac{V_2^{-\gamma+1} - V_1^{-\gamma+1}}{-\gamma+1}\right) = \frac{p_2 V_2^\gamma V_2^{-\gamma+1} - p_1 V_1^\gamma V_1^{-\gamma+1}}{\gamma-1}$$

$$= \frac{P_2 V_2 - P_1 V_1}{\gamma-1}$$

which is Equation 18.12 (note that we have used $p_1 V_1^\gamma = p_2 V_2^\gamma$ to obtain the last equality of the first line).

ASSESS   From the general expression for work done by a change in volume of a gas, we have derived the expression for the work done by an adiabatic change in volume of a gas.

65.   **INTERPRET**   We are to derive the relationship between temperature and volume for adiabatic processes.

DEVELOP   For an adiabatic, the relationship between pressure and volume is $pV^\gamma =$ constant. Use this and the ideal-gas law (Equation 17.2) $pV = nRT$, to derive equation 18.11b, $TV^{\gamma-1} =$ constant.

EVALUATE   From the ideal-gas law,

$$pV = nRT \quad \Rightarrow \quad p = \frac{nRT}{V}$$

Substitute this into $pV^\gamma =$ constant to obtain

$$\frac{nRT}{V}V^\gamma = (nR)TV^{\gamma-1} = \text{constant}$$

$$TV^{\gamma-1} = \frac{\text{constant}}{nR} = \text{a different constant}$$

ASSESS   We have shown what was required. Note that the constants involved in Equations 18.11a and 18.11b are different constants.

67.   **INTERPRET**   We are given the relationship between pressure and volume for compressing a given volume of air, and we are to find the work done during this reversible process.

DEVELOP   The most general equation for work done by a gas is Equation 18.3: $W = -\int_{V_1}^{V_2} pdV$. We are given the initial pressure $p_0 = 1.0$ atm, the initial volume $V_0 = 17$ m^3, the final pressure $p_f = 1.4$ atm, and the formula

$$\left(\frac{p}{p_0}\right)^2 = \frac{V}{V_0}$$

$$p = p_0\sqrt{\frac{V}{V_0}}$$

Insert this into Equation 18.3 and integrate from V_0 to V_f, where V_f is given by

$$V_f = V_0\left(\frac{p_f}{p_0}\right)^2$$

EVALUATE   Performing the integration gives

$$W = -\int_{V_0}^{V_f} p_0 \sqrt{\frac{V}{V_0}} \, dV = -\frac{p_0}{\sqrt{V_0}}\left(\frac{2}{3}\right)V^{\frac{3}{2}}\Big|_{V_0}^{V_f} = -\frac{2p_0}{3\sqrt{V_0}}\left[V_f^{\frac{3}{2}} - V_0^{\frac{3}{2}}\right] = -\frac{2P_0}{3\sqrt{V_0}}\left[V_0^{\frac{3}{2}}\left(\frac{p_f}{p_0}\right)^3 - V_0^{\frac{3}{2}}\right]$$

$$= -\frac{2p_0 V_0}{3}\left[\left(\frac{p_f}{p_0}\right)^3 - 1\right] = -\frac{2(101\times10^3 \text{ Pa})(17 \text{ m}^3)}{3}\left[\left(\frac{1.4 \text{ atm}}{1.0 \text{ atm}}\right) - 1\right] = -2.0 \text{ MJ}$$

ASSESS   Because the work is negative, the gas does 2.0 MJ of work on its environment in this process.

69.   INTERPRET   An ideal gas expands along a given path, and we are to find the work done by the gas. We can do this by using the general expression for work done by a gas that changes pressure and volume, Equation 18.3.
DEVELOP   The gas goes from (p_1, V_1) to (p_2, V_2) where $p_2 = 2p_1$ and $V_2 = 2V_1$. The path it takes is along

$$p = p_1\left[1 + \left(\frac{V - V_1}{V_1^2}\right)^2\right]$$

Insert this expression into the integrand of Equation 18.3 and integrate from V_1 to V_2 to find the work done on the gas. The work done by the gas will be the negative of this result.
EVALUATE   The work done on the gas is

$$W = -\int_{V_1}^{V_2} p \, dV = -\int_{V_1}^{2V_1} p_1\left[1 + \frac{(V - V_1)^2}{V_1^2}\right]dV = -\frac{p_1}{V_1^2}\int_{V_1}^{2V_1}\left[V^2 - 2V_1 V + 2V_1^2\right]dV$$

$$= -\frac{p_1}{V_1^2}\left[\frac{1}{3}V^3 - V_1 V^2 + 2V_1^2 V\right]_{V_1}^{2V_1} = -\frac{p_1}{V_1^2}\left[\frac{1}{3}\left(8V_1^3 - V_1^3\right) - V_1\left(4V_1^2 - V_1^2\right) + 2V_1^2\left(2V_1 - V_1\right)\right]$$

$$= -p_1 V_1\left(\frac{7}{3} - 3 + 2\right) = -\frac{4}{3}p_1 V_1$$

so the work done by the gas is $4p_1 V_1/3$.
ASSESS   The units are pressure times volume, which is joules, as expected.

71.   INTERPRET   We are to calculate the efficiency and waste heat of a power plant, given the heat input and the power produced.
DEVELOP   The efficiency is

$$e = \frac{\text{power out}}{\text{power in}}$$

The waste heat is the difference between the power in and the power out. We need to know that the rate of waste heat produced is less than 400 MW, and the power plant has efficiency is $e \geq 50\%$.
EVALUATE   The power-plant efficiency is

$$e = \frac{360 \text{ MW}}{670 \text{ MW}} = 54\%.$$

Heat is lost at a rate of

$$P_{\text{lost}} = 670 \text{ MW} - 360 \text{ MW} = 310 \text{ MW} < 400 \text{ MW}$$

ASSESS   This power plant meets the requirements.

73.   INTERPRET   We consider the physics behind warm winds called Chinooks.
DEVELOP   We're told that the wind sweeping down from the mountain has no time to exchange heat with the surroundings.
EVALUATE   No heat exchange means the process is adiabatic.
The answer is (d).
ASSESS   Recall that air has one of the lowest thermal conductivities in Table 16.2. So it's perhaps not surprising that a large mass of air might need a lot of time to exchange an appreciable amount of heat with its surroundings.

75.   INTERPRET   We consider the physics behind warm winds called Chinooks.

DEVELOP   From 18.11a, the pressure and volume in the mountains is related to that in the plains by $p_1 V_1^{\gamma} = p_2 V_2^{\gamma}$. We'll assume $\gamma = 1.4$.

EVALUATE   Given the initial and final pressure, we can solve for the relative change in the volume:

$$\frac{\Delta V}{V_1} = \frac{V_2}{V_1} - 1 = \left(\frac{p_1}{p_2} \right)^{1/\gamma} - 1 = \left(\frac{60 \text{ kPa}}{90 \text{ kPa}} \right)^{1/1.4} - 1 = -25\%$$

This implies the volume decreases by less than 50%.

The answer is (d).

ASSESS   It makes sense that the volume of air decreases when exposed to the higher pressures at the base of the mountain.

THE SECOND LAW OF THERMODYNAMICS

19

EXERCISES

Sections 19.2 and 19.3 The Second Law of Thermodynamics and Its Applications

13. **INTERPRET** This problem requires us to calculate the efficiency of reversible heat engines that operate between the given temperatures.

    DEVELOP The maximum efficiency of a reversible engine, operating between two absolute temperatures, $T_h > T_c$, is given by Equation 19.3, $e_{Carnot} = 1 - T_c/T_h$. Apply this to each part of this problem to find the corresponding efficiencies.

    EVALUATE (a) $e = 1 - T_c/T_h = 1 - (273 \text{ K})/(373 \text{ K}) = 26.8\%$.

    (b) $e = 1 - T_c/T_h = \Delta T/T_h = (21 \text{ K})/(298 \text{ K}) = 7.05\%$.

    (c) With room temperature at $e = 1 - T_c/T_h = 1 - (293 \text{ K})/(1273 \text{ K}) = 77.0\%$.

    ASSESS The engine with the largest difference in reservoir temperature has the largest efficiency.

15. **INTERPRET** This problem involves a reversible Carnot engine that runs between the boiling and melting point of He. We are given the engine's efficiency and are asked to find the temperature of its cold reservoir (i.e., the melting point of He).

    DEVELOP Apply Equation 19.3, which gives the efficiency of a Carnot engine. We are given $e_{Carnot} = 0.777$ and $T_h = 4.25$ K, so we can solve for T_c, which will be the melting point of He.

    EVALUATE The melting point of He is

    $$e_{Carnot} = 1 - \frac{T_c}{T_h}$$

    $$T_c = (1 - e_{Carnot})T_h = (1.000 - 0.777)(4.25 \text{ K}) = 0.948 \text{ K}$$

    ASSESS This agrees with the melting point of He found in the literature, so it seems to be a reasonable result.

17. **INTERPRET** We are to find the coefficient of performance of a reversible refrigerator that operates between 0°C and 30°C.

    DEVELOP For a refrigerator, the coefficient of performance is given by Equation 19.4,

    $$COP_{refrigerator} = \frac{Q_c}{Q_h - Q_c}$$

    Use Equation 19.2, $Q_c/Q_h = T_c/T_h$ to convert this to an expression involving temperature:

    $$COP_{refrigerator} = \frac{Q_c}{Q_h - Q_c} = \frac{Q_c}{Q_h}\left(\frac{1}{1 - Q_c/Q_h}\right) = \frac{T_c}{T_h}\left(\frac{1}{1 - T_c/T_h}\right) = \frac{T_c}{T_h - T_c}$$

    which we can solve given that $T_c = 0°C = 273$ K and $T_h = 30°C = 303$ K.

EVALUATE   Inserting the given quantities into the expression for the COP gives

$$COP_{refrigerator} = \frac{T_c}{T_h - T_c} = \frac{273 \text{ K}}{303 \text{ K} - 273 \text{ K}} = 9.10$$

ASSESS   Notice that the temperatures are absolute temperatures (i.e., Kelvin).

19.   INTERPRET   We want to know if the human body can be considered as a heat engine, which has stringent limits on its efficiency.

DEVELOP   If the human body were a heat engine, the maximum efficiency it could attain would be given by the efficiency for a Carnot engine (Equation 19.3): $e = 1 - T_c / T_h$.

EVALUATE   Body temperature is $T_h = 37°C = 310$ K. So if we assume the ambient temperature is $T_c = 20°C = 293$ K, the maximum efficiency would be

$$e = 1 - \frac{293 \text{ K}}{310 \text{ K}} = 5\%$$

So under normal circumstances, the human body is far too efficient at 25% to be a heat engine.

ASSESS   We often say the body "burns" calories, which sounds like it's just releasing random heat from the food we eat. But actually the process is more specific. Energy-storing molecules interact with other molecules to cause precise chemical reactions that result in, for example, a muscle contracting or a neuron producing a current. Not all of the stored energy is converted to useful work, however. Some of it ends up as heat.

Section 19.4 Entropy and Energy Quality

21.   INTERPRET   The problem concerns the entropy increase associated with metabolizing a hamburger.

DEVELOP   We'll assume the energy in the burger, Q, flows into the body as heat. Therefore, the entropy change from state 1 (burger ingested) and state 2 (burger metabolized) is given by Equation 19.6: $\Delta S_{12} = \int_1^2 dQ / T$.

EVALUATE   The body temperature, $T = 37°C = 310$ K, remains constant throughout, so

$$\Delta S_{12} = \frac{1}{T} \int_1^2 dQ = \frac{Q}{T} = \frac{650 \text{ kcal}}{310 \text{ K}} \left(\frac{4.184 \text{ kJ}}{1 \text{ kcal}} \right) = 8.8 \text{ kJ/K}$$

ASSESS   Although we assumed the burger's energy went into heat, the answer would be the same if the body used some of the energy to do work. In either case, the burger's energy is no longer available to do work once it has been metabolized.

23.   INTERPRET   This problem requires us to find the mass of a block of lead given its entropy increase associated with its solid-to-liquid phase transition (i.e., melting it).

DEVELOP   The lead starts at its melting-point temperature, so there is no change in temperature associated with the solid-to-liquid phase change. Therefore, Equation 19.6 for entropy change takes the form

$$\Delta S = \frac{\Delta Q}{T}$$

The heat change is given by Equation 17.5, $\Delta Q = mL_f$, where $L_f = 24.7$ kJ/kg (see Table 17.1). Insert this into the expression for entropy change and solve for the mass m.

EVALUATE   From the above equation, we find the mass of lead to be

$$\Delta S = \frac{mL_f}{T}$$
$$m = \frac{T\Delta S}{L_f} = \frac{(600 \text{ K})(900 \text{ J/K})}{24.7 \text{ kJ/kg}} = 21.9 \text{ kg}$$

ASSESS   As expected, the mass of the block is proportional to the change in entropy.

25. **INTERPRET** We're asked to find the probability that 6 molecules are distributed in different ways inside a box. This has relevance to the statistical interpretation of entropy.

DEVELOP Considering a single molecule, the probability that it is located on the left-side or the right side of the box is ½. Considering 2 molecules, the probability for one particular left-right arrangement (microstate) is ¼. Another way to say this is that there are 4 different ways to sort the molecules between the two sides. For 6 molecules, there are $2^6 = 64$ ways to sort, so the probability for one particular arrangement (microstate) is 1/64. We now have to count how many of these microstates match the following macrostates (see Figures 19.18 and 19.19).

EVALUATE (a) There's only one microstate in which all of the molecules are found on one side of the box, so the probability of this macrostate is 1/64.

(b) It's a bit harder to find the number of microstates with half the molecules on one side, half on the other. So let's label the molecules A, B, C, D, E, and F, and let's identify a microstate by the 3 molecules on the left-hand side. So for example, (ABC) is the microstate with A, B, C on the left-hand side, and the others on the right. We can switch out C in three different ways: (ABD), (ABE) and (ABF). Similarly we can switch out B in three different ways and switch out A in three different ways. That gives us 10 microstates. We count another 10 microstates if we start with (DEF), and switch out D, then E, then F. The total number is 20 microstates.

So the probability of the macrostate with the molecules split evenly between the sides is 20/64.

ASSESS It is 20 times more likely that the 6 molecules will be spread out evenly between the two sides of the box vs. all on one side. In general, if there are n molecules, the probability that k of them will be on one side and $(n-k)$ on the other side is given by the coefficients from the binomial theorem:

$$P = \binom{n}{k} \cdot \frac{1}{2^n} = \left[\frac{n!}{k!(n-k)!} \right] \cdot \frac{1}{2^n}$$

In the case above, $n = 6$ and $k = 3$, so the probability is $P = 6!/(3! \, 3!) \, 2^6 = 20/64$, as we found.

PROBLEMS

27. **INTERPRET** This problem requires us to find the thermodynamic efficiency of a nuclear power plant in winter and in summer, when the temperature of its cold reservoir is 0°C and 25°C, respectively.

DEVELOP From Equation 19.3, the thermodynamic efficiency of a Carnot engine is

$$e_{\text{Carnot}} = 1 - \frac{T_c}{T_h}$$

where the temperatures are in Kelvin.

EVALUATE Inserting the given temperatures for summer and winter gives

$$e_{\text{summer}} = 1 - \frac{298 \text{ K}}{570 \text{ K}} = 47.7\%$$

$$e_{\text{winter}} = 1 - \frac{273 \text{ K}}{570 \text{ K}} = 52.1\%$$

ASSESS The plant is more efficient in winter than in summer because there is a greater heat difference. However, as explained in Section 19.3, irreversible processes, transmission losses, etc., make actual efficiencies less than the theoretical maxima.

29. **INTERPRET** This problem involves a nuclear power plant and asks us to calculate the rate of energy extraction, the efficiency, and the highest temperature the plant attains.

DEVELOP From Equation 16.3, $Q = mc\Delta T$, we see that to raise the temperature of the cooling water by 8.5 K, heat must be exhausted to it at a rate of

$$\frac{dQ_c}{dt} = c\left(\frac{dm}{dt}\right)\Delta T = \left[4.184 \text{ kJ/(kg} \cdot \text{K)}\right]\left(2.8 \times 10^4 \text{ kg/s}\right)\left(8.5 \text{ K}\right) = 996 \text{ MW}$$

We take this to be the rate of all the heat rejected by the power plant. Since the rate of work output dW/dt is also given, the heat input to the plant (extracted from its fuel) is

$$\frac{dQ_h}{dt} = \frac{dQ_c}{dt} + \frac{dW}{dt}$$

where we have used the first law of thermodynamics (see Problem 19.16). In terms of the rates, the efficiency of the plant is

$$e = \frac{dW/dt}{dQ_h/dt}$$

If we consider the plant to operate like a Carnot engine, then its highest temperature can be calculated using $Q_c/Q_h = T_c/T_h$ (from Equation 19.2).

EVALUATE (a) Substituting the values given, we obtain

$$\frac{dQ_h}{dt} = \frac{dQ_c}{dt} + \frac{dW}{dt} = 996 \text{ MW} + 750 \text{ MW} = 1.75 \text{ GW}$$

where the negative sign corresponds to the energy being extracted from the fuel.

(b) The plant's efficiency (from the definition of efficiency in terms of rates) is

$$e = \frac{dW/dt}{dQ_h/dt} = \frac{750 \text{ MW}}{1.75 \text{ GW}} = 43.0\%$$

(c) With the assumption that the plant operates like an ideal Carnot engine, then

$$\frac{T_h}{T_c} = \frac{Q_h}{Q_c} = \frac{dQ_h/dt}{dQ_c/dt} = \frac{1.75 \text{ GW}}{996 \text{ MW}} = 1.75$$

(Note that the energy rate per cycle and the energy rate per second are proportional.) If $T_c = 15°\text{C} = 288 \text{ K}$, then

$$T_h = 1.75 T_c = 1.75(288 \text{ K}) = 505 \text{ K} = 232°\text{C}$$

ASSESS The actual highest temperature would be somewhat greater than this, because the actual efficiency is always less than the Carnot efficiency.

31. **INTERPRET** This problem asks us to find the rate (i.e., kg/s) at which all the power plants in the USA use cooling water. We are given the actual efficiency of the power-plants and the temperature rise in the cooling water.

DEVELOP For a cyclic operation, the change in internal energy is zero, $\Delta U = 0$. From the first law of thermodynamics, we have $W = Q_h - Q_c$, where W is the work done by the system (contrary to the definition of W in Chapter 18), Q_h is the heat absorbed by the system, and Q_c is the heat rejected by the system. Therefore, the total rate at which heat is exhausted by all the power plants is

$$\frac{dQ_c}{dt} = \frac{d}{dt}(Q_h - W) = \frac{dW}{dt}\left(\frac{1}{e} - 1\right) = \left(2 \times 10^{11} \text{ W}\right)\left(\frac{1}{33\%} - 1\right) = 4.06 \times 10^{11} \text{ W}$$

The mass rate of flow at which water could absorb this amount of energy, with only a 5°C temperature rise, is

$$\frac{dQ_c}{dt} = \frac{dm}{dt} c_{water} \Delta T$$

where $c_{water} = 4184 \text{ J/(kg} \cdot \text{K)}$ (see Table 16.1). This equation can be solved to give the mass rate of cooling water used.

EVALUATE Solving the equation above for dm/dt, we obtain

$$\frac{dm}{dt} = \frac{dQ_c}{dt} c_{water} \Delta T = \frac{4.06 \times 10^{11} \text{ W}}{\left[4184 \text{ J/(kg} \cdot \text{K)}\right](5 \text{ K})} = 2 \times 10^7 \text{ kg/s}$$

or about 1 Mississippi (a self-explanatory unit of river flow).

ASSESS To absorb the power output of 2×10^{11} W with only an increase of temperature of 5°C, we expect the mass flow rate to be large.

33. INTERPRET   This problem involves finding the COP of a freezer for which the highest and the lowest temperatures are $T_h = 32°C = 305$ K and $T_c = 0°C = 273$ K . In addition, we are to find how much water at $0°C$ the freezer can freeze in one hour.

DEVELOP   The coefficient of performance (COP) of a reversible freezer is given by Equation 19.4:

$$COP = \frac{Q_c}{W} = \frac{Q_c}{Q_h - Q_c} = \frac{T_c}{T_h - T_c}$$

where we have used Equation 19.2, $T_c/T_h = Q_c/Q_h$ for the last equality. Once the COP is known, we can solve for Q_c and the amount of water the freezer can freeze in one hour, which is $m = Q_c/L_f$ with Lf = 334 kJ/kg (see Equation 17.5 and Table 17.1).

EVALUATE   (a) The COP of the freezer is

$$COP = \frac{T_c}{T_h - T_c} = \frac{273 \text{ K}}{305 \text{ K} - 273 \text{ K}} = 8.53$$

(b) The heat rejected in one hour is $Q_c = COP\ W = 8.53 \times (12 \text{ kWh}) = 369$ MJ , so the water we can freeze is

$$m = \frac{Q_c}{L_f} = \frac{369 \text{ MJ}}{334 \text{ kJ/kg}} = 1.10 \times 10^3 \text{ kg.}$$

ASSESS   Typical freezers have a COP lower than 8.53. Thus, more electrical energy is needed to freeze the same amount of water.

35. INTERPRET   This problem requires us to find the monthly cost of using all the incoming electrical energy to power a heat pump with COP = 3.1 to heat a house. We are given that the electrical energy costs $180 per month in the winter.

DEVELOP   The same electrical energy W used for direct conversion in electric heating would produce heat $Q_h = W + Q_c$. Using Equation 19.4 allows us to express this as

$$Q_h = W + Q_c = W + Q_h\ (1 - 1/COP)$$
$$Q_h = W\ (COP)$$

Thus, the heat pump can produce a factor COP more heat than if the electrical energy is converted directly to heat.

EVALUATE   Because the heat pump is a factor COP more efficient, the cost will be reduced by this same factor, so the monthly heating bill would be

$$\frac{\$180}{COP} = \frac{\$180}{3.1} = \$58$$

ASSESS   The savings are significant, which is why electrical heating is not recommended.

37. INTERPRET   We are to find the minimum COP required to save money if we switch from an oil furnace to an electrically powered heat pump, considering the cost of oil and of electricity. We will do this by calculating the cost of the heat delivered by both the oil-burning heater and the electric heat pump.

DEVELOP   The coefficient of performance (COP) is the relationship between the heat sent to the cold reservoir and the work done. Set the heat Q_c to be the same for both heating mechanisms, and solve for COP. The cost of oil is $\$_{oil} = \frac{\$1.75}{30 \text{ kWh}} = \$0.0583 \text{ kW} \cdot h^{-1}$, and the cost of electricity is $\$_{electric} = \$0.165 \text{ kW} \cdot h^{-1}$. The heat delivered is

$$Q_c = W \times COP$$

(see derivation of Equation 19.4), and we are paying for the work done, so the COP must exceed the ratio of the costs (COP > $\$_{electric}/\$_{oil}$).

EVALUATE   The COP must satisfy

$$COP > \$_{electric}/\$_{oil} = 2.83$$

So in order to be cost-effective, the heat pump must have a COP of greater than 2.83.

ASSESS   Most heat pumps have a COP much higher than this value, so it's probably a good idea to switch.

39. **INTERPRET**   We are asked to find the COP, power usage, and operating cost compared to that of an oil-burning heater of a heat pump. We will assume that the heat pump is a Carnot heat pump.

DEVELOP   The maximum COP of a heat pump (when its heating, not cooling) is given in Equation 19.4b: $COP = T_h / (T_h - T_c)$. In this case, $T_c = 10°C = 283$ K and $T_h = 70°C = 343$ K. In general, the COP for a heat pump is the heat supplied divided by the work input. In terms of rates, that can be written as $COP = H / P$, where H is the supplied heat rate and P is the electric power consumption.

EVALUATE   (a) Assuming the heat pump is maximally efficient,

$$COP = \frac{T_h}{T_h - T_c} = \frac{343 \text{ K}}{343 \text{ K} - 283 \text{ K}} = 5.72 \approx 5.7$$

(b) The power consumption needed to supply heat at 20 kW is

$$P = \frac{H}{COP} = \frac{20 \text{ kW}}{5.72} = 3.5 \text{ kW}$$

(c) Given the utility rate for electric power, the heat pump's hourly operating cost is

$$C_{pump} = (Pt) R_{utility} = (3.5 \text{ kW})(1\text{h})(15.5¢ / \text{kWh}) = 54¢$$

In comparison, an oil furnace, supplying the same heat, would have an hourly operating cost of

$$C_{oil} = \frac{20 \text{ kW}(1\text{h})}{30 \text{ kWh/gal}} ($2.60 / \text{gal}) = $1.73$$

ASSESS   The cost per kWh of oil is actually less than that of electricity: $R_{oil} = 8.7¢ / \text{kWh}$. But the heat pump has such a high COP that it cost three times less to heat the house.

41. **INTERPRET**   Our engine cycle consists of four paths, two of which are isochoric and two of which are isobaric. We are to determine the efficiency, defined as the work done per unit heat absorbed, and compare the result with the efficiency of a Carnot engine operating between the same temperatures. Finally, we need to explain any difference between the two efficiencies.

DEVELOP   Label the states in Fig. 19.22 A, B, C, and D going clockwise from the upper left corner. The work done and the heat absorbed during the isobaric segments AB and CD are

$$W_{AB} = p_A (V_B - V_A) = (6 \text{ atm})(6 \text{ L} - 2 \text{ L}) = 24 \text{ L} \cdot \text{atm}$$
$$W_{CD} = p_C (V_D - V_C) = (3 \text{ atm})(2 \text{ L} - 6 \text{ L}) = -12 \text{ L} \cdot \text{atm}$$

and

$$Q_{AB} = nC_P (T_B - T_A) = n \left(\frac{5}{2} R\right) \left(\frac{p_B V_B}{nR} - \frac{p_A V_A}{nR}\right) = \frac{5}{2}(36 - 12) \text{ L} \cdot \text{atm} = 60 \text{ L} \cdot \text{atm}$$

$$Q_{CD} = nC_P (T_D - T_C) = \frac{5}{2}(p_D V_D - p_C V_C) = \frac{5}{2}(6 - 18) \text{ L} \cdot \text{atm} = -30 \text{ L} \cdot \text{atm}$$

where we have assumed an ideal monatomic gas (see Equation 18.13).
For the isochoric segments, we have

$$Q_{BC} = nC_V (T_C - T_B) = \frac{3}{2}(18 - 36) \text{ L} \cdot \text{atm} = -27 \text{ L} \cdot \text{atm}$$

$$Q_{DA} = nC_V (T_A - T_D) = \frac{3}{2}(12 - 6) \text{ L} \cdot \text{atm} = 9 \text{ L} \cdot \text{atm}$$

and $W_{BC} = W_{DA} = 0$. The net heat added for one cycle is therefore

$$Q_{ABCDA} = Q_{AB} + Q_{BC} + Q_{CD} + Q_{DA} = (60 - 27 - 30 + 9) \text{ L} \cdot \text{atm} = (12 \text{ L} \cdot \text{atm})(101.3 \text{ J/L} \cdot \text{atm})$$
$$= 1.22 \text{ kJ}$$

and the net work done is $W = (24 + 0 - 12 + 0)$ L·atm $= 12$ L·atm $= 1.22$ kJ. Note that the first law of thermodynamics, applied to a cyclic process, requires that $W = Q$ when using the definition that W is the work done by the system (which is opposite to the definition used in Chapter 18).

EVALUATE (a) Since the heat absorbed is $Q_+ = (60 + 9)$ L·atm $= 69$ L·atm, the efficiency is

$$e = \frac{W}{Q_+} = \frac{12 \text{ L·atm}}{69 \text{ L·atm}} = 17.4\%$$

(b) The maximum and minimum temperatures are $T_B = p_B V_B / nR$ and $T_D = p_D V_D / nR$ so the efficiency of a Carnot engine operating between these temperatures is

$$e_{\text{Carnot}} = 1 - \frac{T_D}{T_B} = 1 - \frac{p_D V_D}{p_B V_B} = 1 - \frac{6 \text{ L·atm}}{36 \text{ L·atm}} = 83.3\%$$

This is not a contradiction of Carnot's theorem, because the given engine does not operate between two heat reservoirs at fixed temperatures.

ASSESS The efficiency of a real engine is always less or equal to that of a Carnot engine.

43. **INTERPRET** This problem is about the increase in entropy as the ice is melted and heated up.

DEVELOP The entropy increase is given by Equation 19.6: $\Delta S_{12} = \int_1^2 dQ / T$. We consider the entropy increase in two steps. First, the heat needed to melt the lake ice is $Q = mL_f$, where $L_f = 334$ kJ/kg from Table 17.1. The temperature is constant during the melting, $T_0 = 273$ K. In the second step, the heat input raises the water temperature according to $dQ = mc\,dT$, where $c = 4.184$ kJ/kg·K from Table 16.1. Here, the temperature is not constant, so we will have to integrate.

EVALUATE The entropy increase during melting is

$$\Delta S_{\text{melt}} = \int_1^2 \frac{dQ}{T} = \frac{mL_f}{T_0} = \frac{(94 \text{ Mg})(334 \text{ kJ/kg})}{(273 \text{ K})} = 115 \text{ MJ/K}$$

The entropy increase during warming is

$$\Delta S_{\text{warm}} = \int_{T_0}^{T_1} \frac{mc\,dT}{T} = mc \ln\left(\frac{T_1}{T_0}\right) = (94 \text{ Mg})(4.184 \text{ kJ/kg·K}) \ln\left(\frac{288 \text{ K}}{273 \text{ K}}\right) = 21.0 \text{ MJ/K}$$

So the total entropy increase is

$$\Delta S_{\text{tot}} = \Delta S_{\text{melt}} + \Delta S_{\text{warm}} = 115 \text{ MJ/K} + 21.0 \text{ MJ/K} \approx 140 \text{ MJ/K}$$

ASSESS As expected, the entropy change is positive in both melting and warming processes.

45. **INTERPRET** We are to derive the formula given in the problem statement that describes the entropy change for n moles of an ideal gas that undergoes an isovolumic temperature change from T_1 to T_2.

DEVELOP From the first law of thermodynamics ($dQ = dU + dW$) and the properties of an ideal gas ($dU = nC_V\,dT$ and $pV = nRT$), an infinitesimal entropy change is

$$dS = \frac{dQ}{T} = nC_V \frac{dT}{T} + \frac{p}{T} dV = nC_V \frac{dT}{T} + nR \frac{dV}{V}$$

Integrate from state 1 (T_1, V_1) to state 2 (T_2, V_2), and apply the isovolumic constraint to obtain the given formula.

EVALUATE Integrating the expression above gives

$$\Delta S = nC_V \ln\left(\frac{T_2}{T_1}\right) + nR \ln\left(\frac{V_2}{V_1}\right)$$

For an isovolumic process $V_1 = V_2$ so $\Delta S = nC_V \ln(T_2/T_1)$.

ASSESS Of course, we could have started with $dQ = nC_V\,dT$ at constant volume, but we wanted to display ΔS for a general ideal-gas process, for use in other problems.

47.   **INTERPRET**   This problem involves the entropy change in an ideal diatomic gas heated under three different conditions: constant volume, constant pressure, and adiabatically.

DEVELOP   From Problem 19.45, the entropy change at constant volume is $\Delta S_V = nC_V \ln\left(T_2/T_1\right)$, where $C_V = 5R/2$ for a diatomic gas (see discussion after Equation 18.13). From Problem 19.46, the entropy change at constant pressure is

$$\Delta S_p = nC_p \ln\left(T_2/T_1\right) = n\left(C_V + R\right)\ln\left(T_2/T_1\right)$$

For an adiabatic process, consider the discussion accompanying Figure 19.16. The entropy change is

$$\Delta S = nR \ln\left(\frac{V_2}{V_1}\right) = nR \ln\left[\left(\frac{T_1}{T_2}\right)^{\frac{1}{\gamma-1}}\right] = \frac{nR}{\gamma-1}\ln\left(\frac{T_1}{T_2}\right)$$

where we have used the relation for an adiabatic process $T_1 V_1^{\gamma-1} = T_2 V_2^{\gamma-1}$ (Equation 18.11b). For a diatomic gas,

$$\gamma = \frac{C_P}{C_V} = \frac{C_V + R}{C_V} = \frac{5R/2 + R}{5R/2} = \frac{7}{5}$$

EVALUATE   (a) At constant volume, the entropy change is

$$\Delta S_V = nC_V \ln\left(T_2/T_1\right) = (5.0 \text{ mol})\left(\frac{5}{2}\right)\left[8.314 \text{ J}/(\text{mol·K})\right]\ln\left(\frac{500 \text{ K}}{300 \text{ K}}\right) = 53 \text{ J/K}$$

(b) At constant pressure, the entropy change is

$$\Delta S_p = nC_p \ln\left(T_2/T_1\right) = n\left(C_V + R\right)\ln\left(T_2/T_1\right) = (5.0 \text{ mol})\left(\frac{7}{2}\right)\left[8.314 \text{ J}/(\text{mol·K})\right]\ln\left(\frac{500 \text{ K}}{300 \text{ K}}\right) = 74 \text{ J/K}$$

(c) For an adiabatic process, the entropy change is

$$\Delta S = \frac{nR}{\gamma-1}\ln\left(\frac{T_1}{T_2}\right) = \frac{2(5.0 \text{ mol})\left[8.314 \text{ J}/(\text{mol·K})\right]}{5}\ln\left(\frac{300 \text{ K}}{500 \text{ K}}\right) = -8.5 \text{ J/K}$$

ASSESS   For the adiabatic process, the final volume is less than the initial volume because the final temperature is greater than the initial temperature (see, e.g., Table 18.1), so the entropy decreases, in agreement with the result of the discussion of Figure 19.16.

49.   **INTERPRET**   You want to find the entropy change in going from one state to another state that lie on the same adiabat.

DEVELOP   We're told that the system goes from state p_1, V_1 to state p_2, V_2, where the two states are related by the adiabatic equation: $p_1 V_1^{\gamma} = p_2 V_2^{\gamma}$. Our first inclination would be that the entropy change would be zero, since there is no heat exchange in the adiabatic process that connects these two states. However, we're told that the system goes between the two states in two segments: one a constant pressure process (going from p_1, V_1 to p_1, V_2) and the other a constant volume process (going from p_1, V_2 to p_2, V_2). See the figure below.

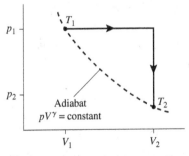

This is an ideal gas, so the temperatures of the three endpoints are: $T_1 = p_1 V_1/nR$, $T_{12} = p_1 V_2/nR$, and $T_2 = p_2 V_2/nR$. From Table 18.1, the differential heat flows for these two processes are:

$$dQ = nC_p dT \quad \text{for constant pressure}$$

$$dQ = nC_V dT \quad \text{for constant volume}$$

Recall that $\gamma = C_p / C_V$.

EVALUATE For the constant pressure process, the entropy change is

$$\Delta S_p = \int_{T_1}^{T_{12}} \frac{nC_p dT}{T} = nC_p \ln\left(\frac{T_{12}}{T_1}\right) = nC_p \ln\left(\frac{V_2}{V_1}\right) = n\gamma C_V \ln\left(\frac{V_2}{V_1}\right)$$

For the constant volume process, the entropy change is

$$\Delta S_V = \int_{T_{12}}^{T_2} \frac{nC_V dT}{T} = nC_V \ln\left(\frac{T_2}{T_{12}}\right) = nC_V \ln\left(\frac{p_2}{p_1}\right) = nC_V \ln\left(\frac{V_1^\gamma}{V_2^\gamma}\right)$$

The total entropy is the sum:

$$\Delta S = \Delta S_p + \Delta S_V = nC_V \left[\gamma \ln\left(\frac{V_2}{V_1}\right) - \gamma \ln\left(\frac{V_2}{V_1}\right) \right] = 0$$

where we have used the mathematical identity: $\ln\left(x^a\right) = a \ln x$. The total entropy change is zero as we expected, since the system could have gone from state 1 to state 2 by an adiabatic process for which $Q = 0$.

ASSESS As explained in the text, the entropy is a state variable, which doesn't depend on how a system arrived at a particular state. Note that this separates entropy from the heat, Q, absorbed or expelled by a system. You can't say that a system contains a particular amount of heat, but you can say that it contains a particular amount of entropy.

51. **INTERPRET** This problem asks for the entropy change of the pan-water system, when thermal equilibrium has been reached after a hot pan has been plunged into the given amount of cold water.

DEVELOP Assume all the heat lost by the pan is gained by the water. The equilibrium temperature is given by Equation 16.4, or

$$T_{eq} = \frac{(2.4 \text{ kg})\left[900 \text{ J}/(\text{kg} \cdot \text{K})\right](428 \text{ K}) + (3.5 \text{ kg})\left[4184 \text{ J}/(\text{kg} \cdot \text{K})\right](288 \text{ K})}{(2.4 \text{ kg})\left[900 \text{ J}/(\text{kg} \cdot \text{K})\right] + (3.5 \text{ kg})\left[4184 \text{ J}/(\text{kg} \cdot \text{K})\right]} = 306 \text{ K}$$

Using the result of Exercise 48, the change in entropy for the pan is

$$\Delta S_{pan} = m_{pan} c_{pan} \ln\left(\frac{T_{eq}}{T_{pan}}\right)$$

Similarly, the change in entropy for the water is

$$\Delta S_{water} = m_{water} c_{water} \ln\left(\frac{T_{eq}}{T_{water}}\right)$$

The sum of these two terms is the change of entropy of the pan-water system.

EVALUATE The entropy change of the pan and water together is

$$\Delta S = \Delta S_{pan} + \Delta S_{water} = m_{pan} c_{pan} \ln\left(\frac{T_{eq}}{T_{pan}}\right) + m_{water} c_{water} \ln\left(\frac{T_{eq}}{T_{water}}\right)$$

$$= (2.4 \text{ kg})\left[900 \text{ J}/(\text{kg} \cdot \text{K})\right] \ln\left(\frac{306 \text{ K}}{428 \text{ K}}\right) + (3.5 \text{ kg})\left[4184 \text{ J}/(\text{kg} \cdot \text{K})\right] \ln\left(\frac{306 \text{ K}}{288 \text{ K}}\right) = 160 \text{ J/K}$$

to two significant figures.

ASSESS The entropy change for the pan is negative, while that of the water is positive. The total entropy change is positive, in accordance with the second law of thermodynamics.

53.   **INTERPRET**   We will calculate the efficiency of the Otto cycle, on which gasoline engines are modeled.

DEVELOP   The engine absorbs heat (Q_h) during combustion, and expels heat to the environment (Q_c) during the exhaust segment. Both these processes are at constant volume, so $Q = nC_V \Delta T$, and the efficiency is:

$$e_{Otto} = \frac{W}{Q_h} = 1 - \frac{Q_c}{Q_h} = 1 - \frac{\Delta T_c}{\Delta T_h}$$

We can find the respective temperature changes assuming the gas mixture in the engine is ideal: $T = pV / nR$.

EVALUATE   (a) The hot temperature change is between point 2 and point 3 in Figure 19.24:

$$\Delta T_h = T_3 - T_2 = \frac{1}{nR}\left[3p_2\left(\tfrac{1}{5}V_1\right) - p_2\left(\tfrac{1}{5}V_1\right)\right] = \frac{2p_2V_1}{5nR}$$

where we use the values for the pressure and volume given in the figure. The cold temperature change is between point 1 and point 4 in the figure, but the pressures aren't given in this case. However, point 1 and point 2 are on the same adiabat $(pV^\gamma = \text{constant})$, so: $p_1 = p_2\left(\tfrac{1}{5}\right)^\gamma$, and similarly for point 4 and point 3: $p_4 = 3p_2\left(\tfrac{1}{5}\right)^\gamma$. Therefore, the cold temperature change can be written:

$$\Delta T_c = T_4 - T_1 = \frac{1}{nR}\left[p_4V_1 - p_1V_1\right] = \frac{2p_2V_1}{5^\gamma nR}$$

From these temperature changes, the efficiency of the Otto cycle is:

$$e_{Otto} = 1 - \frac{\Delta T_c}{\Delta T_h} = 1 - 5^{1-\gamma}$$

(b) The minimum temperature occurs at point 1 at the end of the exhaust segment:

$$T_{min} = T_1 = \frac{p_2V_1}{5^\gamma nR}$$

The maximum temperature occurs at point 3 at the end of combustion:

$$T_{max} = T_3 = \frac{3p_2V_1}{5nR} = 3 \cdot 5^{\gamma-1} T_{min}$$

(c) A Carnot cycle working between minimum and maximum temperatures would have an efficiency of (Equation 19.3):

$$e_{Carnot} = 1 - \frac{T_{min}}{T_{max}} = 1 - \tfrac{1}{3}5^{1-\gamma}$$

So, the Carnot cycle efficiency is greater than that of the Otto cycle: $e_{Carnot} > e_{Otto}$, as we'd expect since the Carnot cycle has the maximum efficiency for an engine.

ASSESS   If we assume $\gamma = 1.4$, just for argument sake, then $e_{Otto} = 47\%$, while $e_{Carnot} = 82\%$. In this light, gasoline engines are woefully inefficient. Much of the combustion energy is lost as exhaust heat.

55.   **INTERPRET**   Find the maximum efficiency of a power plant, given the temperature range of its cycle. We will calculate the Carnot efficiency, and compare this with the actual efficiency.

DEVELOP   The high temperature $T_h = 950°F = 783$ K. The low temperature is $T_c = 90°F = 305$ K. The Carnot efficiency is given by $e = 1 - \frac{T_c}{T_h}$.

EVALUATE   The maximum efficiency is $e = 1 - \frac{T_c}{T_h} = 61\%$.

ASSESS   The actual efficiency of this plant is given as 25%, which is considerably lower due (at least in part) to having to evaporate moisture out of the wood-chip fuel.

57.   **INTERPRET**   We are asked to calculate the entropy change in an object whose heat capacity is inversely proportional to its temperature.

DEVELOP   By definition, the heat capacity relates the heat flowing into an object to the change in its temperature: $dQ = CdT$. In this case, $C = C_0 (T_0 / T)$. We can plug this into Equation 19.6 for the entropy change: $\Delta S = \int C_0 T_0 dT / T^2$.

EVALUATE   Performing the integration from T_0 to T_1:

$$\Delta S_{01} = \int_{T_0}^{T_1} \frac{C_0 T_0 dT}{T^2} = C_0 T_0 \left(\frac{-1}{T} \right) \Bigg|_{T_0}^{T_1} = C_0 \left(1 - \frac{T_0}{T_1} \right)$$

ASSESS   For $T_1 > T_0$, the entropy change is positive. For $T_1 \gg T_0$, the entropy change becomes constant: $\Delta S_{01} \approx C_0$.

59.   **INTERPRET**   You have an infinite heat reservoir, but a finite cool reservoir. The question is how much work can you obtain with an engine placed between the reservoirs before the cool reservoir is "exhausted."

DEVELOP   The infinite heat reservoir will maintain its temperature, T_h, no matter how much heat, Q_h, you extract from it. The cool reservoir, on the other hand, will not maintain its initial temperature, T_{c0}, as heat from the engine is expelled into it. The temperature will rise in the cool reservoir according to $dQ_c = CdT_c$. But once the cool reservoir temperature is equal to T_h, no more work can be extracted.

EVALUATE   You can assume that the engine cycles fast enough that during a single cycle the cool reservoir temperature is approximately constant. To maximize the amount of work that you extract, place a Carnot engine between the reservoirs so that the work extracted during one cycle is:

$$dW = d(Q_h - Q_c) = dQ_c \left[\frac{T_h}{T_c} - 1 \right] = CdT_c \left[\frac{T_h}{T_c} - 1 \right]$$

where $Q_h / Q_c = T_h / T_c$ for a Carnot engine. To find the total work, integrate this expression from T_{c0} to T_h,

$$W = \int_{T_{c0}}^{T_h} CdT_c \left[\frac{T_h}{T_c} - 1 \right] = CT_h \ln (T_c) - CT_c \Big|_{T_{c0}}^{T_h} = CT_h \ln \left(\frac{T_h}{T_{c0}} \right) - C(T_h - T_{c0})$$

If we let $x = T_h / T_{c0}$, then $W = CT_h (\ln x - 1 + 1/x)$.

ASSESS   The work is proportional to the heat capacity, as you might expect. The heat capacity is a measure of how much heat the cool reservoir can accept, so the larger the heat capacity, the more work that can be extracted. You might worry that the work could be negative for some value of $x = T_h / T_{c0}$. For $x \gg 1$, the work is approximately $W \approx CT_h \ln x$, which is positive. For $x \approx 1$, $\ln x \approx x - 1$, and the work is approximately $W \approx CT_h (x - 2 + 1/x)$, which is positive as well. Therefore, the work extracted is positive for all possible temperature differences.

61.   **INTERPRET**   We're asked to find the entropy change when hot and cold water are mixed together. Since the hot and cold water can't be unmixed, Equation 19.6 doesn't apply directly, but you can find a reversible process that mimics this irreversible mixing.

DEVELOP   Before mixing, imagine cooling the hot water and warming the cold water until they are both at T_f, which is the final temperature when the water volumes are irreversible mixed. After the temperatures are equilibrated, the two water samples can simply be added together. This is a reversible process, since we could easily divide the mixed water in half and re-warm one sample and re-cool the other to the original temperatures. But as the final state is the same as in the irreversible mixing case, we claim that the entropy change applies to both system paths.

EVALUATE The differential heat flow in both the warming and cooling processes is $dQ = mc\,dT$. The entropy change, therefore, in cooling the hot water from T_h to T_f is

$$\Delta S_{hf} = \int_{T_h}^{T_f} \frac{dQ}{T} = mc\ln\left(\frac{T_f}{T_h}\right)$$

Similarly, the entropy change in warming the cold water from T_c to T_f is

$$\Delta S_{cf} = \int_{T_c}^{T_f} \frac{dQ}{T} = mc\ln\left(\frac{T_f}{T_c}\right)$$

Adding these entropy changes together for the total gives:

$$\Delta S_{tot} = \Delta S_{hf} + \Delta S_{cf} = mc\ln\left(\frac{T_f^2}{T_h T_c}\right)$$

This is positive when $T_f^2 > T_h T_c$, or equivalently when $T_f^2 - T_h T_c > 0$. Since the sample masses are equal, the final temperature will be at the midpoint between the temperature extremes: $T_f = \frac{1}{2}(T_h + T_c)$. Therefore,

$$T_f^2 - T_h T_c = \tfrac{1}{4}T_h^2 - \tfrac{1}{2}T_h T_c + \tfrac{1}{4}T_c^2 = \tfrac{1}{4}(T_h - T_c)^2 > 0$$

This shows that the total entropy change is positive, as we would expect.

ASSESS Using the final equation above, we can rewrite the total entropy as:

$$\Delta S_{tot} = mc\ln\left(1 + \frac{(T_h - T_c)^2}{4T_h T_c}\right)$$

What this shows is that for a given T_h the entropy change will be greater the smaller that T_c is. In other words, the entropy change is greater when the initial temperature difference is made greater. This is what we would expect.

63. **INTERPRET** We are to find the entropy change for a sample of gas with the given temperature change, where the specific heat of the gas changes with temperature.

DEVELOP We are given an equation for the molar specific heat: $c_p = a + bT + cT^2$, where $a = 33.6$ J/(mol·K), $b = 2.93 \times 10^{-3}$ J/(mol·K^2), and $c = 2.13 \times 10^{-5}$ J/(mol·K^3). The amount of gas is 2 moles, and the temperature changes from $T_1 = 293$ K to $T_2 = 473$ K. From the definition of the molar specific heat (see discussion preceding Equation 18.3), $Q = nC_p \Delta T$, which we use to express the heat change in terms of temperature and specific heat. Insert this into Equation 19.6 to find the entropy change.

EVALUATE

$$dS = \frac{dQ}{T}$$

$$\Delta S = \int_{T_1}^{T_2} \frac{nC_p}{T}\,dT = n\int_{T_1}^{T_2} \frac{(a + bT + cT^2)}{T}\,dT = n\int_{T_1}^{T_2}\left(\frac{a}{T} + b + cT\right)dT$$

$$= n\left[a\ln\left(\frac{T_2}{T_1}\right) + b(T_2 - T_1) + \frac{c}{2}(T_2^2 - T_1^2)\right] = 36.2 \text{ J/K}$$

ASSESS The entropy increases as the temperature increases, as we would expect.

65. **INTERPRET** We will consider the energy consumption of a typical refrigerator.

DEVELOP To get a sense of how the refrigerator works, we can look at Figure 19.6. In the course of a day, an amount of heat, $Q_c = 30$ MJ, is drawn from the fridge's cold interior. But this requires work to be done, $W = 10$ MJ. Specifically, electricity is needed to pump refrigerant through the system.

EVALUATE Both the heat drawn from the fridge interior and the work done by the electrical energy are expelled as heat: $Q_h = Q_c + W$. So, the work effectively ends up as waste heat rejected to the kitchen environment.

The answer is (d).

ASSESS This might sound wasteful: turning high quality electrical energy into heat that gets dumped out of the backside of your fridge. But according to Clausius' statement of the second law of thermodynamics, it's impossible to construct a perfect refrigerator whose sole effect is to transfer heat from a cooler object to a hotter one. An external energy source is needed. However, it's not necessary to use electrical energy. For example, solar refrigerators use sunlight to evaporate water and thus draw heat from the fridge interior.

67. **INTERPRET** We will consider the energy consumption of a typical refrigerator.

DEVELOP We're told the coal-fired power plant has an efficiency of $e = 40\%$. In contrast to the COP, the efficiency is defined as what we want (electrical energy to do work) divided by what we put in (heat from the burning of coal), i.e., $e = W / Q_h$.

EVALUATE To make 10 MJ of electrical energy, the power plant has to burn enough fuel to generate $Q_h = W / e = 25$ MJ.

The answer is (b).

ASSESS Notice what this says: it takes 25 MJ of heat from coal burning to extract 30 MJ of heat from the fridge contents.